T0329581

Control over Communication Networks

IEEE Press
445 Hoes Lane
Piscataway, NJ 08854

Control over Communication Networks

Modeling, Analysis, and Design of Networked Control Systems and Multi-Agent Systems over Imperfect Communication Channels

Jianying Zheng
Beihang University, Beijing, China

Liang Xu
Shanghai University, Shangai, China

Qinglei Hu
Beihang University, Beijing, China

Lihua Xie
Nanyang Technological University, Singapore, Singapore

IEEE Press Series on Control Systems Theory and Applications
Maria Domenica Di Benedetto, Series Editor

Published by John Wiley & Sons, Inc., Hoboken, New Jersey.
Published simultaneously in Canada.

Library of Congress Cataloging-in-Publication Data

Names: Zheng, Jianying, author. | Xu, Liang, author. | Hu, Qinglei, author. | Xie, Lihua, author.
Title: Control over communication networks : modeling, analysis, and design of networked control systems and multi-agent systems over imperfect communication channels / Jianying Zheng, Beihang University, Beijing, China, Liang Xu, Shanghai University, Shanghai, China, Qinglei Hu, Beihang University, Beijing, China, Lihua Xie, Nanyang Technological University, Singapore, Singapore.
Description: Hoboken, New Jersey : Wiley, [2023] | Includes bibliographical references and index.
Identifiers: LCCN 2023002566 (print) | LCCN 2023002567 (ebook) | ISBN 9781119885795 (hardback) | ISBN 9781119885801 (adobe pdf) | ISBN 9781119885818 (epub)
Subjects: LCSH: Supervisory control systems–Reliability. | Remote control. | Robust control. | Uncertainty (Information theory) | Fault tolerance (Engineering) | Multi-agent systems.
Classification: LCC TJ222 .Z46 2023 (print) | LCC TJ222 (ebook) | DDC 620/.46–dc23/eng/20230213
LC record available at https://lccn.loc.gov/2023002566
LC ebook record available at https://lccn.loc.gov/2023002567

Cover Design: Wiley
Cover Image: © ProStockStudio/Shutterstock

Set in 9.5/12.5pt STIXTwoText by Straive, Chennai, India

To my parents, my husband Pengfei, and my daughter Coco (Jianying Zheng)
To my parents, my wife Xiaoxue, and my son Ze (Liang Xu)

Contents

About the Authors

Jianying Zheng received the BE degree in Automation from University of Science and Technology of China in 2010, and the PhD degree in Electronic and Computer Engineering from the Hong Kong University of Science and Technology, Hong Kong, in 2016. Between September 2014 and February 2015, she was a visiting student in the University of Newcastle, Australia. From 2016 to 2018, she worked as a Research Fellow in the School of Electrical and Electronic Engineering, Nanyang Technological University, Singapore. Since 2018, she joined Beihang University, where she is currently an associate professor. Her current research interests include networked control systems, multi-agent systems, and spacecraft formation control.

Liang Xu is a professor at Shanghai University, Shanghai, China. He received his PhD degree in Electrical and Electronic Engineering from Nanyang Technological University, Singapore, in 2018 and the MS and BSc degrees in Automation from Harbin Institute of Technology, China, in 2013 and 2011, respectively. Prior to his current position, he was a postdoctoral scholar at Nanyang Technological University, Singapore from 2018 to 2019, and a postdoctoral scholar at École polytechnique fédérale de Lausanne (EPFL), Switzerland from 2019 to 2022. His research interests include learning to control and networked control systems.

Qinglei Hu obtained his BEng degree in 2001 from the Department of Electrical and Electronic Engineering at the Zhengzhou University, P.R. China, and M. Eng. and PhD degree from the Department of Control Science and Engineering at the Harbin Institute of Technology with specialization in controls, P.R. China, in 2003 and 2006, respectively. From 2003 to 2014, he was with the Department of Control Science and Engineering at Harbin Institute of Technology and was promoted to the rank of professor with tenure in 2012. He worked as a postdoctoral research fellow in Nanyang Technological University from 2006 to 2007 and from 2008 to 2009, he visited University of Bristol as Senior Research Fellow supported by

Royal Society Fellowship, and from 2010 to 2014, he visited Concordia University, Lakehead University, and Nanyang Technological University again as visiting professor. He joined Beihang University in 2014 as "Outstanding Bairen Plan" Professor, and vice director of Institute of Science and Technology.

His research interests include the design and control of spacecraft attitude system and fault tolerant control to aerospace engineering. He has published significantly on the subjects with over 100 technical papers while enjoying the application of the theory through astronautic consulting. He has been actively involved in various technical professional societies such American Institute of Aeronautics and Astronautics (AIAA), Institute of Electrical and Electronics Engineers (IEEE) and American Society of Mechanical Engineers (ASME), as reflected by AIAA Associate Fellow, IEEE Senior Member and general chair of many international conferences. He also previously served as associate editor for Aerospace Science and Technology, etc.

Lihua Xie received the PhD degree in electrical engineering from the University of Newcastle, Australia, in 1992. Since 1992, he has been with the School of Electrical and Electronic Engineering, Nanyang Technological University, Singapore, where he is currently a professor and the director, Centre for Advanced Robotics Innovation Technology (CARTIN). He served as the head of Division of Control and Instrumentation from 2011 to 2014, codirector of Delta-NTU Corporate Laboratory for Cyber-Physical Systems from 2016 to 2021 and director, Center for E-City from July 2011 to June 2013. He is a Fellow of Institute of Electrical and Electronics Engineers (IEEE), International Federation of Automatic Control (IFAC), Chinese Automation Association (CAA), and Academy of Engineering Singapore.

Dr. Xie's research interests include robust control and estimation, networked control systems, multi-agent networks, and unmanned systems. In these areas, he has published 9 books, over 500 journal papers, and over 20 patents/technical disclosures. He was listed as a highly cited researcher by Thomson Routers and Clarivate Analytics. In addition to fundamental research contributions, he has developed a universal navigation system for AGVs in manufacturing and logistics, patented WiFi-based technologies for indoor positioning and human activity recognitions, and reliable and accurate positioning systems for UAV-based structure inspection. He has received many awards for his research.

He is an editor-in-chief for Unmanned Systems and an associate editor for Sciences China – Information Science. He has served as an editor of IET Book Series in Control and an associate editor of a number of journals including IEEE Transactions on Automatic Control, Automatica, IEEE Transactions on Control Systems Technology, IEEE Transactions on Network Control Systems, and IEEE Transactions on Circuits and Systems-II, etc. He was an IEEE Distinguished Lecturer from 2011 to 2014.

Preface

In networked control systems (NCSs), wired or wireless communication channels are used to link components among plants, sensors, and controllers to achieve control objectives. While there are many advantages, NCSs also introduce a series of challenging problems that arise from the limited resources and unreliability of the communication networks used for information transmission. For example, due to congestion, data losses and transmission delays may occur in digital communication channels. Besides, in wireless communication networks, which are widely used in sensor networks and multi-agent systems (MASs), communication channels naturally suffer from inference, fading, and transmission noises. Since control is often used in safety or mission-critical applications, we must take the uncertainties in communication networks into consideration and investigate how they affect the stability and performance of NCSs and MASs.

The book gives a systematical and self-contained description for the analysis and design of NCSs and MASs over imperfect communication networks. Specifically, the book considers fading channels and delayed channels and includes two main parts. In the first part, the stabilization, optimal control, and remote state estimation of linear systems over channels with fading, signal-to-noise constraints, or intermittent measurements are considered. The channel requirements for the mean-square stabilization and optimal control are characterized and the optimal estimator designs and performance analysis are conducted. In the second part, the joint impact of communication channels and network topology on the consensusability of MASs is analyzed. By integrating communication and control theory, we present several fundamental results on the stabilization, optimal control, and estimation of NCSs and the consensus of MASs over imperfect channels. The book intends to provide a unified platform for introducing the analysis and design of NCSs and MASs for researchers working in related areas.

January 2023

Jianying Zheng
Liang Xu
Qinglei Hu
Lihua Xie

Acknowledgments

The contents included in this book are the outgrowth and summary of the authors' academic research achievements on the analysis and design of networked control systems and multi-agent systems over imperfect communication channels in the past several years. Some of the materials contained herein arose from the joint work with our collaborators, and the book would not have been possible without their efforts and support. In particular, the first author is indebted to Prof. Li Qiu from the Hong Kong University of Science and Technology, who led her into the gate of scientific research. We would like to thank our collaborators, including Prof. Jiu-Gang Dong, Prof. Yilin Mo, Prof. Keyou You, Prof. Chao Yang, and Dr. Nan Xiao, for their helpful discussions, and the students Roudan Zhou and Xiao Wang for their assistance in preparing the manuscript.

We gratefully thank the financial support from the National Natural Science Foundation of China under Grants 61903018, 62273017, and 61960206011, the National Key Basic Research Program "Gravitational Wave Detection" Project under Grant 2021YFC2202600, the Beijing Natural Science Foundation under Grant JQ19017, and the Zhejiang Provincial Natural Science Foundation under Grant LD22E050004.

Jianying Zheng
Liang Xu
Qinglei Hu
Lihua Xie

Acronyms

a.e.	almost everywhere
ARE	algebraic Riccati equation
AWGN	additive white Gaussian noise
BMI	bilinear matrix inequality
CEL	compressed edge Laplacian
CIIM	compressed in-incidence matrix
CIM	compressed incidence matrix
DEL	directed edge Laplacian
FSMC	finite-state Markov channel(s)
i.i.d.	independent and identically distributed
IIM	in-incidence matrix
IM	incidence matrix
LMI	linear matrix inequality(-ies)
LTI	linear time-invariant
LQ	linear quadratic
LQG	linear quadratic Gaussian
MARE	modified algebraic Riccati equation(s)
MAS	multi-agent system(s)
MIMO	multi-input-multi-output
MJLS	Markov jump linear system
MSE	mean-square error
MMSE	minimum mean-square error
NCS	networked control system(s)
SNR	signal-to-noise ratio
TCP	transport control protocol
TDMA	time division multiple access

List of Symbols

$\mathbb{N}$	the set of natural numbers
$\mathbb{N}^{+}$	the set of positive natural numbers
$\mathbb{R}(\mathbb{C})$	the set of real (complex) numbers
$\mathbb{R}^{n}(\mathbb{C}^{n})$	the set of n-dimensional real (complex) column vectors
$\mathbb{R}^{m\times n}(\mathbb{C}^{m\times n})$	the set of $m \times n$-dimensional real (complex) matrices
$\mathcal{S}_n$	the set of $n \times n$ symmetric matrices
$\mathcal{P}_n$	the set of $n \times n$ positive semidefinite matrices
$\mathrm{Re}(c)$	the real part of $c \in \mathbb{C}$
$\lvert c\rvert$	the magnitude of $c \in \mathbb{C}$
$\lvert\mathcal{S}\rvert$	the cardinality of set $\mathcal{S}$
$\mathbf{1}$	a column vector of ones
$0_{m\times n}$	an $m \times n$ matrix with all entries being zero
I_N	the N-by-N identity matrix
A'	the transpose of matrix A
A^*	the conjugate transpose of matrix A
$[A]_{ij}$	the ijth element of matrix A
$[A]_{\mathrm{row}\,i}$	the ith row of matrix A
$[A]_{\mathrm{column}\,j}$	the jth column of matrix A
A^{-1}	the inverse of matrix A
$A^{\dagger}$	the Moore–Penrose pseudoinverse of matrix A
$\rho(A)$	the spectral radius of matrix A
$\mathrm{tr}(A)$	the trace of matrix A
$\det(A)$	the determinant of matrix A
$\mathrm{null}(A)$	the null space of matrix A
$\mathrm{diag}(A, B)$	a diagonal matrix with diagonal entries A and B
$\begin{bmatrix} A & * \\ C & B \end{bmatrix}$	the abbreviation of the symmetric matrix $\begin{bmatrix} A & C' \\ C & B \end{bmatrix}$
$\lambda_{\min}(S)$	the minimal eigenvalue of a real symmetric matrix S
$S > 0\ (S \geq 0)$	positive definite (semidefinite) matrix

χ^t	the sequence $\{\chi_i\}_{i=0}^t$
$\otimes$	the Kronecker product
$\odot$	the Hadamard product
$\mathbb{E}\{\cdot\}$	the expectation operator
$\mathbb{E}_x\{\cdot\}$	the expectation conditioned on the event $X = x$
$\mathcal{N}_x(\overline{x}, \Sigma)$	a random variable x with Gaussian distribution of mean $\overline{x}$ and covariance matrix Σ
$f(x)$ $(\Pr(x))$	the probability density function (probability) of the random variable x
$f(x\|y)$ $(\Pr(x\|y))$	the probability density function (probability) of the random variable x conditioned on the event that $Y = y$
e	the Euler's number
$\ln(\cdot)$	the natural logarithm
$\log(\cdot)$	the logarithm to base 2
$\log_x(\cdot)$	the logarithm to base x

1

Introduction

1.1 Introduction and Motivation

1.1.1 Networked Control Systems

Due to the flexible architecture and ease of installation and maintenance, communication networks are widely used in control systems, which result in networked control systems (NCSs), where the plants, actuators, sensors, and controllers are spatially distributed and interconnected by communication channels [Schenato et al., 2007, Hespanha et al.]. NCSs are ubiquitous in industry and daily life, such as teleoperation [Arcara and Melchiorri, 2002], power systems [Wang et al., 2012], and transportation systems [Seiler and Sengupta, 2001].

Even though NCSs have the advantages of low cost, easy implementation, and expansion to large-scale applications, they also introduce new challenging problems arising from the limited resources and unreliability of the communication networks used for information transmission (see Figure 1.1). For example, the time delay may occur in digital communication channels due to data processing and transmission [Tse and Viswanath, 2005, Goldsmith, 2005]. Notably, in wireless communication networks, communication channels naturally suffer from interference, fading, and transmission noises [Tse and Viswanath, 2005, Goldsmith, 2005]. There into, fading is the time variation of channel strengths and is usually caused by two factors: one is the shadowing from obstacles; the other one is the multipath propagation [Tse and Viswanath, 2005, Goldsmith, 2005]. Packet drops can also be modeled as a special case of channel fading. Take Figure 1.2 as an illustration. The wireless signal may transmit through the car and undergo several paths before arriving at the receiver. If the phases of the received signals from different paths are the same, the signal strength is enhanced. Otherwise, the signal strength is reduced as a result of the cancellation of radio waves. Besides, the signal strength at the receiver side might be reduced due to the shadowing from

Control over Communication Networks: Modeling, Analysis, and Design of Networked Control Systems and Multi-Agent Systems over Imperfect Communication Channels, First Edition.
Jianying Zheng, Liang Xu, Qinglei Hu, and Lihua Xie.

the car. Since control is often used in safety- or mission-critical applications, we must take the uncertainties in communication networks into consideration and investigate how they affect the stability and performance of control systems.

The classical control theory mainly deals with the systems with nearly perfect point-to-point connections and focuses on the design of control laws to achieve the given control performance. It can't be applied directly to the NCSs when the uncertainties in the communication network must be considered. A new control paradigm is required to deal with the interplay between control and communication. In this book, one of the main objectives is to study the stabilization, estimation, and optimal control of NCSs over channels with fading, packet drops, or delay.

1.1.2 Multi-Agent Systems

Motivated by the collective behavior in nature, such as schooling fish, flocking birds, and marching locusts, multi-agent systems (MASs) have attracted considerable research interest from the control community [Jadbabaie et al., 2003, Olfati-Saber and Murray, 2004, Olfati-Saber et al., 2007, Bliman and Ferrari-Trecate, 2008, Cao et al., 2008, Ren and Beard, 2008, You and Xie, 2010, Cao et al., 2012, Trentelman et al., 2013, Qi et al., 2016, Qiu et al., 2017, Xu et al., 2018, Zheng et al., 2018]. With the rapid development of wireless communication networks, MASs have been applied in many industrial and military applications. Such systems usually involve large numbers of autonomous agents (e.g. robots, unmanned aerial vehicles, satellites), which share information via local interactions and work together to achieve collective objectives.

For MASs, each agent can have the same or different system dynamics, resulting in different types of MASs, e.g. first- and second-order MASs, linear and nonlinear MASs, homogeneous and heterogeneous MASs. The interactions among the agents form the interaction topology, which can be fixed or time-varying. Then the cooperative control of MASs is based on the system dynamics and the interaction topology to design the control laws, which can be centralized or distributed, to fulfill a task. Typical cooperative control tasks include consensus, formation, swarming/flocking, rendezvous, etc. There into, the consensus problem, which requires all agents to agree on a certain quantity of common interests, builds the foundation of other cooperative tasks.

Existing research on consensus assumes that the communication networks among agents are perfect. However, as mentioned earlier, in practical applications, communication channels naturally suffer from fading, signal-to-noise ratio (SNR) constraints, time delay, etc. Hence, it is of great significance to study how

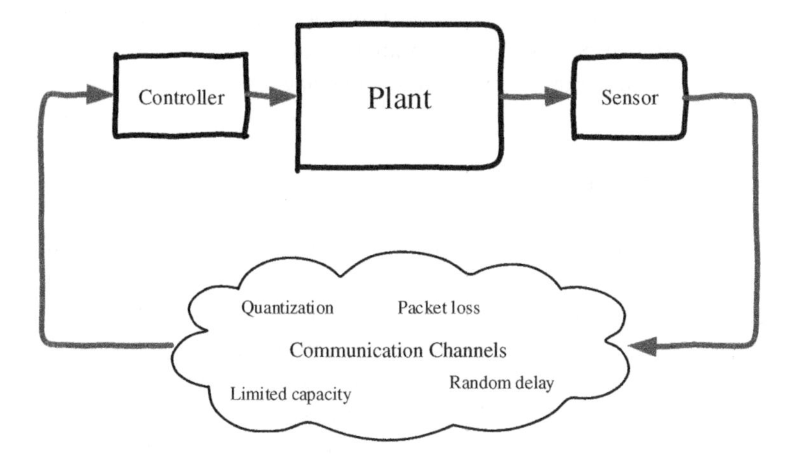

Figure 1.1 Networked control systems.

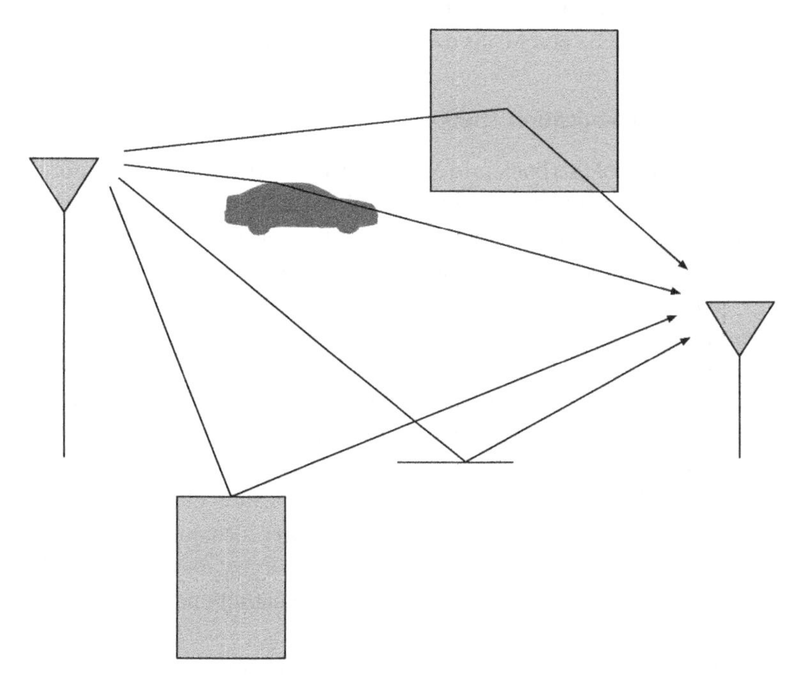

Figure 1.2 Fading phenomenon in wireless communications.

the uncertainties in communication networks influence the consensus of MASs. The other main objective of this book is to analyze the consensus problem of MASs over channels with fading, packet drops, and delay.

1.2 Literature Review

Control over communication channels/networks has been a hot research topic in the past decades [Matveev and Savkin, 2009, Como et al., 2014, You et al., 2015], motivated by the rapid developments of wireless communication technologies that enable the wide connection of geographically distributed devices and systems. However, the inclusion of wireless communication channels/networks also introduces challenges in the analysis and design of control systems due to constraints and uncertainties in wireless communications. We must take the communication channels/networks into consideration and study their impact on the stability and performance of control systems. This section briefly reviews existing results on the analysis and design of NCSs and MASs over imperfect communication channels.

1.2.1 Basics of Communication Theory

One of the main focuses of this book is to characterize the critical channel requirement such that the NCS can be mean-square stabilized. Since the communication channel is used to transmit information about the system state, as illustrated in Figure 1.1, it is expected that if the channel capacity is large enough, the feedback connected system can be mean-square stable. From this perspective, the communication channel capacity might be critical for the mean-square stabilization of control systems.

The channel capacity problem is fundamental in communication theory since it dictates the maximum data rates that can be transmitted over channels with asymptotically small error probability [Tse and Viswanath, 2005, Goldsmith, 2005]. In this subsection, we briefly review the communication channel capacity definitions and discuss why the communication theoretic channel capacity is not the critical characterization of the capacity required for controls. We only discuss discrete memoryless channels, and most of the definitions are borrowed from Cover and Thomas [2006].

A discrete memoryless channel consists of three parts: an input alphabet $\mathcal{X}$, an output alphabet $\mathcal{Y}$, and a probability transition matrix $p(y|x)$ that describes the probability of observing the output symbol y given the input symbol x. The channel is memoryless if the probability distribution of the current channel output conditioned on the current channel input is independent of previous channel inputs or outputs. The configuration of the point-to-point communication system

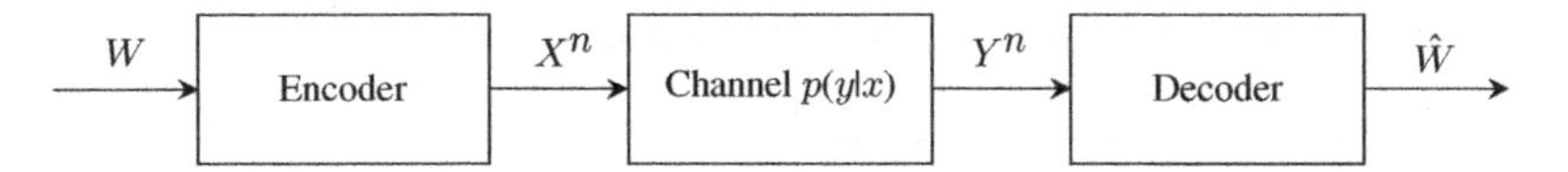

Figure 1.3 Point-to-point communication system.

is depicted in Figure 1.3. We want to transmit a message W reliably through the communication channel with appropriately designed channel encoders and decoders. The (M, n) code in a communication system is defined as follows.

Definition 1.1 ***((M, n) code)*** An (M, n) code for the channel $(\mathcal{X}, p(y|x), \mathcal{Y})$ consists of three parts:

1. A message index set $\{1, 2, \ldots, M\}$.
2. An encoding function $X^n : \{1, 2, \ldots, M\} \rightarrow \mathcal{X}^n$, generating codewords $x^n(1)$, $x^n(2)$, …, $x^n(M)$.
3. A decoding function $g : \mathcal{Y}^n \rightarrow \{1, 2, \ldots, M\}$, generating an estimate for the transmitted message index.

The performance of the code is measured by the decoding error.

Definition 1.2 ***(Decoding error)*** The maximal probability of error for an (M, n) code is defined as $\lambda^{(n)} = \max_{i \in \{1,2,\ldots,M\}} \Pr\left(g(Y^n) \neq i | X^n = x^n(i)\right)$.

The communication channel capacity which measures the maximal capacity for reliably transmitting the information is defined below.

Definition 1.3 ***(Channel capacity)*** The rate R of the (M, n) code is defined as

$$R = \frac{\log M}{n} \text{ bits per transmission.}$$

A rate R is achievable if there exists a sequence of $(\lceil 2^{nR} \rceil, n)$ codes such that $\lambda^{(n)}$ tends to 0 as $n \rightarrow \infty$. The channel capacity C is then defined as the supremum of all achievable rates.

The channel capacity in Definition 1.3 is called the Shannon channel capacity since C. E. Shannon proved in the channel coding theorem that this channel capacity equals the mutual information of the channel maximized over all possible input distributions [Shannon, 2001, Cover and Thomas, 2006]:

$$C = \max_{p(x)} \mathcal{I}(X; Y),$$

where the mutual information $\mathcal{I}(X;Y)$ is defined as

$$\mathcal{I}(X;Y) = \sum_{x\in\mathcal{X},y\in\mathcal{Y}} p(x,y)\log\frac{p(x,y)}{p(x)p(y)}.$$

The Shannon capacity of fading channels has been studied under various scenarios in Goldsmith and Varaiya [1997], Biglieri et al. [1998], Sadeghi et al. [2008], Abou-Faycal et al. [2001], and Caire et al. [1999]. For example it is proved in Goldsmith and Varaiya [1997] that if the channel state information is available at the receiver side, the Shannon channel capacity of a fading channel is

$$C = \int_0^\infty \frac{1}{2}\log\left(1+\frac{\gamma^2 P}{\sigma_\omega^2}\right)p(\gamma)\mathrm{d}\gamma,$$

where $p(\gamma)$ is the probability distribution function of the channel fading γ.

The Shannon channel capacity in Definition 1.3 assumes that the capacity-achieving code can be sufficiently long, which would inevitably result in a large delay. Since delay is critical in control systems, we may expect that the communication theoretic Shannon channel capacity is not the right choice for controls. This has been confirmed by Sahai and Mitter [2006], where another kind of channel capacity is defined, named the anytime capacity, and showed that the anytime capacity should be the critical characterization of channel capacities for controls when moment stability is concerned. However, there is no systemic method to calculate the anytime capacity. In the following, we will briefly review the existing results on the control and estimation of NCSs, and the consensus of MASs over communication networks.

1.2.2 Stabilization of NCSs

1.2.2.1 Control over Noiseless Digital Channels

For control systems, with components connected through noiseless digital communication channels, the celebrated data rate theorem [Nair and Evans, 2004] is an important result in the past decades. The data rate theorem states that to keep the state of a scalar unstable discrete-time linear system

$$x_{t+1} = \lambda x_t + u_t + w_t \tag{1.1}$$

mean-square bounded, the data rate R for the digital communication channel that connects the sensor to the controller should satisfy that

$$R > \log|\lambda|. \tag{1.2}$$

Intuitively, this result has the following explanation, see Figure 1.4. The controller wants to compensate for the expansion of uncertainties in the state estimation during the communication process. To ensure the boundedness of the system state, $\lambda^2/2^{2R}$ should be smaller than one, which gives the data rate theorem.

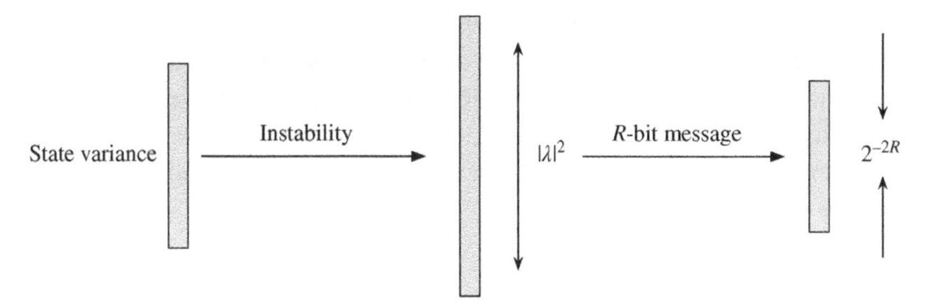

Figure 1.4 Intuitive explanations of the data rate theorem.

The result in (1.2) resembles the Shannon's source-channel coding theorem [Cover and Thomas, 2006], with the left-hand side being the Shannon channel capacity and the right-hand side the source's uncertainty measure. Indeed, the right-hand side of (1.2) denotes the information generating speed of the linear time-invariant (lti) system [Elia, 2004, Nair et al., 2004], which is generating information about the unknown initial system state. This resemblance also motivates the researchers to study the control systems from the perspective of information theory, e.g. see Touchette and Lloyd [2000], Zang and Iglesias [2003], Martins et al. [2007], Martins and Dahleh [2008], Nair [2013], Silva [2013], Ranade and Sahai [2011, 2013, 2015], Ramnarayan et al. [2014], and Ranade [2015].

1.2.2.2 Control over Stochastic Digital Channels

For noisy channels, the stability problem is more complex because different stability definitions require different channel capacities. Matveev and Savkin [2007] prove that for almost sure stability, the Shannon capacity in relation to the unstable dynamics of the system establishes the critical condition for its stabilizability. For moment stability, Sahai and Mitter [2006] show that the Shannon capacity is too optimistic, while the zero-error capacity is too pessimistic, and the anytime capacity is introduced to characterize the stabilizability conditions. Essentially, to keep the η-moment of the state of an unstable scalar plant bounded, it is necessary and sufficient for the feedback channel's anytime capacity corresponding to anytime-reliability $\alpha = \eta \log|\lambda|$ to be greater than $\log|\lambda|$, where λ is the unstable eigenvalue of the plant. The anytime capacity has a more stringent reliability requirement than the Shannon capacity. However, it is worth noting that there exists no systematic method to calculate the anytime capacities of channels. In the control community, the anytime capacity is usually studied under the mean-square stability requirement and is also named the mean-square capacity. In the following, we survey the related results that aim to determine the requirements on noisy channels to ensure that the feedback-connected linear systems can be mean-square stabilized, which, on the other hand, reveals the mean-square capacities for the channels studied.

One important kind of communication channel is the time-varying digital channel. Minero et al. [2009] assumes that the data rate R_t of the time-varying digital channel under consideration is stochastic and iid and gives the mean-square stabilizability condition for a connected discrete-time lti system. For the scalar systems to ensure the mean-square stabilizability, the following condition should be satisfied:

$$\mathbb{E}\left\{\frac{\lambda^2}{2^{2R_t}}\right\} < 1. \tag{1.3}$$

Similar to the explanation of the data rate theorem for noiseless channels, the inequality (1.3) intuitively implies that to ensure mean-square stabilizability, it is necessary and sufficient for the average expanding factor of the system state during one iteration to be smaller than one. For vector systems, necessary and sufficient conditions are provided in the form of stability regions or characterized by rate vectors [Minero et al., 2009].

For a stochastic rate-limited channel, You and Xie [2010] further show that the minimum data rate for the stabilization of a single-input vector system is explicitly given in terms of unstable eigenvalues of the open-loop matrix and the packet dropout rate, which reveals the amount of the additional bit rate required to counter the effect of packet dropouts on stabilization. Sufficient data rate conditions for mean-square stabilization of multiple-input vector systems are also derived there. When the packet drop is correlated over time, the problem becomes much more complicated. You and Xie [2011b] study mean-square stabilization of linear systems over networks with Markovian packet drops. Since the sojourn time of the time-homogeneous Markovian process that models the two-state packet drop process is iid [Xie and Xie, 2009], a randomly sampled system approach is developed in You and Xie [2011b] to derive the mean-square stabilizability condition. The same method is also adopted when deriving the data rate theorem with the additional consideration of system uncertainties in Okano and Ishii [2014]. Borrowing results from Markov jump linear systems, the mean-square stabilizability results for a more general n-state Markovian packet drop process are given in Minero et al. [2013], which contains the two-state Markovian packet drop process as a special case. The existing results in Minero et al. [2009], You and Xie [2010, 2011b], and Minero et al. [2013] are both necessary and sufficient for scalar systems. However, for vector systems, generally, there exists a gap between the derived sufficient conditions and necessary conditions. The main difficulty for deriving conditions that are both necessary and sufficient is how to optimally allocate the bits to each unstable subsystem.

1.2.2.3 Control over Analog Channels

The above results focus on digital channels. As to analog channels, Elia [2005] considers the mean-square stabilization problem over a pure multiplicative noise channel and derives the mean-square capacity of such channels. Since the iid packet drop channel is one special kind of pure multiplicative noise channels, the results obtained in Elia [2005] can be easily used to derive the results for iid packet drop channels. Xiao et al. [2012] further derive sufficient and necessary conditions for mean-square stabilization of multi-input-multi-output (MIMO) systems controlled over parallel multiplicative noise channels. Qiu et al. [2013] propose a channel/controller codesign approach with channel resource allocations to stabilize lti systems controlled with imperfect input channels when the total input channel capacity is fixed. When the subchannel capacities are fixed a priori, Chen et al. [2014] derive the stabilizability condition with a majorization approach. The joint effect of the quantization and multiplicative noise on the mean-square stabilizability is studied in Gu et al. [2015], whereas the case of both time-delay and multiplicative noise is studied in Qi et al. [2017], Su et al. [2017], and Tan et al. [2015]. Chiuso et al. [2014] consider linear quadratic Gaussian (LQG)-like control of scalar systems over communication channels suffering from data losses, delays, and SNR limitations. The stability of the closed-loop system depends on a tradeoff among the snr constraint, packet loss probability, and time delay.

Braslavsky et al. [2007] study the mean-square stabilization problem over an additive white Gaussian noise (awgn) channel and characterize the critical capacity to ensure mean-square stabilizability. To ensure the mean-square stabilization of a networked scalar system, the channel parameters should satisfy the following relation:

$$\log|\lambda| < \frac{1}{2}\log\left(1 + \frac{\mathcal{P}}{\sigma_\omega^2}\right) \tag{1.4}$$

with $\mathcal{P}/\sigma_\omega^2$ denoting the SNR of the awgn channel. They also show that for the output feedback case, the capacity required for the awgn channel is generally larger than that of the state feedback case, unless the plant is minimum phase. They further show that the extension from linear encoders/decoders to more general causal encoders/decoders cannot provide additional benefits of increasing the channel capacity [Freudenberg et al., 2010].

Specifically, the results stated above deal with multiplicative noise channels or awgn channels separately. While in wireless communications, it is practical to consider them as a whole. Xiao and Xie [2011] have derived the necessary and sufficient conditions for such kinds of channels to ensure the mean-square stabilizability under a linear encoder/decoder. It is still unknown whether we can achieve a larger stabilizability region with a more general causal encoder/decoder. We provide a positive answer to this question in Chapters 2 and 3.

1.2.3 LQ Optimal Control of NCSs over Fading Channels

As one of the most fundamental problems in control theory, linear-quadratic (LQ) optimal control has attracted great attention and has been extensively studied for deterministic and stochastic linear systems. See Dragan et al. [2010], Fragoso et al. [1998], Freiling and Hochhaus [2003], Kalman [1960], Wonham [1968], and Zhou et al. [1996] and the references therein. It aims to design an optimal state-feedback controller such that the closed-loop system is stable, and the quadratic cost function is minimized. When the channel fading is characterized by a random process, the associated NCS can be treated as a stochastic system and the LQ optimal control problem over fading channels can be tackled in virtue of some results from the stochastic case. Motivated by the control of some macroeconomic models, Katayama [1976] considers the LQ optimal control for LTI systems with only a scalar gain. In Huang et al. [2006], the LQ optimal control for a general stochastic system with both state- and control-dependent scalar noise is studied. The optimal control law was obtained under certain assumptions of mean-square stabilizability and exact observability. Much of the research treats the LQ optimal control as part of the LQG control problem. The LQG control of a MIMO system with a single packet dropping input channel and a single packet dropping output channel is considered in Imer et al. [2006] and Sinopoli et al. [2005]

It is well known that the solvability of a deterministic LQ optimal control problem depends on the existence of a stabilizing solution to the associated algebraic Riccati equation (ARE). Not surprisingly, as shown in the aforementioned literature, a class of modified algebraic Riccati equations (MAREs) play a vital role in these stochastic optimal control problems, as well as the stochastic optimal filtering problem. Specifically, the solvability of the stochastic LQ/LQG optimal control boils down to the existence problem of a mean-square stabilizing solution to the associated MARE. However, unlike AREs, which have been studied extensively (see Lancaster and Rodman [1995], Qiu [1999], and Zhou et al. [1996] and the references therein), the solutions and properties of MAREs are still under investigation. In most of the existing research, e.g. Wonham [1968], Katayama [1976], Freiling and Hochhaus [2003], Imer et al. [2006], Huang et al. [2008], and Garone et al. [2012], only sufficient conditions are obtained to ensure the existence of a mean-square stabilizing solution. In Garone et al. [2012], a sufficient condition for the existence of a stabilizing solution to a MARE, which is associated with LQG control over erasure channels with perfect acknowledgment, is given in terms of the loss probabilities as well as the classical stabilizability and detectability. A similar condition is provided by Wonham [1968]. In Zhang et al. [2008], by assuming stabilizability and exact detectability, a MARE is shown to have a stabilizing solution. The sufficient condition provided by Freiling and Hochhaus [2003] is given in terms of mean-square stabilizability and another definition of detectability, which is dual to the mean-square stabilizability. Note that the stabilizability of stochastic systems or NCSs over fading channels is usually defined in the mean-square sense, while there are several ways to define the stochastic observability and detectability from different perspectives in these papers. As seen above, detectability for stochastic systems is always assumed. Is it necessary? Does there exist a necessary and sufficient condition to ensure the existence of a mean-square stabilizing solution? In Dragan et al. [2010], one numerical necessary and sufficient condition for a discrete-time MARE is given in terms of the feasibility of some linear matrix inequalities (LMIs). However, such a condition does not have explicit interpretations with respect to the dynamical properties of the underlying stochastic system. It is of great significance to obtain an explicit necessary and sufficient condition for the existence of a mean-square stabilizing solution to the MAREs and thus solve the LQ optimal control problem for NCSs over fading channels. We will thoroughly investigate the solvability of MAREs in Chapter 4.

1.2.4 Estimation of NCSs with Intermittent Communication

In NCSs, intermittent communication is frequently involved, which may result from unreliable channels or the stochastic manner of data transmission. The intermittent communication may influence the performance of the components in a NCS; for example if the estimator receives measurements from the sensors intermittently, it may be unstable. In this subsection, we summarize existing results on remote state estimation with intermittent communication. We mainly focus on the stability issue of Kalman filtering over intermittent observations and the optimal state estimator design problem in the presence of packet drops and sensor scheduling.

1.2.4.1 Stability of Kalman Filtering with Intermittent Observations

The stability issue of remote state estimation caused by intermittent measurements has been investigated by many researchers. Particularly, the stability problem of Kalman filtering caused by intermittent measurements from one single sensor is well studied in Sinopoli et al. [2004]. They show the existence of a critical arrival rate below which the estimation error may diverge; they also provide lower and upper bounds of the critical arrival rate. This result is further developed by Mo and Sinopoli [2008, 2012]. Researchers also work on the stability issue of multisensor cases. Liu and Goldsmith [2004] consider a system with two sensors and provide a form of the lower bound of the expected estimation error covariance. Rong [2012] extends the result of the lower bound in Liu and Goldsmith [2004] to the system with multiple sensors and also proposes an explicit form of the upper bound of the critical arrival rate for the system with a single sensor, which is an improvement of Sinopoli et al. [2004]. When just one sensor is chosen at each time, the upper bound of the expected estimation error covariance is proposed in Gupta et al. [2006]. In Chapter 5, a comprehensive study of the stability of multisensor Kalman filtering with intermittent observations is studied, which provides lower and upper bounds of the expected estimation error covariance and gives conditions on the divergence and convergence of them.

1.2.4.2 Remote State Estimation with Sensor Scheduling

In a distributed sensor network, a large number of sensors are deployed on a vast terrain and take measurements of some processes of interest and then transmit the sensed information to a remote center. The sensors are often supplied by battery power. Since it is usually difficult to replace the battery when a sensor runs out of power, to make sufficient use of each sensor and prolong its lifetime, one needs to design an appropriate sensor communication schedule. Sensor scheduling algorithms can be roughly categorized as off-line schedules and event-triggered schedules. The off-line schedules are designed based on the communication frequency requirement and the statistics of the systems [Yang and Shi, 2011, Shi et al., 2011, Mo et al., 2014]. Compared with off-line schedules, event-triggered schedules depend on both the statistics and the realization of the system, which is expected to achieve better performance than off-line ones. Many triggering rules have been proposed in the literature based on the conditions that the estimation error [Xia et al., 2017], error in predicated output [Trimpe, 2014], functions of the estimation error [Wu et al., 2013, Han et al., 2015], or the error covariance [Trimpe and D'Andrea, 2014], exceed a given threshold. For example, a measurement innovation-based, event-triggered sensor-scheduling scheme is proposed to reduce the communication rate in the remote state estimation problem in Wu et al. [2013]. However, since the innovation is not Gaussian, only suboptimal estimators can be obtained. Stochastic event-triggered sensor-scheduling algorithms are further proposed in Han et al. [2015] to handle the non-Gaussian problem. Both open-loop and closed-loop schedules are proposed, and it is shown that the conditional distributions of the system state are Gaussian. As a result, closed-form minimum mean-square error (MMSE) estimators are obtained. A similar non-Gaussianity phenomenon could appear when transmit power control is used in sensor networks [Li et al., 2018]. To overcome the problem, a transmit power controller based on a specific quadratic form of measurement innovations is carefully designed in Li et al. [2018] to preserve the Gaussianity of posterior state distribution, which facilities the MMSE estimator design and performance analysis.

Wireless communications are mostly utilized in sensor networks, and packet drops are inevitable in wireless communications. Therefore, it is necessary to study how packet drops affect sensor-scheduling algorithms [Leong et al., 2017, Mo et al., 2014]. It should be noted that, for off-line schedulers and estimation error covariance-based event-triggered schedulers, there is no need to distinguish between the channel loss event and the hold of transmission event when designing estimators. As long as the estimator receives the packet, it can conduct the measurement update to improve the estimate and vice versa. However, the case is different for the stochastic event-triggered sensor-scheduling algorithms in Han et al. [2015] where the sensor measurement is used as the trigger criterion and the hold of transmission event contains information about the sensor measurement. In the presence of possible channel losses, the estimator cannot decide whether the nonreception of the packet can be attributed to the sensor measurement or the channel loss. If it is due to that the sensor measurement lies below the given threshold, then this information can be leveraged to improve the estimate. However, if it is caused by the channel loss, the estimator will have no information about the sensor measurement and no update will be carried out. This fact complicates the optimal estimator design. Furthermore, it is proved in Kung et al. [2017] that, in the presence of channel losses, the Gaussian properties with the stochastic event-triggered sensor-scheduling algorithms in Han et al. [2015] no longer hold. Chapter 6 is devoted to providing solutions to the remote state estimation problem of LTI systems with stochastic event-triggered sensor schedules in the presence of packet drops.

1.2.5 Distributed Consensus of MASs

In many applications, single-agent systems are incapable of dealing with complex tasks. Cooperation among mass becomes necessary, and thus they have attracted great attention from various research fields, ranging from biology [Zhu et al., 2017], to robotics [Hu et al., 2013], control theory to applied mathematics, among many others. Among various cooperative tasks, consensus, which requires all agents to reach an agreement on a certain quantity of common interest, builds the foundation of others [Olfati-Saber et al., 2007, Ren and Beard, 2008, Cao et al., 2012]. One question arises before control synthesis: whether there exist distributed controllers such that the mas can achieve consensus. This problem is usually referred to as the consensusability of mass. Several important results have been derived to answer this question, under an undirected/directed communication topology [Ma and Zhang, 2010, You and Xie, 2011a, Li et al., 2010, Trentelman et al., 2013]. In Ma and Zhang [2010], it is shown that to ensure the consensus of a continuous-time linear mas, the lti dynamics should be stabilizable and detectable, and the undirected communication topology should be connected. Furthermore, You and Xie [2011a] and Gu et al. [2012] show that for a discrete-time linear mas, the product of the unstable eigenvalues of the system matrix should additionally be upper bounded by a function of the eigenratio of the undirected graph. Extensions to directed graphs and robust consensus can be found in Li et al. [2010] and Trentelman et al. [2013]. Most of the consensusability results discussed above are derived assuming perfect communications. However, this is not the case in practical applications, where communication channels naturally suffer from limited data rate constraints, SNR constraints, time delay, and so on. Therefore, it is necessary to study the consensusability problem of mass under communication channel constraints.

Li et al. [2011] consider the average consensus problem for discrete-time first-order mass over rate-limited channels with undirected graphs. A distributed consensus protocol based on dynamic encoding and decoding is proposed, and the average consensus can be achieved with only one-bit information exchange between each pair of adjacent agents at every time step. The extensions to the case with bounded time-delay and time-varying graphs for first-order mass can be found in Liu et al. [2011] and Li and Xie [2011], respectively. Li and Xie [2012] consider the distributed coordination problem of second-order MASs with partially measurable states under rate-limited communication channels. A quantized-observer-based encoding-decoding scheme and a distributed coordinated control law are proposed. The two bits quantization is sufficient for the asymptotic synchronization of agent states. Determining the critical data rate for distributed consensus of general nth order mass can be challenging. Only limited results exist for special kinds of nth order systems; see Qiu et al. [2016, 2017].

Consensusability of MASs over communication channels with time delays has been studied in Olfati-Saber and Murray [2004], Bliman and Ferrari-Trecate [2008], Yu et al. [2010], Zhou and Lin [2014], and Qi et al. [2013, 2016, 2014]. Specifically, a necessary and sufficient condition for consensusability under an undirected communication graph with constant time delays is given in Olfati-Saber and Murray [2004], while sufficient conditions for the existence of average consensus under an undirected communication graph with bounded communication delays are provided in Bliman and Ferrari-Trecate [2008]. A truncated predictor feedback approach is established in Zhou and Lin [2014] to solve the consensus problem with time delays in both the communication network and control inputs. Utilizing techniques from robust control, Qi et al. [2013, 2014, 2016] characterize the maximal tolerable time-delay for the existence of a linear distributed consensus controller for discrete-time linear mass over undirected graphs and directed graphs, respectively. The results show that the consensusability is related to the time delay, unstable poles, and nonminimum phase zeros of the system dynamics.

There are also results on the consensusability of MASs over communication channels corrupted by packet dropouts. In Hatano and Mesbahi [2005], consensusability with probability 1 is addressed under the assumptions that each link has an equal packet dropout rate and the mean of the random graph is complete. Moreover, sufficient conditions based on the second smallest eigenvalue of the mean Laplacian matrix are provided in Kar and Moura [2009] for mean-square consensusability in undirected networks with nonuniform packet dropout rates.

In Tahbaz-Salehi and Jadbabaie [2008], a necessary and sufficient condition for almost sure asymptotic consensus over directed networks with stochastic i.i.d. weight matrices is obtained.

In this book, we are interested in the consensusability problems of discrete-time linear mass over fading networks and delay. The framework considered in Xiao et al. [2014] deals with identical fading networks with undirected communication topologies only. It is still unknown how the directed communication topology and nonidentical fading networks affect the consensusability of mass, and this problem will be analyzed in Chapters 7, 8 and 10. For the consensus of discrete-time MASs with delay and packet dropouts, the approaches given in Zong et al. [2016] depend heavily on the order and structure of the system matrices and cannot be extended to general MASs. We will discuss the consensus of general discrete-time MASs with constant communication delay and packet dropouts in Chapter 9. The synchronization problem of a special class of MASs, named Vicsek model, with delay is also studied in Chapter 11.

1.3 Preview of the Book

The rest of this book is organized as follows.

In Chapters 2–4, we study the stabilization and optimal control problem for NCSs.

In Chapter 2, we consider the mean-square stabilization problem of discrete-time LTI systems over a power-constrained fading channel. Fundamental limitations on the mean-square stabilizability are obtained via information-theoretic arguments. For scalar and two-dimensional systems, necessary and sufficient conditions for the mean-square stabilizability are provided, respectively. Moreover, an adaptive time division multiple access (TDMA) communication scheme is designed for high-dimensional systems, which achieves a larger stabilizability region than the conventional TDMA communication scheme and is proved to be optimal under certain situations.

Chapter 3 studies the mean-square stabilization problem of discrete-time LTI systems over Gaussian finite-state Markov channels, which suffer from both SNR constraint and correlated channel fading modeled by a Markov process. The existence of a fundamental limitation for mean-square stabilization is first established. Then, sufficient stabilization conditions under a TDMA communication scheme are derived in terms of the stability of a Markov jump linear system. Moreover, we present a necessary and sufficient condition for mean-square stabilization of two-dimensional systems controlled over power-constrained Markov lossy channels. Furthermore, improved sufficient stabilization conditions are derived based on an adaptive TDMA communication scheme for general high-dimensional systems, which achieve a larger stabilization region than the TDMA communication scheme.

In Chapter 4, LQ optimal control problem for a discrete-time LTI system with random input gains is studied. The finite-horizon case can be solved by dynamic programming, while the solvability of the infinite-horizon case is equivalent to the existence of a mean-square stabilizing solution to the associated MARE. By virtue of the theory of cone-invariant operators, some properties of the associated MARE are obtained and an explicit necessary and sufficient condition ensuring the existence of the mean-square stabilizing solution to the MARE is derived. Such a condition is compatible with the one ensuring the stabilizing solution to the standard ARE, and it indicates that the common condition of observability or detectability of certain stochastic systems is unnecessary. With this result, the LQ optimal control problem of NCSs with random input gains can be well solved under the framework of channel/controller codesign.

In Chapters 5 and 6, we study the remote state estimation problem with intermittent communication.

In Chapter 5, we extend the stability theory on Kalman filtering with intermittent measurements from the scenario of one single sensor to one of multiple sensors. Based on the measurements received intermittently from a group of sensors, the estimator computes the estimates of the process states by multisensor Kalman filtering. Because of the intermittent measurements, the estimator may be unstable. A notion of transmission capacity, which is related to the communication rates of sensors, is proposed. It is shown that the expected estimation error covariance diverges for all feasible communication rates collections of the sensors when the transmission capacity is below a certain value; meanwhile, when the transmission capacity is above another certain value, there exists a feasible communication rates collection such that the expected estimation error covariance is bounded.

In Chapter 6, we study the remote state estimation problem of linear time-invariant systems with stochastic event-triggered sensor schedules in the presence of packet drops between the sensor and the estimator. Due to the existence of packet drops, the Gaussianity at the estimator side no longer holds. It is proved that the system state conditioned on the available information at the estimator side is Gaussian mixture distributed. The MMSE estimator can be obtained from a bank of Kalman filters. Since the optimal estimators require exponentially increasing computation and memory with time, suboptimal estimators to reduce computational complexities by limiting the length and number of hypotheses are further provided. In the end, simulations are conducted to illustrate the performance of the optimal and suboptimal estimators.

Starting with Chapter 7, we turn the attention from NCSs to MASs.

In Chapter 7, we investigate the consensus problem of MASs over undirected fading networks. That is, each agent can only receive corrupted information about its neighborhoods' states, which make it difficult to reach consensus. How the channel fading affects the consensus of MASs over undirected networks is the essential problem to address in this chapter. For consensus over identical fading networks, a decomposition method is used and the mean-square consensus problem is transformed to a simultaneous mean-square stabilization problem. For consensus over nonidentical fading networks, the edge Laplacian defined for undirected graphs is introduced to model the consensus error dynamics. Then sufficient consensus conditions are derived, which demonstrate how the system dynamics, the communication quality, and the network topological structure interplay with each other to allow the existence of a linear distributed consensus controller.

In Chapter 8, the mean-square consensus problem of discrete-time linear MASs over analog fading networks with directed graphs is further studied. However, since the graph Laplacian for directed graphs may contain complex eigenvalues, and there is no appropriate definition of edge Laplacian for directed graphs, the method used in Chapter 7 for the consensus of MASs over undirected fading networks cannot be applied to directed graph cases, which complicate the consensusability analysis due to the coupling between the channel fading and the network topology. In this chapter, we introduce the definitions of compressed in-incidence matrix (CIIM), compressed incidence matrix (CIM), and compressed edge Laplacian (CEL) to facilitate the modeling and consensus analysis. It is then shown that the mean-square consensusability is solely determined by the edge state dynamics on a directed spanning tree. As a result, sufficient conditions are provided for mean-square consensus over fading networks with directed graphs in terms of fading parameters, the network topology, and the agent dynamics.

In Chapter 9, a more practical and complex scenario is taken into account, i.e. the multi-agents communicate through the networks with both delay and packet dropouts. The approaches given in Chapter 8 cannot be directly applied to the case when both communication delays and packet dropouts exist in the communication channels. By proposing some novel mean-square stability criteria of NCSs with delay and multiplicative noise and employing the notion of CEL from Chapter 8 to build the dynamics over edges to separate the packet dropouts from the network topology, a sufficient consensusability condition in terms of the communication delay, the packet dropout rates, the communication network topology, and the agent dynamics is provided. Closed-form consensusability conditions can be obtained under some specific configurations.

In Chapters 7, 8, and 9, the fading or packet dropouts are assumed to be i.i.d., which cannot capture the temporal correlation of channel fadings. In Chapter 10, it is assumed that the multi-agents communicate with each other through undirected Markovian packet loss channels. For the case with identical Markovian packet losses, a necessary and sufficient consensus condition is derived based on the stability of Markov jump linear systems, together with a numerically verifiable consensus criterion in terms of the feasibility of LMIs. For the case with nonidentical packet loss, the edge Laplacian is used to model the consensus error on edges rather than on vertexes and a sufficient consensus condition in terms of LMIs is given.

In Chapter 11, the synchronization problem of a special class of MASs, named Vicsek model, is studied. Simulation results show that all agents, using only the local rule, might eventually move in the same direction, exhibiting the synchronization behavior. A theoretical analysis of the synchronization of both linear and nonlinear Vicsek models with bounded time-varying delays is given in this chapter. Some delay-dependent synchronization conditions, imposed only on the initial state, are derived. Compared to the synchronization conditions based on some connectivity conditions on the dynamically changing neighbor graphs in the literature, the derived results can be easily checked.

1.4 Preliminaries

1.4.1 Graph Theory

In this subsection, we introduce the basis of graph theory used to model the interactions among agents in MASs. For detailed reference to graph theory, please refer to Godsil and Royle [2001] and Mesbahi and Egerstedt [2010].

A directed graph $\mathcal{G} = (\mathcal{V}, \mathcal{E})$ is used to characterize the interaction among agents, where $\mathcal{V} = \{1, 2, \ldots, N\}$ is the node set representing N agents and $\mathcal{E} \subseteq \mathcal{V} \times \mathcal{V}$ is the edge set with ordered pairs of nodes denoting the information transmission among agents. An edge $(i, j) \in \mathcal{E}$ means that the ith agent can send information to the jth agent, where node i and node j are called the initial node and terminal node of this edge, respectively. Self-edge (i, i) is not allowed. The neighborhood set $\mathcal{N}_i$ of agent i is defined as $\mathcal{N}_i = \{j \in \mathcal{V} \mid (j, i) \in \mathcal{E}\}$. A sequence of edges $\{(i_{j-1}, i_j)\}_{j=1}^{l}$ with $(i_{j-1}, i_j) \in \mathcal{E}$ and $\{i_j\}_{j=0}^{l}$ being distinct is called a directed path from agent i_0 to agent i_l. A directed cycle is a directed path starting and ending at the same node. A graph contains a directed spanning tree if it has at least one node with directed paths to all other nodes. A graph is called strongly connected if, for any two agents i and j, there is a path in each direction between them. A graph is complete if there exists an edge for each pair of nodes. The underlying graph of $\mathcal{G}$ is the graph obtained by treating the edges of $\mathcal{G}$ as unordered pairs.

The adjacency matrix A_{adj} is defined as $[A_{\text{adj}}]_{ii} = 0$, $[A_{\text{adj}}]_{ij} = 1$ if $(j, i) \in \mathcal{E}$ and $[A_{\text{adj}}]_{ij} = 0$, otherwise. When A_{adj} is symmetric, $\mathcal{G}$ is called an undirected graph.

An undirected graph is connected if there is a path between every pair of distinct nodes. The in-degree and out-degree of agent i are given by $\deg_i = \sum_{j=1}^{N} [A_{\text{adj}}]_{ij}$ and $\deg_i^{\text{out}} = \sum_{j=1}^{N} [A_{\text{adj}}]_{ji}$, respectively. A graph $\mathcal{G}$ is called balanced if and only if $\deg_i = \deg_i^{\text{out}}$ for any $i = 1, \dots, N$. Denote the Laplacian matrix of $\mathcal{G}$ by $\mathcal{L} = \mathcal{D} - \mathcal{A}$, where $\mathcal{D} = \text{diag}\{\deg_1, \dots, \deg_N\}$. The eigenvalues of $\mathcal{L}$ are denoted by $\lambda_i \in \mathbb{C}, i = 1, \dots, N$ with an ascending order in magnitude, i.e. $0 = |\lambda_1| \le |\lambda_2| \le \cdots \le |\lambda_N|$. Then the graph Laplacian $\mathcal{L}$ has the following property.

Lemma 1.1 ***(Ren and Beard [2008])***

1. *For a directed graph $\mathcal{G}$, all the eigenvalues of $\mathcal{L}$ have nonnegative real parts. Then $\mathcal{G}$ contains a directed spanning tree if and only if zero is a simple eigenvalue of $\mathcal{L}$ with a right eigenvector* **1**, *equivalently to say, $\lambda_2 \neq 0$.*
2. *For an undirected graph $\mathcal{G}$, $\lambda_i \ge 0$ for all $i \in \mathcal{N}$. Then $\mathcal{G}$ is connected if and only if $\lambda_2 > 0$. Moreover, $\mathcal{G}$ is complete if and only if $\lambda_2 = \lambda_N$.*

1.4.2 Hadamard Product and Kronecker Product

The Hadamard product (also known as Schur product) refers to the component-wise multiplication of matrices with the same dimension. That is, given matrices $A, B \in \mathbb{C}^{m\times n}$, the Hadamard product, denoted by $A \odot B \in \mathbb{C}^{m\times n}$, is defined as

$$[A \odot B]_{ij} = [A]_{ij}[B]_{ij}.$$

For example, the Hadamard product of a 2×2 matrix A and a 2×2 matrix B is

$$\begin{bmatrix} a_{11} & a_{12} \\ a_{21} & a_{22} \end{bmatrix} \odot \begin{bmatrix} b_{11} & b_{12} \\ b_{21} & b_{22} \end{bmatrix} = \begin{bmatrix} a_{11}b_{11} & a_{12}b_{12} \\ a_{21}b_{21} & a_{22}b_{22} \end{bmatrix}.$$

The Hadamard product is commutative, associative, and distributive over addition. That is, given matrices A, B, and C with the same size, and a scalar k, it holds that

$$A \odot B = B \odot A,$$

$$(kA) \odot B = A \circ (kB) = k\,(A \odot B),$$

$$A \odot (B \odot C) = (A \odot B) \odot C,$$

$$A \odot (B + C) = A \odot B + A \odot C.$$

If D and E are diagonal matrices, then

$$D(A \odot B)E = (DAE) \odot B = (DA) \odot (BE) = (AE) \odot (DB) = A \odot (DBE).$$

The Kronecker product is an operation that maps two matrices of arbitrary size into a larger block matrix. That is, given matrices $A \in \mathbb{C}^{m\times n}$ and $B \in \mathbb{C}^{p\times q}$

$$A = \begin{bmatrix} a_{11} & \cdots & a_{1n} \\ \vdots & \ddots & \vdots \\ a_{m1} & \cdots & a_{mn} \end{bmatrix}, \quad B = \begin{bmatrix} b_{11} & \cdots & b_{1q} \\ \vdots & \ddots & \vdots \\ b_{p1} & \cdots & a_{pq} \end{bmatrix},$$

the Kronecker product, denoted by $A \otimes B \in \mathbb{C}^{mp\times nq}$, is defined as

$$A \otimes B = \begin{bmatrix} a_{11}B & \cdots & a_{1n}B \\ \vdots & \ddots & \vdots \\ a_{m1}B & \cdots & a_{mn}B \end{bmatrix}.$$

The Kronecker product is associative and distributive over addition. That is, given matrices A, B, and C, it holds that

$$A \otimes (B + C) = A \otimes B + A \otimes C,$$
$$(A \otimes B) \otimes C = A \otimes (B \otimes C).$$

Unlike the matrix product or Hadamard product, it is in general noncommutative. That is for arbitrary matrices A and B, usually $A \otimes B \neq B \otimes A$.

When $A \in \mathbb{C}^{n\times n}$ and $B \in \mathbb{C}^{m\times m}$, let $\lambda_1, \dots, \lambda_n$ be the eigenvalues of A and $\mu_1, \dots, \mu_m$ be those of B. Then the eigenvalues of $A \otimes B$ are $\{\lambda_i \mu_j, i = 1, \dots, n, j = 1, \dots, m\}$. It also holds that $A \otimes B$ is invertible if and only if both A and B are invertible, i.e. $(A \otimes B)^{-1} = A^{-1} \otimes B^{-1}$.

If A and C are matrices of the same size, B and D are matrices of the same size, then the following mixed-product property holds

$$(A \otimes B) \odot (C \otimes D) = (A \odot C) \otimes (B \odot D).$$

The Kronecker product is also closely related to the vectorization operator (denoted by vec(·)), which stacks the column vectors of a matrix into one column vector. To be specific,

$$\left(B' \otimes A\right) \ \text{vec}(X) = \text{vec}(AXB) = \text{vec}(C).$$

Thus, by using the Kronecker product, we can see that the equation $AXB = C$, where A, B, and C are given matrices, has a unique solution X, if and only if A and B are invertible.

For a detailed reference to Hadamard product and Kronecker product, please refer to Horn and Johnson [1985, 1991].

Bibliography

I. C. Abou-Faycal, M. D. Trott, and S. Shamai. The capacity of discrete-time memoryless Rayleigh-fading channels. *IEEE Transactions on Information Theory*, 47(4):1290–1301, 2001.

P. Arcara and C. Melchiorri. Control schemes for teleoperation with time delay: A comparative study. *Robotics and Autonomous Systems*, 38(1):49–64, 2002.

E. Biglieri, J. Proakis, and S. Shamai. Fading channels: Information-theoretic and communications aspects. *IEEE Transactions on Information Theory*, 44(6):2619–2692, 1998.

P. A. Bliman and G. Ferrari-Trecate. Average consensus problems in networks of agents with delayed communications. *Automatica*, 44(8):1985–1995, 2008.

J. H. Braslavsky, R. H. Middleton, and J. S. Freudenberg. Feedback stabilization over signal-to-noise ratio constrained channels. *IEEE Transactions on Automatic Control*, 52(8):1391–1403, 2007.

G. Caire, G. Taricco, and E. Biglieri. Optimum power control over fading channels. *IEEE Transactions on Information Theory*, 45(5):1468–1489, 1999.

M. Cao, A. S. Morse, and B. D. O. Anderson. Reaching a consensus in a dynamically changing environment: A graphical approach. *SIAM Journal on Control and Optimization*, 47(2):575–600, 2008.

Y. Cao, W. Yu, W. Ren, and G. Chen. An overview of recent progress in the study of distributed multi-agent coordination. *IEEE Transactions on Industrial informatics*, 9(1):427–438, 2012.

W. Chen, S. Wang, and L. Qiu. When MIMO control meets MIMO communication: A majorization condition for networked stabilizability. *ArXiv e-prints*, 1408:3500, 2014.

A. Chiuso, N. Laurenti, L. Schenato, and A. Zanella. LQG-like control of scalar systems over communication channels: The role of data losses, delays and SNR limitations. *Automatica*, 50(12):3155–3163, 2014.

G. Como, B. Bernhardsson, and A. Rantzer. *Information and control in networks*. Springer International Publishing, Cham, 2014.

T. M. Cover and J. A. Thomas. *Elements of information theory*. Wiley-Interscience, Hoboken, NJ, 2006.

V. Dragan, T. Morozan, and A. M. Stoica. *Mathematical methods in robust control of discrete-time linear stochastic systems*. Springer, New York, 2010.

N. Elia. When Bode meets Shannon: Control-oriented feedback communication schemes. *IEEE Transactions on Automatic Control*, 49(9):1477–1488, 2004.

N. Elia. Remote stabilization over fading channels. *Systems & Control Letters*, 54(3):237–249, 2005.

M. D. Fragoso, O. L. V. Costa, and C. E. De Souza. A new approach to linearly perturbed Riccati equations arising in stochastic control. *Applied Mathematics and Optimization*, 37(1):99–126, 1998.

G. Freiling and A. Hochhaus. Properties of the solutions of rational matrix difference equations. *Computers & Mathematics with Applications*, 45(6):1137–1154, 2003.

J. S. Freudenberg, R. H. Middleton, and V. Solo. Stabilization and disturbance attenuation over a Gaussian communication channel. *IEEE Transactions on Automatic Control*, 55(3):795–799, 2010.

E. Garone, B. Sinopoli, A. Goldsmith, and A. Casavola. LQG control for MIMO systems over multiple erasure channels with perfect acknowledgment. *IEEE Transactions on Automatic Control*, 57(2):450–456, 2012.

C. Godsil and G. Royle. *Algebraic graph theory*. Springer, New York, 2001.

A. Goldsmith. *Wireless communications*. Cambridge University Press, Cambridge, 2005.

A. J. Goldsmith and P. P. Varaiya. Capacity of fading channels with channel side information. *IEEE Transactions on Information Theory*, 43(6):1986–1992, 1997.

G. Gu, L. Marinovici, and F. L. Lewis. Consensusability of discrete-time dynamic multiagent systems. *IEEE Transactions on Automatic Control*, 57(8):2085–2089, 2012.

G. Gu, S. Wan, and L. Qiu. Networked stabilization for multi-input systems over quantized fading channels. *Automatica*, 61:1–8, 2015.

V. Gupta, T. H. Chung, B. Hassibi, and R. M. Murray. On a stochastic sensor selection algorithm with applications in sensor scheduling and sensor coverage. *Automatica*, 42(2):251–260, 2006.

D. Han, Y. Mo, J. Wu, S. Weerakkody, B. Sinopoli, and L. Shi. Stochastic event-triggered sensor schedule for remote state estimation. *IEEE Transactions on Automatic Control*, 60(10):2661–2675, 2015.

Y. Hatano and M. Mesbahi. Agreement over random networks. *IEEE Transactions on Automatic Control*, 50(11):1867–1872, 2005.

J. P. Hespanha, P. Naghshtabrizi, and X. Yonggang. A survey of recent results in networked control systems. *Proceedings of the IEEE*, 95(1):138–162.

R. A. Horn and C. R. Johnson. *Matrix analysis*. Cambridge University Press, 1985.

R. A. Horn and C. R. Johnson. *Topics in matrix analysis*. Cambridge University Press, New York, 1991.

J. Hu, J. Xu, and L. Xie. Cooperative search and exploration in robotic networks. *Unmanned Systems*, 1(01):121–142, 2013.

Y. Huang, W. Zhang, and H. Zhang. Infinite horizon LQ optimal control for discrete-time stochastic systems. In *Proceedings of the 6th World Congress on Control and Automation*, pages 252–256, 2006.

Y. Huang, W. Zhang, and H. Zhang. Infinite horizon linear quadratic optimal control for discrete-time stochastic systems. *Asian Journal of Control*, 10(5):608–615, 2008.

O. C. Imer, S. Yüksel, and T. Başar. Optimal control of LTI systems over unreliable communication links. *Automatica*, 42(9):1429–1439, 2006.

A. Jadbabaie, J. Lin, and A. S. Morse. Coordination of groups of mobile autonomous agents using nearest neighbor rules. *IEEE Transactions on automatic control*, 48(6):988–1001, 2003.

R. E. Kalman. Contributions to the theory of optimal control. *Bulletin de la Societe Mathematique de Mexicana*, 5(2):102–119, 1960.

S. Kar and J. M. F. Moura. Distributed consensus algorithms in sensor networks with imperfect communication: Link failures and channel noise. *IEEE Transactions on Signal Processing*, 57(1):355–369, 2009.

T. Katayama. On the matrix Riccati equation for linear systems with random gain. *IEEE Transactions on Automatic Control*, 21(5):770–771, 1976.

E. Kung, J. Wu, D. Shi, and L. Shi. On the nonexistence of event-based triggers that preserve Gaussian state in presence of package-drop. In *Proceedings of the 2017 American Control Conference*, pages 1233–1237, 2017.

P. Lancaster and L. Rodman. *Algebraic Riccati equations*. Oxford University Press, Oxford, 1995.

A. S. Leong, S. Dey, and D. E. Quevedo. Sensor scheduling in variance based event triggered estimation with packet drops. *IEEE Transactions on Automatic Control*, 62(4):1880–1895, 2017.

T. Li and L. Xie. Distributed consensus over digital networks with limited bandwidth and time-varying topologies. *Automatica*, 47(9):2006–2015, 2011.

T. Li and L. Xie. Distributed coordination of multi-agent systems with quantized-observer based encoding-decoding. *IEEE Transactions on Automatic Control*, 57(12):3023–3037, 2012.

T. Li, M. Fu, L. Xie, and J. F. Zhang. Distributed consensus with limited communication data rate. *IEEE Transactions on Automatic Control*, 56(2):279–292, 2011.

Y. Li, J. Wu, and T. Chen. Transmit power control and remote state estimation with sensor networks: A Bayesian inference approach. *Automatica*, 97:292–300, 2018.

Z. Li, Z. Duan, G. Chen, and L. Huang. Consensus of multiagent systems and synchronization of complex networks: A unified viewpoint. *IEEE Transactions on Circuits and Systems I: Regular Papers*, 57(1):213–224, 2010.

X. Liu and A. Goldsmith. Kalman filtering with partial observation losses. In *Proceedings of the 43rd IEEE Conference on Decision and Control*, volume 4, pages 4180–4186. IEEE, 2004.

S. Liu, T. Li, and L. Xie. Distributed consensus for multiagent systems with communication delays and limited data rate. *SIAM Journal on Control and Optimization*, 49(6):2239–2262, 2011.

C. Q. Ma and J. F. Zhang. Necessary and sufficient conditions for consensusability of linear multi-agent systems. *IEEE Transactions on Automatic Control*, 55(5):1263–1268, 2010.

N. C. Martins and M. A. Dahleh. Feedback control in the presence of noisy channels: Bode-like fundamental limitations of performance. *IEEE Transactions on Automatic Control*, 53(7):1604–1615, 2008.

N. C. Martins, M. A. Dahleh, and J. C. Doyle. Fundamental limitations of disturbance attenuation in the presence of side information. *IEEE Transactions on Automatic Control*, 52(1):56–66, 2007.

A. Matveev and A. Savkin. An analogue of Shannon information theory for detection and stabilization via noisy discrete communication channels. *SIAM Journal on Control and Optimization*, 46(4):1323–1367, 2007.

A. S. Matveev and A. V. Savkin. *Estimation and control over communication networks.* Birkhäuser, Boston, MA, 2009.

M. Mesbahi and M. Egerstedt. *Graph theoretic methods in multiagent networks.* Princeton University Press, Princeton, NJ, 2010.

P. Minero, M. Franceschetti, S. Dey, and G. N. Nair. Data rate theorem for stabilization over time-varying feedback channels. *IEEE Transactions on Automatic Control*, 54(2):243–255, 2009.

P. Minero, L. Coviello, and M. Franceschetti. Stabilization over Markov feedback channels: The general case. *IEEE Transactions on Automatic Control*, 58(2):349–362, 2013.

Y. Mo and B. Sinopoli. A characterization of the critical value for Kalman filtering with intermittent observations. In *Proceedings of the 47th IEEE Conference on Decision and Control*, pages 2692–2697, 2008.

Y. Mo and B. Sinopoli. Kalman filtering with intermittent observations: Tail distribution and critical value. *IEEE Transactions on Automatic Control*, 57(3):677–689, 2012.

Y. Mo, E. Garone, and B. Sinopoli. On infinite-horizon sensor scheduling. *Systems & control letters*, 67:65–70, 2014.

G. N. Nair. A nonstochastic information theory for communication and state estimation. *IEEE Transactions on Automatic Control*, 58(6):1497–1510, 2013.

G. N. Nair and R. J. Evans. Stabilizability of stochastic linear systems with finite feedback data rates. *SIAM Journal on Control and Optimization*, 43(2):413–436, 2004.

G. N. Nair, R. J. Evans, I. M. Y. Mareels, and W. Moran. Topological feedback entropy and nonlinear stabilization. *IEEE Transactions on Automatic Control*, 49(9):1585–1597, 2004.

K. Okano and H. Ishii. Stabilization of uncertain systems with finite data rates and Markovian packet losses. *IEEE Transactions on Control of Network Systems*, 1(4):298–307, 2014.

R. Olfati-Saber and R. M. Murray. Consensus problems in networks of agents with switching topology and time-delays. *IEEE Transactions on Automatic Control*, 49(9):1520–1533, 2004.

R. Olfati-Saber, J. A. Fax, and R. M. Murray. Consensus and cooperation in networked multi-agent systems. *Proceedings of the IEEE*, 95(1):215–233, 2007.

T. Qi, L. Qiu, and J. Chen. Multi-agent consensus under delayed feedback: Fundamental constraint on graph and fundamental bound on delay. In *Proceedings of the 2013 American Control Conference*, pages 952–957, 2013.

T. Qi, L. Qiu, and J. Chen. Topological constraints on consensus via delayed output feedback over directed graph. In *Proceedings of the 33rd Chinese Control Conference*, pages 1360–1365, 2014.

T. Qi, L. Qiu, and J. Chen. MAS consensus and delay limits under delayed output feedback. *IEEE Transactions on Automatic Control*, 62(9):4660–4666, 2016.

T. Qi, J. Chen, W. Su, and M. Fu. Control under stochastic multiplicative uncertainties: Part I, fundamental conditions of stabilizability. *IEEE Transactions on Automatic Control*, 62(3):1269–1284, 2017.

L. Qiu. On the generalized eigenspace approach for solving Riccati equations. In *Reprints of the 14th IFAC World Congress*, volume D, pages 281–286, 1999.

L. Qiu, G. Gu, and W. Chen. Stabilization of networked multi-input systems with channel resource allocation. *IEEE Transactions on Automatic Control*, 58(3):554–568, 2013.

Z. Qiu, L. Xie, and Y. Hong. Quantized leaderless and leader-following consensus of high-order multi-agent systems with limited data rate. *IEEE Transactions on Automatic Control*, 61(9):2432–2447, 2016.

Z. Qiu, L. Xie, and Y. Hong. Data rate for distributed consensus of multi-agent systems with high-order oscillator dynamics. *IEEE Transactions on Automatic Control*, 62(11):6065–6072, 2017.

G. Ramnarayan, G. Ranade, and A. Sahai. Side-information in control and estimation. In *Proceedings of the 2014 IEEE International Symposium on Information Theory*, pages 171–175, 2014.

G. Ranade. Active systems with uncertain parameters: An information-theoretic perspective. PhD Thesis, EECS Department, University of California, Berkeley, 2014.

G. Ranade and A. Sahai. Implicit communication in multiple-access settings. In *Proceedings of the 2011 IEEE International Symposium on Information Theory*, pages 998–1002, 2011.

G. Ranade and A. Sahai. Non-coherence in estimation and control. In *Proceedings of the 51st Annual Allerton Conference on Communication, Control, and Computing*, pages 189–196, 2013.

G. Ranade and A. Sahai. Control capacity. In *Proceedings of the 2015 IEEE International Symposium on Information Theory*, pages 2221–2225, 2015.

W. Ren and R. W. Beard. *Distributed consensus in multi-vehicle cooperative control: theory and applications*. Springer-Verlag London Limited, London, 2008.

B. Rong. State estimation over packet-dropping channels. Master's thesis, Hong Kong University of Science and Technology, 2012.

P. Sadeghi, R. A. Kennedy, P. B. Rapajic, and R. Shams. Finite-state Markov modeling of fading channels - a survey of principles and applications. *IEEE Signal Processing Magazine*, 25(5):57–80, 2008.

A. Sahai and S. Mitter. The necessity and sufficiency of anytime capacity for stabilization of a linear system over a noisy communication link - Part I: Scalar systems. *IEEE Transactions on Information Theory*, 52(8):3369–3395, 2006.

L. Schenato, B. Sinopoli, M. Franceschetti, K. Poolla, and S. S. Sastry. Foundations of control and estimation over lossy networks. *Proceedings of the IEEE*, 95(1):163–187, 2007.

P. Seiler and R. Sengupta. Analysis of communication losses in vehicle control problems. In *American Control Conference, 2001. Proceedings of the 2001*, volume 2, pages 1491–1496. IEEE, 2001.

C. E. Shannon. A mathematical theory of communication. *ACM SIGMOBILE Mobile Computing and Communications Review*, 5(1):3–55, 2001.

L. Shi, P. Cheng, and J. Chen. Sensor data scheduling for optimal state estimation with communication energy constraint. *Automatica*, 47(8):1693–1698, 2011.

A. Silva. Invariance entropy for random control systems. *Mathematics of Control, Signals, and Systems*, 25(4):491–516, 2013.

B. Sinopoli, L. Schenato, M. Franceschetti, K. Poolla, M. I. Jordan, and S. S. Sastry. Kalman filtering with intermittent observations. *IEEE Transactions on Automatic Control*, 49(9):1453–1464, 2004.

B. Sinopoli, L. Schenato, M. Franceschetti, K. Poolla, and S. S. Sastry. Optimal control with unreliable communication: The TCP case. In *Proceedings of the American Control Conference*, pages 3354–3359, 2005.

W. Su, J. Chen, M. Fu, and T. Qi. Control under stochastic multiplicative uncertainties: Part II, optimal design for performance. *IEEE Transactions on Automatic Control*, 62(3):1285–1300, 2017.

A. Tahbaz-Salehi and A. Jadbabaie. A necessary and sufficient condition for consensus over random networks. *IEEE Transactions on Automatic Control*, 53(3):791–795, 2008.

C. Tan, L. Li, and H. Zhang. Stabilization of networked control systems with both network-induced delay and packet dropout. *Automatica*, 59:194–199, 2015.

H. Touchette and S. Lloyd. Information-theoretic limits of control. *Physical Review Letters*, 84(6):1156–1159, 2000.

H. L. Trentelman, K. Takaba, and N. Monshizadeh. Robust synchronization of uncertain linear multi-agent systems. *IEEE Transactions on Automatic Control*, 58(6):1511–1523, 2013.

S. Trimpe. Stability analysis of distributed event-based state estimation. In *Proceedings of the 53rd Annual Conference on Decision and Control*, pages 2013–2019, 2014.

S. Trimpe and R. D'Andrea. Event-based state estimation with variance-based triggering. *IEEE Transactions on Automatic Control*, 59(12):3266–3281, 2014.

D. Tse and P. Viswanath. *Fundamentals of wireless communication*. Cambridge University Press, Cambridge, 2005.

S. Wang, X. Meng, and T. Chen. Wide-area control of power systems through delayed network communication. *IEEE Transactions on Control Systems Technology*, 20(2):495–503, 2012.

W. M. Wonham. On a matrix Riccati equation of stochastic control. *SIAM Journal on Control*, 6(4):681–697, 1968.

J. Wu, Q. Jia, K. H. Johansson, and L. Shi. Event-based sensor data scheduling: Trade-off between communication rate and estimation quality. *IEEE Transactions on Automatic Control*, 58(4):1041–1046, 2013.

M. Xia, V. Gupta, and P. J. Antsaklis. Networked state estimation over a shared communication medium. *IEEE Transactions on Automatic Control*, 62(4):1729–1741, 2017.

N. Xiao and L. Xie. Analysis and design of discrete-time networked systems over fading channels. In *Proceedings of the 30th Chinese Control Conference*, pages 6562–6567, 2011.

N. Xiao, L. Xie, and L. Qiu. Feedback stabilization of discrete-time networked systems over fading channels. *IEEE Transactions on Automatic Control*, 57(9):2176–2189, 2012.

N. Xiao, L. Xie, Y. Niu, and Y. Hong. Distributed estimation over analog fading channels using constant-gain estimators. In *Proceedings of the 19th IFAC World Congress*, pages 2872–2877, 2014.

L. Xie and L. Xie. Stability analysis of networked sampled-data linear systems with Markovian packet losses. *IEEE Transactions on Automatic Control*, 54(6):1375–1381, 2009.

L. Xu, J. Zheng, N. Xiao, and L. Xie. Mean square consensus of multi-agent systems over fading networks with directed graphs. *Automatica*, 95:503–510, 2018.

C. Yang and L. Shi. Deterministic sensor data scheduling under limited communication resource. *IEEE Transactions on Signal Processing*, 59(10):5050–5056, 2011.

K. You and L. Xie. Minimum data rate for mean square stabilization of discrete LTI systems over lossy channels. *IEEE Transactions on Automatic Control*, 55(10):2373–2378, 2010.

K. You and L. Xie. Network topology and communication data rate for consensusability of discrete-time multi-agent systems. *IEEE Transactions on Automatic Control*, 56(10):2262–2275, 2011a.

K. You and L. Xie. Minimum data rate for mean square stabilizability of linear systems with Markovian packet losses. *IEEE Transactions on Automatic Control*, 56(4):772–785, 2011b.

K. You, N. Xiao, and L. Xie. *Analysis and design of networked control systems*. Springer London, London, 2015.

W. Yu, G. Chen, and M. Cao. Some necessary and sufficient conditions for second-order consensus in multi-agent dynamical systems. *Automatica*, 46(6):1089–1095, 2010.

G. Zang and P. A. Iglesias. Nonlinear extension of Bode's integral based on an information-theoretic interpretation. *Systems & Control Letters*, 50(1):11–19, 2003.

W. Zhang, H. Zhang, and B. S. Chen. Generalized Lyapunov equation approach to state-dependent stochastic stabilization/detectability criterion. *IEEE Transactions on Automatic Control*, 53(7):1630–1642, 2008.

J. Zheng, L. Xu, L. Xie, and K. You. Consensusability of discrete-time multiagent systems with communication delay and packet dropouts. *IEEE Transactions on Automatic Control*, 64(3):1185–1192, 2018.

B. Zhou and Z. Lin. Consensus of high-order multi-agent systems with large input and communication delays. *Automatica*, 50(2):452–464, 2014.

K. Zhou, J. C. Doyle, and K. Glover. *Robust and optimal control*, volume 40. Prentice Hall, Upper Saddle River, NJ, 1996.

B. Zhu, L. Xie, D. Han, X. Meng, and R. Teo. A survey on recent progress in control of swarm systems. *Science China Information Sciences*, 60(7):1–24, 2017.

X. Zong, T. Li, and J. Zhang. Consensus control of discrete-time multi-agent systems with time-delays and multiplicative measurement noises. *Scientia Sinica Mathematica*, 46(10):1617–1636, 2016.

2

Stabilization over Power Constrained Fading Channels

2.1 Introduction

Traditionally, control over multiplicative noise communication channels and additive noise communication channels are studied separately, see Elia [2005], Braslavsky et al. [2007], and Freudenberg et al. [2010]. While in wireless communications, since the signal-to-noise ratio (SNR) constraint and the channel fading are both unavoidable [Goldsmith, 2005, Tse and Viswanath, 2005], it is practical to consider them as a whole. In this chapter, we are interested in a power-constrained fading channel which is subject to both fading and SNR constraints. We aim to characterize the conditions on the communication channel to ensure the mean-square stabilization of discrete-time linear time-invariant (LTI) systems. Note that Xiao and Xie [2011] have derived the necessary and sufficient condition for such kind of channels to ensure mean-square stabilizability under a linear encoder/decoder. It is still unknown whether we can achieve a larger stabilizability region with a more general causal encoder/decoder. This chapter provides a positive answer to this question. While this chapter only studies the state feedback case, the techniques proposed in Chapter 3 can be used to address the output feedback case.

This chapter is organized as follows: the problem formulation is provided in Section 2.2. The fundamental limitation of stabilizability over a power-constrained fading channel is studied in Section 2.3. In Section 2.4, conditions for the mean-square stabilizability are provided. Section 2.5 provides numerical illustrations. This chapter ends with concluding remarks in Section 2.6.

2.2 Problem Formulation

This chapter studies the following discrete-time linear system

$$x_{t+1} = Ax_t + Bu_t, \tag{2.1}$$

Control over Communication Networks: Modeling, Analysis, and Design of Networked Control Systems and Multi-Agent Systems over Imperfect Communication Channels, First Edition.
Jianying Zheng, Liang Xu, Qinglei Hu, and Lihua Xie.

where $x \in \mathbb{R}^n$ is the system state, $u \in \mathbb{R}$ is the control input, and (A, B) is controllable. The initial state $x_0 = [x_{1,0}, \dots, x_{n,0}]'$ is randomly generated from a Gaussian distribution with zero mean and bounded covariance matrix. Without loss of generality, the following assumption is made as in Zaidi et al. [2014] and Minero et al. [2009].

Assumption 2.1 All the eigenvalues of A are either on or outside the unit circle.

The configuration of the networked control system is depicted in Figure 2.1. The system state x_t is observed and encoded by the sensor/encoder $\mathscr{E}_t(\cdot)$ and transmitted to the controller/decoder $\mathscr{D}_t(\cdot)$ through a slow fading channel. The sensor/encoder $\mathscr{E}_t(\cdot)$ and the controller/decoder $\mathscr{D}_t(\cdot)$ are allowed to be of any causal form and can use all the available information till time t to generate their output. The fading channel is modeled as

$$r_t = \gamma_t s_t + \omega_t, \tag{2.2}$$

where s_t denotes the channel input, which has an average power constraint, i.e. $\mathbb{E}\{s_t^2\} \leq P$; r_t represents the channel output; $\{\gamma_t\}_{t\geq 0}$ is the iid channel fading with bounded mean and variance; $\{\omega_t\}_{t\geq 0}$ is an additive white Gaussian noise (AWGN) with zero-mean and variance σ_ω^2. We also assume that x_0, $\{\gamma_t\}_{t\geq 0}$ and $\{\omega_t\}_{t\geq 0}$ are independent; after each transmission, the instantaneous fading γ_t is known at the decoder side at every step and there exists a channel feedback that transmits one-step delayed information of r_t and γ_t from the decoder to the encoder.

In this chapter, for the given plant (2.1), we try to characterize requirements on the power-constrained fading channel (2.2), such that there exist coding and controlling strategies $\{\mathscr{E}_t(\cdot)\}_{t\geq 0}$, $\{\mathscr{D}_t(\cdot)\}_{t\geq 0}$ that can mean-square stabilize the system, i.e. to render $\lim_{t\to\infty} \mathbb{E}\{x_t x_t'\} = 0$.

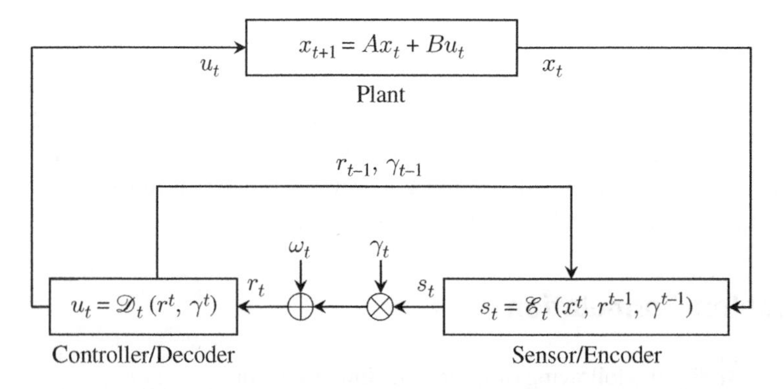

Figure 2.1 Networked control over a power-constrained fading channel. Source: L. Xu et al. 2017/with permission of IEEE.

Remark 2.1 The knowledge of the fading level at the decoder side can be obtained for slow fading channels via receiver estimation in each sampling interval [Tse and Viswanath, 2005]. In a pilot-based channel estimation scheme, a known sequence is first transmitted and used for the receiver to estimate the channel state. Since the fading is slow varying and approximately constant in each sampling interval, the channel fading can be estimated with reasonable accuracy [Goldsmith, 2005, Tse and Viswanath, 2005]. Thus to simplify the study, we assume the perfect knowledge of the channel fading as in Dey et al. [2009].

Remark 2.2 Noiseless channel feedback may not be available in some settings. However, there are situations where this assumption is natural [Silva et al., 2010, Schalkwijk and Kailath, 1966]. A good example is the communication with a satellite. The power in the ground-to-satellite direction can be much larger than in the reverse direction that the first link can be considered as an (essentially) noiseless link [Schalkwijk and Kailath, 1966]. Besides, fading can be used to model quantization effects in digital channels [Su et al., 2011], where knowing the channel input equally means knowing the channel output. Fading channels can also be used to model channels suffering from the packet loss [Elia, 2005], where the use of acknowledgment is equivalent to having a noiseless channel feedback. In some scenarios, the channel feedback can be realized through the plant with suitable designed control policies [Tatikonda and Mitter, 2004]. Thus, the assumption of noiseless channel feedback is widely used in network control research; see Sahai and Mitter [2006], Martins and Dahleh [2008], Charalambous and Farhadi [2008], Silva et al. [2010], Silva and Pulgar [2013], Kumar et al. [2014], and Zaidi et al. [2016].

Remark 2.3 The remote control setting in Figure 2.1 has been widely adopted in networked control research (e.g. You and Xie [2011] and Minero et al. [2009, 2013]). The aerial robotics research platform in Lupashin et al. [2014] is one example of our feedback control configuration. The attitude and position of the aerial robot are observed via a sensing system such as a motion capture system. The observed value is processed on one or more standard computers and then transmitted to the aerial robot over wireless channels to implement the control algorithm.

2.3 Fundamental Limitations

Since the entropy power provides a lower bound for the mean-square value of the system state [Freudenberg et al., 2010], we can treat the entropy power as a measure of the uncertainty of the system state and analyze its update, which poses a fundamental limitation of networked control over fading channels. The result is formalized in the following lemma: the proof essentially follows the same steps as in Minero et al. [2009], Freudenberg et al. [2010], and Kumar et al. [2014]; however, with some differences due to the channel structure.

Lemma 2.1 *There exist coding and controlling strategies $\{\mathscr{E}_t(\cdot)\}_{t\geq 0}$, $\{\mathscr{D}_t(\cdot)\}_{t\geq 0}$, such that the system (2.1) can be mean-square stabilized over the channel (2.2), only if*

$$(\det A)^{\frac{2}{n}}\mathbb{E}\left\{e^{-\frac{2}{n}c_t}\right\} < 1, \tag{2.3}$$

where $c_t = \frac{1}{2}\ln\left(1+\frac{\gamma_t^2 P}{\sigma_\omega^2}\right)$ is the instantaneous Shannon channel capacity of (2.2).

The following definitions are needed in the proof of Lemma 2.1 and are stated first, which are borrowed from Freudenberg et al. [2010]. Let f_X and $f_{X|y}$ denote the probability density of a random variable X, and the probability density of X conditioned on the event $Y = y$, respectively. The differential entropy of X is defined as $\mathscr{H}(X) = -\mathbb{E}\{\ln f_X\}$. The entropy of X conditioned on the event $Y = y$ is defined by $\mathscr{H}_y(X) = \mathscr{H}(X|Y = y) = -\mathbb{E}_y\{\ln f_{X|y}\}$. The random variable associated with $\mathscr{H}_y(X)$ is denoted by $\mathscr{H}_Y(X)$. The conditional entropy of X given the event $Y = y$ and averaged over Y is defined by $\mathscr{H}(X|Y) = \mathbb{E}\{\mathscr{H}_Y(X)\}$, and the conditional entropy of X given the events $Y = y$ and $Z = z$ and averaged only over Y by $\mathscr{H}_z(X|Y) = \mathbb{E}_z\{\mathscr{H}_{Y,Z}(X)\}$. The mutual information between two random variables X and Y conditioned on the event $Z = z$ is defined by $\mathscr{I}_z(X; Y) = \mathscr{H}_z(X) - \mathscr{H}_z(X|Y)$. Given a random variable $X \in \mathbb{R}^n$, the entropy power of X is defined by $\mathscr{N}(X) = \frac{1}{2\pi e}e^{\frac{2}{n}\mathscr{H}(X)}$. Denote the entropy power of X given the event $Y = y$ by $\mathscr{N}_y(X) = \frac{1}{2\pi e}e^{\frac{2}{n}\mathscr{H}_y(X)}$, and the random variable associated with $\mathscr{N}_y(X)$ by $\mathscr{N}_Y(X)$. The conditional entropy power of X given the event $Y = y$ and averaged over Y is defined by $\mathscr{N}(X|Y) = \mathbb{E}\{\mathscr{N}_Y(X)\}$. For any encoding strategy, the following lemma shows that the amount of information that the channel output contains about the source equals that the channel output contains about the channel input.

Lemma 2.2 *Let X be a random variable, $f(X)$ be a function of X, and $Y = f(X) + N$ with N being a random variable that is independent of X. Then $\mathscr{I}(X; Y) = \mathscr{I}(f(X); Y)$.*

Proof: Since $\mathscr{H}(Y|X) = \mathscr{H}(Y|X, f(X)) \leq \mathscr{H}(Y|f(X))$, we have $\mathscr{H}(Y) = \mathscr{I}(X; Y) + \mathscr{H}(Y|X) \leq \mathscr{I}(X; Y) + \mathscr{H}(Y|f(X))$. Thus,

$$\mathscr{H}(Y) - \mathscr{H}(Y|f(X)) = \mathscr{I}(Y; f(X)) \leq \mathscr{I}(X; Y).$$

Besides, since $X \to f(X) \to Y$ forms a Markov chain, $Y \to f(X) \to X$ also forms a Markov chain. The data processing inequality [Cover and Thomas, 2006] then implies that $\mathcal{I}(X;Y) \leq \mathcal{I}(f(X);Y)$. Combining the two facts, we have $\mathcal{I}(X;Y) = \mathcal{I}(f(X);Y)$. □

Proof of Lemma 2.1: Here we use the uppercase letters $\mathcal{X}, \mathcal{S}, \mathcal{R}, \Gamma$ to denote random variables of the system state, the channel input, the channel output, and the channel fading. We use the lowercase letters x, s, r, γ to denote their realizations. The notation χ^t denotes a sequence $\{\chi_i\}_{i=0}^t$. The average entropy power of $\mathcal{X}_t$ conditioned on $(\mathcal{R}^t, \Gamma^t)$ is

$$\begin{aligned}\mathcal{N}(\mathcal{X}_t|\mathcal{R}^t,\Gamma^t) &= \mathbb{E}\{\mathcal{N}_{\mathcal{R}^t,\Gamma^t}(\mathcal{X}_t)\}\\ &\overset{(a)}{=} \mathbb{E}\{\mathbb{E}_{\mathcal{R}^{t-1},\Gamma^t}\{\mathcal{N}_{\mathcal{R}^t,\Gamma^t}(\mathcal{X}_t)\}\}\\ &\overset{(b)}{=} \frac{1}{2\pi e}\mathbb{E}\{\mathbb{E}_{\mathcal{R}^{t-1},\Gamma^t}\{e^{\frac{2}{n}\mathcal{H}_{\mathcal{R}^t,\Gamma^t}(\mathcal{X}_t)}\}\},\end{aligned}$$

where (a) follows from the law of total expectation and (b) from the definition of entropy power. Since

$$\begin{aligned}\mathbb{E}_{r^{t-1},\gamma^t}\{e^{\frac{2}{n}\mathcal{H}_{\mathcal{R}^t,\Gamma^t}(\mathcal{X}_t)}\} &\overset{(c)}{\geq} e^{\frac{2}{n}\mathbb{E}_{r^{t-1},\gamma^t}\{\mathcal{H}_{\mathcal{R}^t,\Gamma^t}(\mathcal{X}_t)\}}\\ &\overset{(d)}{=} e^{\frac{2}{n}\mathcal{H}_{r^{t-1},\gamma^t}(\mathcal{X}_t|\mathcal{R}_t)}\\ &= e^{\frac{2}{n}(\mathcal{H}_{r^{t-1},\gamma^t}(\mathcal{X}_t)-\mathcal{I}_{r^{t-1},\gamma^t}(\mathcal{X}_t;\mathcal{R}_t))}\\ &\geq e^{\frac{2}{n}(\mathcal{H}_{r^{t-1},\gamma^t}(\mathcal{X}_t)-\mathcal{I}_{r^{t-1},\gamma^t}(\mathcal{X}^t;\mathcal{R}_t))}\\ &\overset{(e)}{=} e^{\frac{2}{n}(\mathcal{H}_{r^{t-1},\gamma^t}(\mathcal{X}_t)-\mathcal{I}_{r^{t-1},\gamma^t}(\mathcal{S}_t;\mathcal{R}_t))}\\ &\overset{(f)}{\geq} e^{\frac{2}{n}(\mathcal{H}_{r^{t-1},\gamma^t}(\mathcal{X}_t)-c_t)}\\ &\overset{(g)}{=} e^{-\frac{2}{n}c_t}e^{\frac{2}{n}\mathcal{H}_{r^{t-1},\gamma^{t-1}}(\mathcal{X}_t)},\end{aligned}$$

where (c) follows from Jensen's inequality; (*d*) from the definition of conditional entropy; (*e*) from Lemma 2.2; (*f*) from the definition of channel capacity, i.e. $\mathcal{I}_{r^{t-1},\gamma^t}(\mathcal{S}_t;\mathcal{R}_t) \leq c_t$; and (g) from the fact that $\mathcal{X}_t$ is independent of Γ_t, we have

$$\mathcal{N}(\mathcal{X}_t|\mathcal{R}^t,\Gamma^t) \geq \frac{1}{2\pi e}\mathbb{E}\{e^{-\frac{2}{n}c_t}e^{\frac{2}{n}\mathcal{H}_{\mathcal{R}^{t-1},\Gamma^{t-1}}(\mathcal{X}_t)}\} = \mathbb{E}\{e^{-\frac{2}{n}c_t}\}\mathcal{N}(\mathcal{X}_t|\mathcal{R}^{t-1},\Gamma^{t-1}).$$

Since

$$\begin{aligned}e^{\frac{2}{n}\mathcal{H}_{r^t,\gamma^t}(\mathcal{X}_{t+1})} &= e^{\frac{2}{n}\mathcal{H}_{r^t,\gamma^t}(A\mathcal{X}_t+BU_t)} \overset{(h)}{=} e^{\frac{2}{n}\mathcal{H}_{r^t,\gamma^t}(A\mathcal{X}_t)}\\ &\overset{(i)}{=} e^{\frac{2}{n}\mathcal{H}_{r^t,\gamma^t}(\mathcal{X}_t)+\frac{2}{n}\ln|\det A|}\\ &= (\det A)^{\frac{2}{n}}e^{\frac{2}{n}\mathcal{H}_{r^t,\gamma^t}(\mathcal{X}_t)},\end{aligned}$$

where (h) follows from the fact that $u_t = \mathscr{D}_t(r^t, \gamma^t)$ and (i) from Theorem 8.6.4 in Cover and Thomas [2006], we have

$$\mathcal{N}(\mathcal{X}_{t+1}|\mathcal{R}^t, \Gamma^t) = \mathbb{E}\{\frac{1}{2\pi e}(\det A)^{\frac{2}{n}} e^{\frac{2}{n}\mathscr{H}_{R^t,\Gamma^t}(\mathcal{X}_t)}\} = (\det A)^{\frac{2}{n}} \mathcal{N}(\mathcal{X}_t|\mathcal{R}^t, \Gamma^t).$$

In view of the above results, we have

$$\mathcal{N}(\mathcal{X}_{t+1}|\mathcal{R}^t, \Gamma^t) \geq (\det A)^{\frac{2}{n}} \mathbb{E}\{e^{-\frac{2}{n}c_t}\} \mathcal{N}(\mathcal{X}_t|\mathcal{R}^{t-1}, \Gamma^{t-1}).$$

In light of Proposition II.1 in Freudenberg et al. [2010], to ensure mean-square stability, $\mathcal{N}(\mathcal{X}_{t+1}|\mathcal{R}^t, \Gamma^t)$ should converge to zero asymptotically, which requires $(\det A)^{\frac{2}{n}} \mathbb{E}\{e^{-\frac{2}{n}c_t}\} < 1$. The proof is completed. □

Let $\lambda_1, \ldots, \lambda_d$ denote the distinct unstable eigenvalues (if λ_i is complex, we exclude from this list its complex conjugate) of A in (2.1) with $|\lambda_1| \geq |\lambda_2| \geq \cdots \geq |\lambda_d|$. Let m_i represent the algebraic multiplicity of each λ_i. The real Jordan canonical form J of A then has a form that $J = \text{diag}(J_1, \ldots, J_d) \in \mathbb{R}^{n \times n}$ [Nair and Evans, 2004, Minero et al., 2009], where $J_i \in \mathbb{R}^{\nu_i \times \nu_i}$ and $|\det J_i| = |\lambda_i|^{\nu_i}$, with $\nu_i = m_i$ if $\lambda_i \in \mathbb{R}$, and $\nu_i = 2m_i$ otherwise. We can equivalently study the following dynamical system instead of (2.1):

$$x_{t+1} = Jx_t + OBu_t, \tag{2.4}$$

for some transformation matrix O. Each block J_i has an invariant real subspace $\mathcal{A}_{o_i}$ of dimension $\varrho_i o_i$, for any $o_i \in \{0, \ldots, m_i\}$, where $\varrho_i = 1$ if $\lambda_i \in \mathbb{R}$, and $\varrho_i = 2$ otherwise. Consider the subspace $\mathcal{A}$ formed by taking the product of $\mathcal{A}_{o_i}$, $i = 1, \ldots, d$. The total dimension of $\mathcal{A}$ is $\sum_{i=1}^d \varrho_i o_i$ and the real Jordan form for the dynamics in the subspace $\mathcal{A}$ is $J^\mathcal{V}$ with $|\det J^\mathcal{V}| = \prod_{i=1}^d |\lambda_i|^{\varrho_i o_i}$. Since (2.1) is mean-square stabilizable, the dynamics in the subspace $\mathcal{A}$ is also mean-square stabilizable. In view of Lemma 2.1, the following fundamental limitations can be obtained.

Theorem 2.1 *There exist coding and controlling strategies* $\{\mathscr{E}_t(\cdot)\}_{t\geq 0}, \{\mathscr{D}_t(\cdot)\}_{t\geq 0}$, *such that the system (2.1) can be mean-square stabilized over the channel (2.2) only if* $[\ln|\lambda_1|, \ldots, \ln|\lambda_d|]' \in \mathbb{R}^d$ *satisfy that for all* $o_i \in \{0, \ldots, m_i\}$, $i = 1, \ldots, d$ *with* $o = \sum_{i=1}^d \varrho_i o_i$,

$$\sum_{i=1}^d \varrho_i o_i \ln|\lambda_i| < -\frac{o}{2} \ln \mathbb{E}\left\{\left(\frac{\sigma_\omega^2}{\sigma_\omega^2 + \gamma_t^2 P}\right)^{\frac{1}{o}}\right\}. \tag{2.5}$$

Theorem 2.1 implies that even in the presence of a noiseless channel feedback, there still exists a fundamental limitation for the stabilizability of networked control over power-constrained fading channels. Besides, for scalar systems, where

$A = \lambda_1$, $\ln|\lambda_1|$ should satisfy the following constraint to ensure mean-square stabilizability

$$\ln|\lambda_1| < -\frac{1}{2}\ln\mathbb{E}\left\{\frac{\sigma_\omega^2}{\sigma_\omega^2 + \gamma_t^2 P}\right\}. \tag{2.6}$$

Moreover, for two-dimensional systems with distinct eigenvalues λ_1, λ_2, the following requirement in addition to (2.6) should be satisfied

$$\ln|\lambda_1| + \ln|\lambda_2| < -\ln\mathbb{E}\left\{\left(\frac{\sigma_\omega^2}{\sigma_\omega^2 + \gamma_t^2 P}\right)^{\frac{1}{2}}\right\}. \tag{2.7}$$

2.4 Mean-Square Stabilizability

The existence of a noiseless channel feedback implies that there is no dual effect of control [Yuksel and Basar, 2013], i.e. separation between estimation and control holds, which will simplify the coding design. Indeed, we have the following lemma.

Lemma 2.3 ***(Kumar et al. [2014])*** *If (A, B) is controllable, and there exists an estimate $\hat{x}_t$ for the initial system state x_0, such that the estimation error $e_t = \hat{x}_t - x_0$ satisfies the following property:*

$$\mathbb{E}\left\{e_t\right\} = 0, \tag{2.8}$$

$$\lim_{t\to\infty} A^t\mathbb{E}\left\{e_t e_t'\right\}(A')^t = 0, \tag{2.9}$$

the system (2.1) can be mean-square stabilized by the controller

$$u_t = K\left(A^t\hat{x}_t + \sum_{i=1}^{t} A^{t-i}Bu_{i-1}\right)$$

with K being selected such that $A + BK$ is stable.

Remark 2.4 Assumption 2.1 can be justified from Lemma 2.3. Suppose that the system matrix A contains eigenvalues that are within the unit circle. Then, the real Jordan canonical form J of A has the diagonal structure $J = \text{diag}(J_u, J_s)$, where J_u contains eigenvalues that are either on or outside the unit circle and J_s contains eigenvalues that are within the unit circle. The initial system state x_0 can be partitioned correspondingly as $x_0 = [x_{u,0}', x_{s,0}']'$, with $x_{u,0}$ being the initial system state that corresponds to eigenvalues either on or outside the unit circle and $x_{s,0}$ corresponding to eigenvalues within the unit circle. In view of Lemma 2.3, if there exists an estimate $\hat{x}_t = [\hat{x}'_{u,t}, \hat{x}'_{s,t}]'$ for the initial system state x_0, such that the estimation error $e_t = \hat{x}_t - x_0 = [e'_{u,t}, e'_{s,t}]'$ satisfies (2.8) and (2.9), the system can be mean-square stabilized. The conditions (2.8) and (2.9) are equivalent to the following requirements:

$$\mathbb{E}\left\{e_{u,t}\right\} = 0, \tag{2.10}$$

$$\lim_{t\to\infty} J_u^t \mathbb{E}\left\{e_{u,t}e'_{u,t}\right\} (J'_u)^t = 0, \tag{2.11}$$

$$\mathbb{E}\left\{e_{s,t}\right\} = 0, \tag{2.12}$$

$$\lim_{t\to\infty} J_s^t \mathbb{E}\left\{e_{s,t}e'_{s,t}\right\} (J'_s)^t = 0. \tag{2.13}$$

Simply let $\hat{x}_{s,t} = 0$. Since $\lim_{t\to\infty} J_s^t = 0$ and $\mathbb{E}\left\{x_{s,0}\right\} = 0$, we know that (2.12) and (2.13) hold. Thus, we only need to use the channel to transmit the information about $x_{u,0}$ and design communication schemes to satisfy (2.10) and (2.11). Therefore, we can ignore the stable part of the system dynamics without loss of generality.

In the sequel, we shall focus on the construction of communication/estimation algorithms which can achieve (2.8) and (2.9). To better convey our ideas, we start with scalar systems.

2.4.1 Scalar Systems

Theorem 2.2 *Suppose $A = \lambda_1 \in \mathbb{R}$. There exist coding and controlling strategies $\{\mathscr{E}_t(\cdot)\}_{t\geq 0}, \{\mathscr{D}_t(\cdot)\}_{t\geq 0}$, such that the system (2.1) can be mean-square stabilized over the channel (2.2) if and only if (2.6) holds.*

The necessity follows directly from Theorem 2.1. For the sufficiency, we can show that a variation of the Schalkwijk coding scheme [Schalkwijk and Kailath, 1966] can stabilize the scalar system if (2.6) holds. The proof is similar to that of the awgn case in Kumar et al. [2014] with some differences due to the existence of channel fading.

Proof: Suppose the estimate of x_0 given by the decoder is $\hat{x}_t$ at time t and the estimation error is $e_t = \hat{x}_t - x_0$. The encoder is designed as

$$s_0 = \sqrt{\frac{P}{\sigma_{x_0}^2}} x_0, \quad s_t = \sqrt{\frac{P}{\sigma_{e_{t-1}}^2}} \left(\hat{x}_{t-1} - x_0\right), \ t \geq 1, \tag{2.14}$$

with $\sigma_{x_0}^2, \sigma_{e_{t-1}}^2$ representing the variances of x_0 and e_{t-1}, respectively. The decoder is designed as

$$\hat{x}_0 = \sqrt{\frac{\sigma_{x_0}^2}{P}} r_0, \quad \hat{x}_t = \hat{x}_{t-1} - \frac{\mathbb{E}_{\gamma_t}\{r_t e_{t-1}\}}{\mathbb{E}_{\gamma_t}\{r_t^2\}} r_t, \ t \geq 1. \tag{2.15}$$

Since at time t, the encoder knows the one-step delayed channel output r_{t-1}, the fading γ_{t-1}, and the decoding law, it can thus simulate the decoder to obtain the estimate $\hat{x}_{t-1}$. With the designed encoder (2.14) and decoder (2.15), it is easy to show that $\mathbb{E}\left\{e_0\right\} = 0$ and $\mathbb{E}\left\{e_0^2\right\}$ is bounded. When $t \geq 1$, we have from (2.15) that

$$e_t = e_{t-1} - \frac{\mathbb{E}_{\gamma_t}\{r_t e_{t-1}\}}{\mathbb{E}_{\gamma_t}\{r_t^2\}} r_t. \tag{2.16}$$

By induction arguments, we have $\mathbb{E}\left\{e_t\right\} = 0$ for all $t \geq 1$. Thus, (2.8) is satisfied. Denote $\hat{e}_{t-1} = \mathbb{E}_{\gamma_t}\{r_t e_{t-1}\}/\mathbb{E}_{\gamma_t}\{r_t^2\} r_t$. Since $\hat{e}_{t-1}$ is the minimal mean-square error (MMSE) estimate of e_{t-1} based on r_t, from (2.16), we have

$$\begin{aligned}
\mathbb{E}\{e_t^2\} &= \mathbb{E}\{\mathbb{E}_{\gamma_t}\{(e_{t-1} - \hat{e}_{t-1})^2\}\} \\
&\overset{(a)}{=} \mathbb{E}\left\{\frac{\sigma_\omega^2}{\sigma_\omega^2 + \gamma_t^2 \mathcal{P}} \mathbb{E}\{e_{t-1}^2\}\right\} \\
&= \mathbb{E}\left\{\frac{\sigma_\omega^2}{\sigma_\omega^2 + \gamma_t^2 \mathcal{P}}\right\}^t \mathbb{E}\{e_0^2\},
\end{aligned}$$

where (a) is a direct consequence of the MMSE. Thus, if $\lambda_1^2 \mathbb{E}\left\{\frac{\sigma_\omega^2}{\sigma_\omega^2 + \gamma_t^2 \mathcal{P}}\right\} < 1$, the designed encoder/decoder pair (2.14) and (2.15) can guarantee (2.9). In view of Lemma 2.3, the sufficiency is proved. □

Remark 2.5 Since γ_t is known at the decoder side, we can show that a slight modification of the coding scheme in Zaidi et al. [2014], where the expectation is replaced with the conditional expectation with respect to γ_t, can stabilize the closed-loop system without channel feedback if (2.6) holds.

Remark 2.6 Theorem 2.2 indicates that the anytime capacity of the power constrained fading channel (2.2) corresponding to the anytime-reliability $2 \ln |\lambda_1|$ is $C_a = -\frac{1}{2} \ln \mathbb{E}\left\{\frac{\sigma_\omega^2}{\sigma_\omega^2 + \gamma_t^2 \mathcal{P}}\right\}$. From Jensen's inequality, we know that $\mathbb{E}\left\{e^{-2c_t}\right\} \geq e^{-2\mathbb{E}\{c_t\}}$ and the equality holds if and only if c_t is a constant. Thus, it follows that $C_a = \frac{1}{2} \ln \frac{1}{\mathbb{E}\{e^{-2c_t}\}} \leq \frac{1}{2} \ln \frac{1}{e^{-2\mathbb{E}\{c_t\}}} = \mathbb{E}\left\{c_t\right\} = C_{\text{Shannon}}$, which means that the anytime capacity of the power constrained fading channel is no greater than its Shannon capacity. Besides, for awgn channels, where c_t is a constant, we have that the anytime capacity is equal to its Shannon capacity, which coincides with the results in Sahai and Mitter [2006].

2.4.2 Two-Dimensional Systems

The stabilizability condition for two-dimensional systems is stated in Theorem 2.3.

Theorem 2.3 *Suppose $n = 2$. There exist coding and controlling strategies $\{\mathscr{E}_t(\cdot)\}_{t\geq 0}$, $\{\mathscr{D}_t(\cdot)\}_{t\geq 0}$, such that the system (2.1) can be mean-square stabilized over the channel (2.2) if and only if (2.5) holds.*

In this subsection, we only provide the optimal communication scheme for two-dimensional systems with unstable eigenvalues having different magnitudes, i.e. $A = \begin{bmatrix} \lambda_1 & 0 \\ 0 & \lambda_2 \end{bmatrix}$ with $\lambda_1, \lambda_2 \in \mathbb{R}$ and $|\lambda_1| > |\lambda_2| \geq 1$, and in view of Theorem 2.1, it suffices to show that a sufficient stabilizability condition is (2.6) and (2.7). For the case of two-dimensional systems with eigenvalues of equal magnitude, the communication scheme designed in the Section 2.4.4 is shown to be optimal; see Corollary 2.1.

2.4.2.1 Communication Structure

Since there are two sources $x_{1,0}$ and $x_{2,0}$, we design two encoder/decoder pairs in the communication scheme and also a scheduler to multiplex the channel use. The ith encoder/decoder pair is used to transmit the information of $x_{i,0}$. The scheduler determines which encoder/decoder pair should use the channel. Suppose at time t, the ith encoder/decoder pair has access to the channel. The encoder i first generates a symbol $s_{i,t}$ and transmits it to the decoder through the communication channel. The decoder i then forms an estimate $\hat{x}_{i,t}$ based on the channel output $r_{i,t}$. The controller maintains an array $\hat{x}_t = [\hat{x}_{1,t}, \hat{x}_{2,t}]'$ that represents the most recent estimate of x_0, which is set to 0 at $t = 0$. When the information about $x_{i,0}$ is transmitted, only $\hat{x}_{i,t}$ is updated at the controller side. The controller applies the control law in Lemma 2.3 to the plant at every step.

The structure of the communication protocol is illustrated in Figure 2.2, where t_k^i is the time when the ith encoder/decoder pair is scheduled to use the channel for its kth transmission.

2.4.2.2 Encoder/Decoder Design

The following encoding/decoding strategy is used, which is modified from (2.14) and (2.15). The encoder i is designed as

$$s_{i,t_0^i} = \sqrt{\frac{P}{\sigma^2_{x_{i,0}}}} x_{i,0},$$

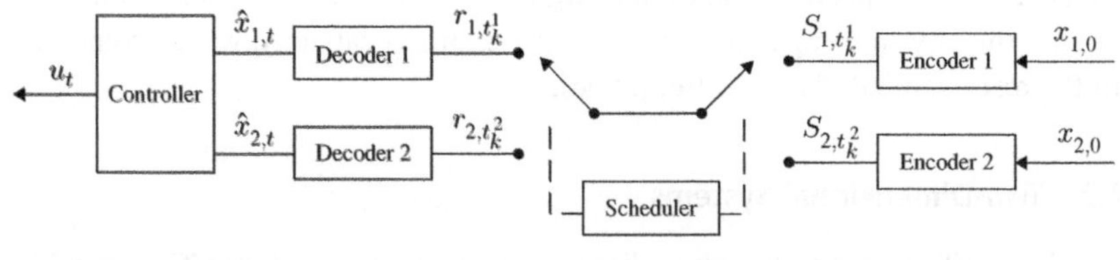

Figure 2.2 Transmission protocol configuration. Source: L. Xu et al. 2017/with permission of IEEE.

$$s_{i,t_k^i} = \sqrt{\frac{\mathcal{P}}{\sigma^2_{e_{i,t_{k-1}^i}}}}(\hat{x}_{i,t_{k-1}^i} - x_{i,0}), \quad k \geq 1, \tag{2.17}$$

where $\sigma^2_{x_{i,0}}$ and $\sigma^2_{e_{i,t}}$ represent the variance of $x_{i,0}$ and $e_{i,t}$, respectively, with $e_{i,t}$ being the ith component of the estimation error e_t. The decoder i satisfies

$$\begin{aligned}
\hat{x}_{i,t_0^i} &= \sqrt{\frac{\sigma^2_{x_{i,0}}}{\mathcal{P}}} r_{i,t_0^i}, \\
\hat{x}_{i,t_k^i} &= \hat{x}_{i,t_{k-1}^i} - \frac{\mathbb{E}_{\gamma_{t_k^i}}\{r_{i,t_k^i} e_{i,t_{k-1}^i}\}}{\mathbb{E}_{\gamma_{t_k^i}}\{r^2_{i,t_k^i}\}} r_{i,t_k^i}, \quad k \geq 1.
\end{aligned} \tag{2.18}$$

2.4.2.3 Scheduler Design

Let $\delta = \frac{\sigma^2_\omega}{\sigma^2_\omega + \mathcal{P}}$. Define the scheduling indication vector as $\Phi(t) = [\phi_1(t), \phi_2(t)]'$ with $\phi_1(t), \phi_2(t) \in \{0, 1\}$, and $\phi_1(t) + \phi_2(t) = 1$. When the ith encoder/decoder pair is scheduled to use the channel at time t, the variable $\phi_i(t)$ is set to 1, otherwise, it is set to 0. Let $\Psi_l(i,j) = \prod_{k=i}^{j} (\frac{\sigma^2_\omega}{\sigma^2_\omega + \gamma_k^2 \mathcal{P}})^{\phi_l(k)}$ with $l = 1, 2$, $i, j \in \mathbb{N}^+$ and $i \leq j$. Similar to the analysis for scalar systems, we can show that with the encoder (2.17) and the decoder (2.18), (2.8) always holds and $\mathbb{E}\left\{e^2_{i,t}\right\} = \mathbb{E}\left\{\Psi_i(t_0^i + 1, t)\right\} \mathbb{E}\left\{e^2_{i,t_0^i}\right\}$ for $i = 1, 2$. Since $\phi_i(t) = 0$ when $t < t_0^i$, to guarantee (2.9), we should design schedulers to ensure that, under the stochastic channel fading, $\lim_{t\to\infty} \mathbb{E}\left\{\lambda_1^{2t}\Psi_1(1,t)\right\} = 0$ and $\lim_{t\to\infty} \mathbb{E}\left\{\lambda_2^{2t}\Psi_2(1,t)\right\} = 0$, or equivalently $\lim_{t\to\infty} \mathbb{E}\{\lambda_1^{2t}\Psi_1(1,t) + \lambda_2^{2t}\Psi_2(1,t)\} = 0$. Thus, the scheduler should be designed to optimally allocate ϕ_1 and ϕ_2 to minimize $\lambda_1^{2t}\Psi_1(1,t) + \lambda_2^{2t}\Psi_2(1,t)$. The optimal allocation should satisfy

$$\sum_{j=1}^{t} \phi_2(j) \ln \frac{\sigma^2_\omega}{\sigma^2_\omega + \gamma_j^2 \mathcal{P}} = 2t \ln \frac{|\lambda_1|}{|\lambda_2|} + \sum_{j=1}^{t} \phi_1(j) \ln \frac{\sigma^2_\omega}{\sigma^2_\omega + \gamma_j^2 \mathcal{P}},$$

which is obtained by requiring $\lambda_1^{2t}\Psi_1(1,t) = \lambda_2^{2t}\Psi_2(1,t)$. To this end, Algorithm 2.1 is designed, which enforces ϕ_1 and ϕ_2 to meet the above requirement when t is sufficiently large.

In Algorithm 2.1, τ_1 is the scheduler parameter to be defined latter; $\check{T}_k = \sum_{j=1}^{k} \overline{T}_j$, $k \in \mathbb{N}^+$ is the time when k rounds of transmissions are completed and $\check{T}_0 = 0$; $\overline{T}_k$ denotes the total time period to complete the kth round of transmissions, i.e. $\overline{T}_k = T_k^1 + T_k^2$. Here we assume that both the encoder and the decoder know the scheduling algorithm. Since the switching among transmissions in Algorithm 2.1 relies on the fading process, which is known to the encoder and the decoder, they are both aware of when to switch transmissions and which encoder/decoder pair is currently using the channel. Thus, we do not need to consider the coordination among the encoders and the decoders. The scheduled transmission periods are depicted in Figure 2.3.

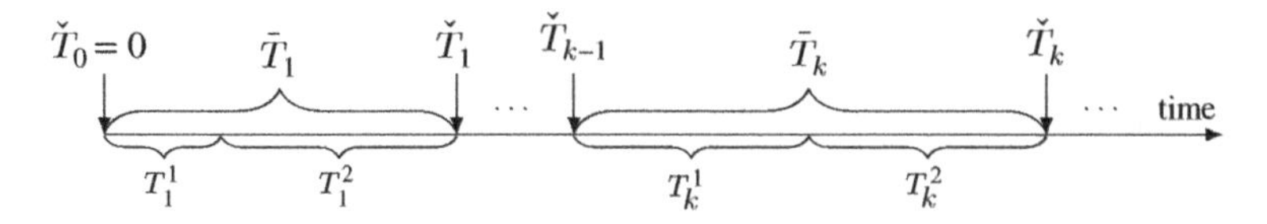

Figure 2.3 Scheduled transmissions with Algorithm 2.1. Source: L. Xu et al. 2017/with permission of IEEE.

Algorithm 2.1: Chasing and Optimal Stopping Scheduler for Power Constrained Fading Channels.

In the kth round of transmissions

- The first encoder/decoder pair is scheduled to use the channel until

$$\sum_{t=\check{T}_{k-1}+1}^{\check{T}_{k-1}+T_k^1} \ln \frac{\sigma_\omega^2}{\sigma_\omega^2 + \gamma_t^2 P} < \tau_1 \ln \delta \tag{2.19}$$

  with T_k^1 being the minimal time period satisfying (2.19).
  - If

$$\tau_1 \ln \delta + 2T_k^1 \ln \frac{|\lambda_1|}{|\lambda_2|} < 0 \tag{2.20}$$

    the second encoder/decoder pair is scheduled to use the channel, until

$$\sum_{t=\check{T}_{k-1}+T_k^1+1}^{\check{T}_{k-1}+T_k^1+T_k^2} \ln \frac{\sigma_\omega^2}{\sigma_\omega^2 + \gamma_t^2 P} < 2(T_k^1 + T_k^2) \ln \frac{|\lambda_1|}{|\lambda_2|} + \tau_1 \ln \delta \tag{2.21}$$

    with T_k^2 being the minimal time period satisfying (2.21).
  - Otherwise, set $T_k^2 = 0$, and no transmission is carried out.
- Repeat this process.

It is clear from Algorithm 2.1 that $\overline{T}_i$ is independent of $\overline{T}_j$, and T_i^2 is independent of T_j^2 for any $i \neq j$, $i, j \in \mathbb{N}^+$. The switching condition (2.20) of Algorithm 2.1 implies that if $T_k^1 < T^c := \frac{\tau_1 \ln \delta}{2(\ln|\lambda_2| - \ln|\lambda_1|)}$ and after the first encoder/decoder pair completes its transmission, the second encoder/decoder pair can use the channel. Otherwise, the first encoder/decoder pair continues to use the channel.

2.4.2.4 Scheduler Parameter Selection

If (2.6) holds, there exists θ_b with $0 < \theta_b < 1$ such that $\mathbb{E}\left\{\left(\frac{\sigma_\omega^2}{\sigma_\omega^2+\gamma_t^2 P}\right)^{\theta_b}\right\} = \lambda_1^{-2}$. Let $f(\theta_a) = 2\theta_a \ln \frac{|\lambda_1|}{|\lambda_2|} - \ln\mathbb{E}\left\{\left(\frac{\sigma_\omega^2}{\sigma_\omega^2+\gamma_t^2 P}\right)^{\theta_a}\right\} - 2\ln|\lambda_1|$. If (2.7) holds, since

$f(0) = -2\ln|\lambda_1| < 0$, $f(\frac{1}{2}) = -\ln\mathbb{E}\left\{\left(\frac{\sigma_\omega^2}{\sigma_\omega^2+\gamma_l^2 P}\right)^{\frac{1}{2}}\right\} - \ln|\lambda_1| - \ln|\lambda_2| > 0$ and $f(\theta_a)$ is increasing in θ_a, there exists θ_a with $0 < \theta_a < \frac{1}{2}$ such that $f(\theta_a) = 0$, i.e. $\mathbb{E}\left\{\left(\frac{\sigma_\omega^2}{\sigma_\omega^2+\gamma_l^2 P}\right)^{\theta_a}\right\} = \lambda_1^{2(\theta_a-1)}\lambda_2^{-2\theta_a}$. The positive constant τ_1 is then selected to satisfy

$$\tau_1 > \max\left\{\frac{-\ln(\lambda_1^{2(2-\theta_a)}\lambda_2^{2\theta_a}) - \ln 4}{(1-2\theta_a)\ln\delta}, \frac{-\ln\lambda_1^2 - \ln 2}{(1-\theta_b)\ln\delta}\right\}. \tag{2.22}$$

2.4.2.5 Proof of Theorem 2.3

The necessity follows from Theorem 2.1. The gist of the sufficiency proof is to show that under Algorithm 2.1, $\mathbb{E}\{\lambda_l^{2\overline{T}_1}\Psi_l(1,\overline{T}_1)\} < 1$ for $l = 1, 2$. Since the transmission is scheduled periodically and $\{\overline{T}_k\}$ is iid, we may expect that $\lim_{t\to\infty}\mathbb{E}\left\{\lambda_l^{2t}\Psi_l(1,t)\right\} = 0$ holds, which together with Lemma 2.3 can guarantee the mean-square stabilizability. The following lemma is needed in the proof of Theorem 2.3.

Lemma 2.4 *Suppose $\{W_i\}$ with $W_i \leq 0$ is iid with bounded nonzero mean, define $B_t = \sum_{i=1}^t W_i$ and let T be the first time such that $B_T < \varphi T + \Theta$ with given $\varphi \geq 0$, $\Theta < 0$. If there exists $\theta \geq 0$ such that $\mathbb{E}\{e^{\theta(W_i-\varphi)}\} = \lambda^{-2}$, then $\mathbb{E}\{\lambda^{2T}\} \leq \lambda^2 e^{-\theta\Theta}$.*

Proof: When $\varphi > 0$, since B_t is nonincreasing and $\varphi t + \Theta$ is increasing, the stopping time T is bounded. When $\varphi = 0$, T is unbounded if and only $\Theta \leq \lim_{t\to\infty}\sum_{i=1}^t W_i \leq 0$. Since $\{W_i\}$ is iid, in view of the law of large numbers, we have

$$\Pr\left(\lim_{t\to\infty}\sum_{i=1}^{t} W_i/t = \mathbb{E}\left\{W_i\right\}\right) = 1.$$

Thus, $\Pr(\lim_{t\to\infty}\sum_{i=1}^t W_i = \infty) = 1$, which implies $\Pr(\Theta \leq \lim_{t\to\infty}\sum_{i=1}^t W_i \leq 0) = 0$. Thus, T is almost surely bounded.

Define $Y_t = e^{\theta B_t + bt}$ with $b = 2\ln|\lambda| - \theta\varphi$, then $\mathbb{E}\left\{Y_{t+1}|Y_t,\dots,Y_1\right\} = Y_t\mathbb{E}\left\{e^{\theta W_{t+1}+b}\right\} = Y_t$. Thus, Y_t is a martingale. Since T is either a bounded or an almost surely bounded stopping time, in view of the optional stopping theorem [Ash and Doléans-Dade, 2000], we have $\mathbb{E}\left\{Y_T\right\} = \mathbb{E}\left\{Y_1\right\} = 1$.

Define $\eta = \varphi T + \Theta - B_T$. Since $B_T < \varphi T + \Theta$ and $B_{T-1} \geq \varphi(T-1) + \Theta$, we have $\eta > 0$. When $\varphi = 0$, since $B_{T-1} \geq \Theta$, we have

$$\eta = \Theta - B_T = \Theta - B_{T-1} - W_T \leq -W_T.$$

When $\varphi > 0$ and $\varphi(T-1) + \Theta \leq B_{T-1} \leq \varphi T + \Theta$, we have

$$\eta = \varphi T + \Theta - B_T = \varphi(T-1) + \Theta - B_{T-1} + \varphi - W_T \leq \varphi - W_T.$$

When $\varphi > 0$ and $\varphi T + \Theta < \mathcal{B}_{T-1} \leq 0$, we have

$$\eta = \varphi T + \Theta - \mathcal{B}_T = \varphi T + \Theta - \mathcal{B}_{T-1} - W_T < -W_T.$$

Thus in general, $\eta \leq \varphi - W_T$.

Since

$$\mathbb{E}\{Y_T\} = \mathbb{E}\{e^{\theta(\varphi T+\Theta-\eta)+bT}\} = e^{\theta\Theta}\mathbb{E}\{e^{(\theta\varphi+b)T}e^{-\theta\eta}\} = e^{\theta\Theta}\mathbb{E}\{\lambda^{2T}e^{-\theta\eta}\} = 1,$$

and

$$\mathbb{E}\{\lambda^{2T}e^{-\theta\eta}\} \geq \mathbb{E}\{\lambda^{2T}e^{\theta(W_T-\varphi)}\}$$
$$= \mathbb{E}\{\mathbb{E}_T\{\lambda^{2T}e^{\theta(W_T-\varphi)}\}\} = \mathbb{E}\{\lambda^{2T}\mathbb{E}_T\{e^{\theta(W_T-\varphi)}\}\} \overset{(a)}{=} \lambda^{-2}\mathbb{E}\{\lambda^{2T}\},$$

where (a) follows from the definition of θ, we have $\mathbb{E}\left\{\lambda^{2T}\right\} \leq \lambda^2 e^{-\theta\Theta}$. □

Proof of Theorem 2.3: Let $W_k = \ln \frac{\sigma_\omega^2}{\sigma_\omega^2+\gamma_k^2 P}$. Then it is immediate from (2.19) that T_1^1 is the first time such that $\mathcal{B}_{T_1^1} < \varphi_1 T_1^1 + \Theta_1$ with $\varphi_1 = 0$ and $\Theta_1 = \tau_1 \ln \delta$. Since there exist $0 < \theta_a < \frac{1}{2}$, $0 < \theta_b < 1$ such that $\mathbb{E}\{e^{\theta_a(W_k-\varphi_1)}\} = \lambda_1^{2(\theta_a-1)}\lambda_2^{-2\theta_a}$, $\mathbb{E}\{e^{\theta_b(W_k-\varphi_1)}\} = \lambda_1^{-2}$, from Lemma 2.4, we have

$$\mathbb{E}\left\{\lambda_1^{2(1-\theta_a)T_1^1}\lambda_2^{2\theta_a T_1^1}\right\} \leq \lambda_1^{2(1-\theta_a)}\lambda_2^{2\theta_a}\delta^{-\tau_1\theta_a}, \tag{2.23}$$

$$\mathbb{E}\left\{\lambda_1^{2T_1^1}\right\} \leq \lambda_1^2\delta^{-\tau_1\theta_b}. \tag{2.24}$$

Suppose $T_1^1 < T^c$. Let $\mathcal{B}_t = \sum_{k=1}^t W_{T_1^1+k}$. In view of the stopping condition (2.21), we know that T_1^2 is the first time instant after T_1^1 satisfying that $\mathcal{B}_{T_1^2} < \varphi_2 T_1^2 + \Theta_2$ with $\varphi_2 = 2\ln\frac{|\lambda_1|}{|\lambda_2|}$ and $\Theta_2 = 2T_1^1 \ln\frac{|\lambda_1|}{|\lambda_2|} + \tau_1 \ln\delta$. Since $\mathbb{E}\{e^{\theta_a(W_k-\varphi_2)}\} = \lambda_1^{-2}$, in view of Lemma 2.4, we have

$$\mathbb{E}_\zeta\{\lambda_1^{2T_1^2}\} < \lambda_1^2 e^{-\theta_a\Theta_2}, \tag{2.25}$$

where ζ denote the event $T_1^1 < T^c$. Since $\theta_a < 1$, when $T_1^1 \geq T^c$, we have $2T_1^1(\theta_a-1)\ln\frac{|\lambda_1|}{|\lambda_2|} \leq 2T^c(\theta_a-1)\ln\frac{|\lambda_1|}{|\lambda_2|} < \tau_1(1-\theta_a)\ln\delta + \ln 2 + 2\ln|\lambda_1|$. Rearranging both sides and applying the natural exponential function, we have $\Omega := \lambda_2^{2T_1^1} - 2\lambda_1^{2(1+T_1^1)}e^{-\theta_a\Theta_2}\delta^{\tau_1} < 0$. In view of the conditional expectation, we have

$$\mathbb{E}\left\{\sum_{i=1}^{2}\lambda_i^{2\overline{T}_1}\Psi_i(1,\overline{T}_1)\right\} \leq \mathbb{E}\left\{\lambda_1^{2\overline{T}_1}\delta^{\tau_1} + \lambda_2^{2\overline{T}_1}\Psi_2(1,\overline{T}_1)\right\}$$
$$= \mathbb{E}\left\{\mathbb{E}_\zeta\left\{\lambda_1^{2\overline{T}_1}\delta^{\tau_1} + \lambda_2^{2\overline{T}_1}\Psi_2(1,\overline{T}_1)\right\}\right\} + \mathbb{E}\left\{\mathbb{E}_\xi\left\{\lambda_1^{2\overline{T}_1}\delta^{\tau_1} + \lambda_2^{2\overline{T}_1}\Psi_2(1,\overline{T}_1)\right\}\right\}$$
$$\overset{(a)}{\leq} \mathbb{E}\left\{\mathbb{E}_\zeta\left\{2\lambda_1^{2(T_1^1+T_1^2)}\delta^{\tau_1}\right\}\right\} + \mathbb{E}\left\{\mathbb{E}_\xi\left\{\lambda_1^{2T_1^1}\delta^{\tau_1} + \lambda_2^{2T_1^1}\right\}\right\}$$

$$\overset{(b)}{\leq} \mathbb{E}\left\{\mathbb{E}_{\zeta}\left\{2\lambda_1^{2(1+T_1^1)}e^{-\theta_a\Theta_2}\delta^{\tau_1}\right\}\right\} + \mathbb{E}\left\{\mathbb{E}_{\xi}\left\{\lambda_1^{2T_1^1}\delta^{\tau_1} + \lambda_2^{2T_1^1}\right\}\right\}$$
$$= \mathbb{E}\left\{2\lambda_1^{2(1+T_1^1)}e^{-\theta_a\Theta_2}\delta^{\tau_1}\right\} + \mathbb{E}\left\{\mathbb{E}_{\xi}\left\{\lambda_1^{2T_1^1}\delta^{\tau_1} + \Omega\right\}\right\}$$
$$\overset{(c)}{\leq} 2\lambda_1^2\delta^{\tau_1(1-\theta_a)}\mathbb{E}\left\{\lambda_1^{2(1-\theta_a)T_1^1}\lambda_2^{2\theta_a T_1^1}\right\} + \mathbb{E}\left\{\mathbb{E}_{\xi}\left\{\lambda_1^{2T_1^1}\delta^{\tau_1}\right\}\right\}$$
$$\overset{(d)}{\leq} 2\lambda_1^{2(2-\theta_a)}\lambda_2^{2\theta_a}\delta^{(1-2\theta_a)\tau_1} + \lambda_1^2\delta^{(1-\theta_b)\tau_1},$$

where ξ denotes the event $T_1^1 \geq T^c$; (a) follows from (2.21); (b) follows from (2.25); (c) follows from the fact that when $T_1^1 \geq T^c$, $\Omega < 0$; (d) follows from (2.23) and (2.24). Since $\delta^{1-2\theta_a} < 1$ and $\delta^{1-\theta_b} < 1$, if τ_1 is selected to satisfy (2.22), we have that $\lambda_1^2\delta^{(1-\theta_b)\tau_1} < \frac{1}{2}$, $2\lambda_1^{2(2-\theta_a)}\lambda_2^{2\theta_a}\delta^{(1-2\theta_a)\tau_1} < \frac{1}{2}$, which guarantees

$$\mathbb{E}\left\{\lambda_1^{2\overline{T}_1}\Psi_1(1,\overline{T}_1) + \lambda_2^{2\overline{T}_1}\Psi_2(1,\overline{T}_1)\right\} < 1.$$

Thus, we have

$$\mathbb{E}\left\{\lambda_1^{2\overline{T}_1}\Psi_1(1,\overline{T}_1)\right\} < 1, \quad \mathbb{E}\left\{\lambda_2^{2\overline{T}_1}\Psi_2(1,\overline{T}_1)\right\} < 1. \tag{2.26}$$

Since

$$\Psi_l(1,\check{T}_k) = \prod_{j=1}^{k}\Psi_l(\check{T}_{j-1}+1,\check{T}_{j-1}+\overline{T}_j)$$

and $\{\Psi_l(\check{T}_{j-1}+1,\check{T}_{j-1}+\overline{T}_j)\}_{j=1}^k$ are iid, we have

$$\sum_{k=0}^{\infty}\mathbb{E}\left\{\sum_{j=1}^{\overline{T}_{k+1}}\lambda_l^{\check{T}_k+j}\Psi_l(1,\check{T}_k)\right\} \tag{2.27}$$

$$= \sum_{k=0}^{\infty}\mathbb{E}\left\{\sum_{j=1}^{\overline{T}_{k+1}}\lambda_l^{\overline{T}_0+\cdots+\overline{T}_k+j}\prod_{j=1}^{k}\Psi_l(\check{T}_{j-1}+1,\check{T}_{j-1}+\overline{T}_j)\right\}$$

$$= \sum_{k=0}^{\infty}\mathbb{E}\left\{\frac{\lambda_l^{\overline{T}_{k+1}+2}-\lambda_l^2}{\lambda_l^2-1}\right\}\mathbb{E}\left\{\lambda_l^{\overline{T}_1}\Psi_l(1,\overline{T}_1)\right\}^k, \tag{2.28}$$

for $l = 1, 2$. In view of (2.26), we further have that

$$\mathbb{E}\left\{\sum_{t=1}^{\infty}(\lambda_1^{2t}\Psi_1(1,t) + \lambda_2^{2t}\Psi_2(1,t))\right\}$$
$$= \sum_{k=0}^{\infty}\mathbb{E}\left\{\sum_{j=1}^{\overline{T}_{k+1}}(\lambda_1^{\check{T}_k+j}\Psi_1(1,\check{T}_k+j) + \lambda_2^{\check{T}_k+j}\Psi_2(1,\check{T}_k+j))\right\}$$

$$\leq \sum_{k=0}^{\infty} \mathbb{E}\left\{ \sum_{j=1}^{\overline{T}_{k+1}} \left(\lambda_1^{\check{T}_k+j} \Psi_1(1,\check{T}_k) + \lambda_2^{\check{T}_k+j} \Psi_2(1,\check{T}_k) \right) \right\} < \infty,$$

which implies that $\lim_{t\to\infty} \mathbb{E}\left\{\lambda_1^{2t}\Psi_1(1,t) + \lambda_2^{2t}\Psi_2(1,t)\right\} = 0$. The proof of sufficiency is completed. □

2.4.3 High-Dimensional Systems: TDMA Scheduler

For general n-dimensional systems, the communication structure is designed similarly to that of the two-dimensional systems. There are n encoder/decoder pairs of the form (2.17) and (2.18) to transmit the information of $x_{i,0}$, $i = 1, \dots, n$. A scheduler is designed to multiplex the channel use. Define ϕ_i, $\Psi_i(\cdot,\cdot)$, $i = 1, \dots, n$ analogously to the two-dimensional case. Similarly, we can prove that with such communication structure, (2.8) always holds and to guarantee (2.9), we only need to ensure that, $\lim_{t\to\infty} \mathbb{E}\left\{\lambda_i^{2t}\Psi_i(1,t)\right\} = 0$ for all $i = 1, \dots, n$, or equivalently, $\lim_{t\to\infty} \mathbb{E}\left\{\sum_{i=1}^{n} \lambda_i^{2t}\Psi_i(1,t)\right\} = 0$. Thus, the schedulers should be designed to optimally allocate ϕ_i to minimize $\sum_{i=1}^{n} \lambda_i^{2t}\Psi_i(1,t)$. The optimal choice of ϕ_i^* should satisfy

$$\sum_{j=1}^{t} \phi_i^*(j) \ln \frac{\sigma_\omega^2}{\sigma_\omega^2 + \gamma_j^2 P} = \left(\sum_{j=1}^{t} \ln \frac{\sigma_\omega^2}{\sigma_\omega^2 + \gamma_j^2 P} + 2t \sum_{i=1}^{n} \ln|\lambda_i| \right) \Big/ n - 2t \ln|\lambda_i|.$$

However ϕ_i^* is determined by $\sum_{j=1}^{t} \ln \frac{\sigma_\omega^2}{\sigma_\omega^2+\gamma_j^2 P}$, which is not causally available when transmitting $x_{i,0}$ at any time $k < t$. When $n = 2$, we can achieve the desired optimal allocation by first fixing ϕ_1 to be such that $\sum_{j=1}^{T_1^1} \phi_1(j) \ln\left(\frac{\sigma_\omega^2}{\sigma_\omega^2+\gamma_j^2 P}\right) < \tau_1 \ln \delta$ and then requiring ϕ_2 to achieve (2.21). However, this method is not applicable to the case of $n \geq 3$. In the following, we propose Algorithm 2.2 based on the time division multiple access (TDMA) principle and analyze the corresponding stability regions for general high-dimensional systems, where τ_i, $i = 1, \dots, n$ are scheduler parameters and their existence are shown in the proof of Theorem 2.4.

In conjunction with the scheduling Algorithm 2.2, the following sufficient condition can be obtained.

Theorem 2.4 *There exist coding and controlling strategies $\{\mathscr{E}_t(\cdot)\}_{t\geq 0}$, $\{\mathscr{D}_t(\cdot)\}_{t\geq 0}$, such that the system (2.1) can be mean-square stabilized over the channel (2.2) if*

$$\sum_{i=1}^{d} v_i \ln|\lambda_i| < -\frac{1}{2} \ln \mathbb{E}\left\{ \frac{\sigma_\omega^2}{\sigma_\omega^2 + \gamma_t^2 P} \right\}. \tag{2.29}$$

Algorithm 2.2: TDMA Scheduler for Power-Constrained Fading Channels

In the kth round of transmissions

- The first encoder/decoder pair is scheduled to use the channel for a duration of τ_1.
- ...
- The jth encoder/decoder pair is scheduled to use the channel for a duration of τ_j.
- ...
- The nth encoder/decoder pair is scheduled to use the channel for a duration of τ_n.
- Repeat this process.

Proof: Without loss of generality, here we assume that $\lambda_1, \ldots, \lambda_d$ are real and $m_i = 1$. For other cases, readers can refer to the analysis discussed in Chapter 2 of Como et al. [2014]. Specifically, under this assumption, J is a diagonal matrix and $d = n$. In Algorithm 2.2, the sensor transmits periodically with a period of $\overline{\tau} = \sum_{i=1}^{n} \tau_i$. The relative transmission frequency for $x_{j,0}$ is $\alpha_j = \frac{\tau_j}{\overline{\tau}}$ among the period of $\overline{\tau}$ with $\sum_{j=1}^{n} \alpha_j = 1$. Similar to the analysis in Section 2.4.2, we can show that (2.8) always holds and

$$\mathbb{E}\left\{e_{i,k\overline{\tau}}^2\right\} = \mathbb{E}\left\{\frac{\sigma_\omega^2}{\sigma_\omega^2 + \gamma_t^2 \mathcal{P}}\right\}^{\alpha_i k\overline{\tau}} \mathbb{E}\left\{e_{i,0}^2\right\}$$

under the designed communication scheme. If $\lambda_i^2 \mathbb{E}\left\{\frac{\sigma_\omega^2}{\sigma_\omega^2+\gamma_t^2\mathcal{P}}\right\}^{\alpha_i} < 1$ for all $i = 1, \ldots, n$, the sufficient condition in Lemma 2.3 can be satisfied. To complete the proof, we only need to show the equivalence between the requirement $\lambda_i^2 \mathbb{E}\left\{\frac{\sigma_\omega^2}{\sigma_\omega^2+\gamma_t^2\mathcal{P}}\right\}^{\alpha_i} < 1$ for all $i = 1, \ldots, n$ and (2.29). On the one hand, since $\sum_{i=1}^{n} \alpha_i = 1$, if $\lambda_i^2 \mathbb{E}\left\{\frac{\sigma_\omega^2}{\sigma_\omega^2+\gamma_t^2\mathcal{P}}\right\}^{\alpha_i} < 1$ for all $i = 1, \ldots, n$, we know that (2.29) holds. On the other hand, if (2.29) holds, we can simply choose $\alpha_i = \frac{\ln|\lambda_i|}{\sum_i \ln|\lambda_i|}$, which satisfies the requirement that $\sum_{i=1}^{n} \alpha_i = 1$ and $\lambda_i^2 \mathbb{E}\left\{\frac{\sigma_\omega^2}{\sigma_\omega^2+\gamma_t^2\mathcal{P}}\right\}^{\alpha_i} < 1$ for all $i = 1, \ldots, n$. The sufficiency is proved. □

2.4.4 High-Dimensional Systems: Adaptive TDMA Scheduler

The tdma scheduler only allocates transmissions based on time. Since we also have the channel state information at the receiver side, we may utilize this information to achieve better control performance. Moreover, we have the following stabilization result.

Theorem 2.5 *There exist coding and controlling strategies* $\{\mathscr{E}_t(\cdot)\}_{t\geq 0}, \{\mathscr{D}_t(\cdot)\}_{t\geq 0}$, *such that the system* (2.1) *can be mean-square stabilized over the channel* (2.2) *if there exist* α_i, $i = 1, \ldots, d$, *with* $0 < \alpha_i \leq 1$ *and* $\sum_{i=1}^{d} \alpha_i = 1$, *such that for all* $i = 1, \ldots, d$,

$$\ln|\lambda_i| < -\frac{1}{2}\ln\mathbb{E}\left\{\left(\frac{\sigma_\omega^2}{\sigma_\omega^2 + \gamma_t^2 \mathcal{P}}\right)^{\frac{\alpha_i}{v_i}}\right\}. \tag{2.30}$$

The above stabilizability result is achieved via an adaptive tdma scheduler. Different from the tdma scheduler, the adaptive tdma scheduler used here is adapted to the fading process. It switches the transmission only if certain stopping conditions are satisfied. By incorporating the information of the fading process, a larger stabilizability region is achieved. The detailed scheduler design and stability analysis are given as follows.

2.4.4.1 Scheduling Algorithm

The scheduler is described in Algorithm 2.3, where the parameters τ_i, $i = 1, \ldots, n$ are defined in the sequel; $\check{T}_k = \sum_{j=1}^{k} \overline{T}_j$, $k \in \mathbb{N}^+$ is the time when k rounds of transmissions are completed and $\check{T}_0 = 0$, and $\overline{T}_k$ denotes the total time period to complete the kth round of transmissions, i.e. $\overline{T}_k = \sum_{i=1}^{n} T_k^i$. Since the fading $\{\gamma_t\}$ is iid, it is clear from Algorithm 2.3 that T_k^i is independent of T_k^j, for any $i \neq j$, $i, j \in \{1, 2, \ldots, n\}$, $k \in \mathbb{N}^+$ and the random variables $\{\overline{T}_1, \overline{T}_2, \ldots\}$ are iid.

2.4.4.2 Scheduler Parameter Selection

If (2.30) holds, there exist θ_i, $i = 1, \ldots, d$ with $0 \leq \theta_i < \frac{\alpha_i}{v_i}$, such that $\mathbb{E}\left\{\left(\frac{\sigma_\omega^2}{\sigma_\omega^2+\gamma_t^2\mathcal{P}}\right)^{\theta_i}\right\} = |\lambda_i|^{-2}$. The positive constants τ_j, $j = 1, \ldots, n$ are selected as follows: if $x_{j,0}$ is the jth component of x_0 in (2.4) that corresponds to the eigenvalue λ_i, $i = 1, \ldots, d$, then τ_j is selected to be

$$\tau_j = -\frac{2n\alpha_i}{v_i \ln\delta}\left(\max_{k\in\{1,\ldots,d\}} \frac{\ln|\lambda_k|}{\alpha_k/v_k - \theta_k} + \iota\right), \quad j = 1, \ldots, n, \tag{2.34}$$

with ι being an arbitrary positive constant.

2.4.4.3 Proof of Theorem 2.5

Here we only consider the case that $\lambda_1, \ldots, \lambda_d$ are real and $m_i = v_i = 1$. We can easily extend the analysis to other cases by combining the following analysis with the argument used in Chapter 2 of Como et al. [2014]. The sufficiency proof is focused on showing that $\lim_{t\to\infty}\mathbb{E}\left\{\lambda_i^{2t}\Psi_i(1,t)\right\} = 0$ for all $i = 1, \ldots, n$ under Algorithm 2.3. Similar to the derivation of (2.24), with Algorithm 2.3, we can show that $\mathbb{E}\left\{\lambda_i^{2T_1^j}\right\} \leq \delta^{-\tau_j\theta_i}\lambda_i^2$. Since $T_1^1, T_1^2, \ldots, T_1^n$ are independent of each

---

Algorithm 2.3: Adaptive TDMA Scheduler for Power-Constrained Fading Channels

---

In the kth round of transmissions

- The first encoder/decoder pair is scheduled to use the channel, until

$$\sum_{t=\check{T}_{k-1}+1}^{\check{T}_{k-1}+T_k^1} \ln \frac{\sigma_\omega^2}{\sigma_\omega^2+\gamma_t^2\mathcal{P}} < \tau_1 \ln \delta, \tag{2.31}$$

  with T_k^1 being the minimal time period satisfying (2.31).
- ...
- The jth encoder/decoder pair is scheduled to use the channel, until

$$\sum_{t=\check{T}_{k-1}+T_k^1+\cdots+T_k^{j-1}+1}^{\check{T}_{k-1}+T_k^1+\cdots+T_k^{j-1}+T_k^j} \ln \frac{\sigma_\omega^2}{\sigma_\omega^2+\gamma_t^2\mathcal{P}} < \tau_j \ln \delta, \tag{2.32}$$

  with T_k^j being the minimal time period satisfying (2.32).
- ...
- The nth encoder/decoder pair is scheduled to use thechannel, until

$$\sum_{t=\check{T}_{k-1}+T_k^1+\cdots+T_k^{n-1}+1}^{\check{T}_{k-1}+T_k^1+\cdots+T_k^{n-1}+T_k^n} \ln \frac{\sigma_\omega^2}{\sigma_\omega^2+\gamma_t^2\mathcal{P}} < \tau_n \ln \delta, \tag{2.33}$$

  with T_k^n being the minimal time period satisfying (2.33).
- Repeat this process.

---

other, we further have $\mathbb{E}\left\{\lambda_i^{2\sum_{j=1}^n T_1^j}\delta^{\tau_i}\right\} \le \delta^{\tau_i-\theta_i\sum_{j=1}^n \tau_j}\lambda_i^{2n}$. If τ_i is selected as (2.34), then $\sum_{i=1}^n \tau_i = -\frac{2n}{\ln\delta}\left(\max_j \frac{\ln|\lambda_j|}{\alpha_j-\theta_j}+\iota\right)$ and $\tau_i/\left(\sum_{j=1}^n \tau_j\right) = \alpha_i$ for all $i = 1,\ldots,n$. Thus, we have

$$\begin{aligned}
\mathbb{E}\left\{\lambda_i^{2\sum_{j=1}^n T_1^j}\delta^{\tau_i}\right\} &\le (\delta^{\alpha_i-\theta_i})^{\sum_{j=1}^n \tau_j}\lambda_i^{2n} \\
&= (\delta^{\alpha_i-\theta_i})^{-\frac{2n}{\ln\delta}\left(\max_j \frac{\ln|\lambda_j|}{\alpha_j-\theta_j}+\iota\right)}(\delta^{\alpha_i-\theta_i})^{\frac{2n}{\ln\delta}\frac{\ln|\lambda_i|}{\alpha_i-\theta_i}} \\
&= (\delta^{\alpha_i-\theta_i})^{\frac{2n}{\ln\delta}\left(\frac{\ln|\lambda_i|}{\alpha_i-\theta_i}-\max_j \frac{\ln|\lambda_j|}{\alpha_j-\theta_j}-\iota\right)}.
\end{aligned}$$

Since $\theta_i < \alpha_i$ and $0 < \delta < 1$, we have

$$\mathbb{E}\left\{\lambda_i^{2\overline{T}_1}\delta^{\tau_i}\right\} = \mathbb{E}\left\{\lambda_i^{2\sum_{j=1}^n T_1^j}\delta^{\tau_i}\right\} < 1, \tag{2.35}$$

for all $i = 1, \ldots, n$. Since $\Psi_i(1, \check{T}_k) = \prod_{j=1}^{k} \Psi_i(\check{T}_{j-1} + 1, \check{T}_{j-1} + \overline{T}_j)$ and $\Psi_i(\check{T}_{j-1} + 1, \check{T}_{j-1} + \overline{T}_j) < \delta^{\tau_i}$ for any $j \in \mathbb{N}^+$, in view of (2.35), we have

$$\begin{aligned}
\mathbb{E}\left\{\sum_{t=1}^{\infty} \lambda_i^{2t} \Psi_i(1, t)\right\} &= \sum_{k=0}^{\infty} \mathbb{E}\left\{\sum_{j=1}^{\overline{T}_{k+1}} \lambda_i^{2(\check{T}_k+j)} \Psi_i(1, \check{T}_k + j)\right\} \\
&< \sum_{k=0}^{\infty} \mathbb{E}\left\{\sum_{j=1}^{\overline{T}_{k+1}} \lambda_i^{2(\check{T}_k+j)} \prod_{j=1}^{k} \Psi_i(\check{T}_{j-1} + 1, \check{T}_{j-1} + \overline{T}_j)\right\} \\
&< \sum_{k=0}^{\infty} \mathbb{E}\left\{\sum_{j=1}^{\overline{T}_{k+1}} \lambda_i^{2(\check{T}_k+j)} \delta^{k\tau_i}\right\} \\
&= \sum_{k=0}^{\infty} \mathbb{E}\left\{\lambda_i^{2\overline{T}_1} \delta^{\tau_i}\right\}^k \mathbb{E}\{(\lambda_i^{2\overline{T}_{k+1}+2} - \lambda_i^2)/(\lambda_i^2 - 1)\} < \infty.
\end{aligned}$$

Thus, $\lim_{t\to\infty} \mathbb{E}\left\{\lambda_i^{2t} \Psi_i(1, t)\right\} = 0$ for all $i = 1, \ldots, n$. The proof of sufficiency is completed.

Remark 2.7 The stabilizability conditions in the derived theorems above involve the calculation of the expectation $\mathbb{E}\left\{\left(\frac{\sigma_\omega^2}{\sigma_\omega^2 + \gamma_t^2 \mathcal{P}}\right)^\alpha\right\}$ for some α. For some fading distributions, we can give the closed form of this term. For example, when $\gamma_t \sim$ Bernoulli(ϵ),[1] this term is given by $(1 - \epsilon)\left(\frac{\sigma_\omega^2}{\sigma_\omega^2 + \mathcal{P}}\right)^\alpha + \epsilon$. For other fading distributions that are not possible to calculate the closed forms, this term can be evaluated numerically via MATLAB or Mathematica.

Remark 2.8 In Theorem 2.5, the stabilizability condition is expressed in terms of parameters α_is. α_i has the physical interpretation that it represents the fraction of channel resources that is allocated to the subdynamics corresponding to the eigenvalue λ_i. For the given communication channel and system matrix, the existence of α_is can be checked via the following feasibility problem

$$\exists \alpha_i > 0, i = 1, \ldots, d$$

$$\text{s.t.} \sum_{i=1}^{d} \alpha_i = 1 \tag{2.36}$$

$$|\lambda_i|^2 < f_i(\alpha_i), \quad i = 1, \ldots, d \tag{2.37}$$

---

1 $\Pr(\gamma_t = 0) = \epsilon$, $\Pr(\gamma_t = 1) = 1 - \epsilon$, where $\gamma_t = 0$ represents the appearance of fading and $\gamma_t = 1$ means that the channel is free of fading.

with $f_i(\alpha_i) := \mathbb{E}\left\{\left(\frac{\sigma_\omega^2}{\sigma_\omega^2+\gamma_t^2\mathcal{P}}\right)^{\frac{\alpha_i}{\nu_i}}\right\}^{-1}$. Since $f_i(\alpha_i)$ is increasing in α_i and $f_i(0) = 1 \le |\lambda_i|^2$, there exists $\alpha_i^* \ge 0$ such that $f_i(\alpha_i^*) = |\lambda_i|^2$ (binary search can be used to find equation roots to obtain α_i^*). In view of (2.37), any feasible α_i must satisfy that $\alpha_i > \alpha_i^*$. If $\sum_i \alpha_i^* \ge 1$, there exists no feasible solution since (2.36) is violated. Otherwise, one feasible solution is given by $\alpha_i = \frac{\alpha_i^*}{\sum_i \alpha_i^*}$.

Remark 2.9 Theorem 2.5 indicates that the stabilzable region of

$$[\ln|\lambda_1|, \dots, \ln|\lambda_d|]' \in \mathbb{R}^d$$

for a given power-constrained fading channel achieved with Algorithm 2.3 is

$$\mathcal{O} = \cup_{\alpha_i>0, \sum_i \alpha_i = 1} \times_{i\in\{1,\dots,d\}} \left[0, -\frac{1}{2}\ln\mathbb{E}\left\{\left(\frac{\sigma_\omega^2}{\sigma_\omega^2+\gamma_t^2\mathcal{P}}\right)^{\frac{\alpha_i}{\nu_i}}\right\}\right),$$

where $\times$ denotes the Cartesian product. We can prove that $\mathcal{O}$ is convex. Suppose $\mathsf{x} = [\mathsf{x}_1, \dots, \mathsf{x}_d]' \in \mathcal{O}$ and $\mathsf{y} = [\mathsf{y}_1, \dots, \mathsf{y}_d]' \in \mathcal{O}$. Then there exist $[\vartheta_1, \dots, \vartheta_d]'$ with $\vartheta_i > 0$, $\sum_i \vartheta_i = 1$ and $[\eta_1, \dots, \eta_d]'$ with $\eta_i > 0$, $\sum_i \eta_i = 1$ such that $\mathsf{x}_i < -\frac{1}{2}\ln\mathbb{E}\left\{\left(\frac{\sigma_\omega^2}{\sigma_\omega^2+\gamma_t^2\mathcal{P}}\right)^{\frac{\vartheta_i}{\nu_i}}\right\}$, $\mathsf{y}_i < -\frac{1}{2}\ln\mathbb{E}\left\{\left(\frac{\sigma_\omega^2}{\sigma_\omega^2+\gamma_t^2\mathcal{P}}\right)^{\frac{\eta_i}{\nu_i}}\right\}$ for $i = 1, \dots, d$. Let $\mathsf{z} = [\mathsf{z}_1, \dots, \mathsf{z}_d]' = c\mathsf{x} + (1-c)\mathsf{y}$ with $0 < c < 1$, then $\mathsf{z}_i = c\mathsf{x}_i + (1-c)\mathsf{y}_i$ and

$$\begin{aligned}
\mathsf{z}_i &< -\frac{c}{2}\ln\mathbb{E}\left\{\left(\frac{\sigma_\omega^2}{\sigma_\omega^2+\gamma_t^2\mathcal{P}}\right)^{\frac{\vartheta_i}{\nu_i}}\right\} - \frac{1-c}{2}\ln\mathbb{E}\left\{\left(\frac{\sigma_\omega^2}{\sigma_\omega^2+\gamma_t^2\mathcal{P}}\right)^{\frac{\eta_i}{\nu_i}}\right\} \\
&= -\frac{1}{2}\ln\mathbb{E}\left\{\left(\frac{\sigma_\omega^2}{\sigma_\omega^2+\gamma_t^2\mathcal{P}}\right)^{\frac{\vartheta_i}{\nu_i}}\right\}^c \mathbb{E}\left\{\left(\frac{\sigma_\omega^2}{\sigma_\omega^2+\gamma_t^2\mathcal{P}}\right)^{\frac{\eta_i}{\nu_i}}\right\}^{1-c} \\
&\overset{(a)}{\le} -\frac{1}{2}\ln\mathbb{E}\left\{\left(\frac{\sigma_\omega^2}{\sigma_\omega^2+\gamma_t^2\mathcal{P}}\right)^{\frac{c\vartheta_i+(1-c)\eta_i}{\nu_i}}\right\},
\end{aligned}$$

where (a) follows from the Hölder's inequality. Thus, there exist α_is with $\alpha_i = c\vartheta_i + (1-c)\eta_i > 0$ and $\sum_i \alpha_i = 1$ such that $\mathsf{z}_i < -\frac{1}{2}\ln\mathbb{E}\left\{\left(\frac{\sigma_\omega^2}{\sigma_\omega^2+\gamma_t^2\mathcal{P}}\right)^{\frac{\alpha_i}{\nu_i}}\right\}$ for all $i = 1, \dots, d$, which means $\mathsf{z} \in \mathcal{O}$. Thus, $\mathcal{O}$ is convex.

Remark 2.10 The sufficiency achieved via the tdma scheduler in Algorithm 2.2 can be alternatively formulated as follows: if there exist α_is with $0 < \alpha_i \leq 1$ and $\sum_{i=1}^{d} \alpha_i = 1$, such that

$$\ln|\lambda_i| < -\frac{\alpha_i}{2\nu_i}\ln\mathbb{E}\left\{\frac{\sigma_\omega^2}{\sigma_\omega^2+\gamma_t^2\mathcal{P}}\right\}, \tag{2.38}$$

for all $i = 1, 2, \ldots, d$, the system (2.1) can be mean-square stabilized. Since $f(z) = z^{\frac{\alpha_i}{\nu_i}}$ with $0 < \frac{\alpha_i}{\nu_i} \leq 1$ is concave, from the Jensen's inequality, we have

$$-\frac{\alpha_i}{2\nu_i}\ln\mathbb{E}\left\{\frac{\sigma_\omega^2}{\sigma_\omega^2+\gamma_t^2\mathcal{P}}\right\} \leq -\frac{1}{2}\ln\mathbb{E}\left\{\left(\frac{\sigma_\omega^2}{\sigma_\omega^2+\gamma_t^2\mathcal{P}}\right)^{\frac{\alpha_i}{\nu_i}}\right\}.$$

Thus, any λ_i that satisfies (2.38) must also satisfy (2.30) with the same α_i, which implies that the adaptive tdma scheduler in this chapter achieves a stabilizability region no smaller than the tdma scheduler.

Remark 2.11 If $\gamma_t = 1$, channel (2.2) degenerates to an awgn channel and the necessary and sufficient condition to ensure mean-square stabilizability, following from (2.5) and (2.30), is $\sum_{i=1}^{d} \nu_i \ln|\lambda_i| < \frac{1}{2}\ln\left(1+\frac{\mathcal{P}}{\sigma_\omega^2}\right)$, which recovers the results in Braslavsky et al. [2007] and Freudenberg et al. [2010]. If $\gamma_t \sim$ Bernoulli(ϵ), by taking the limit $\sigma_\omega^2 \to 0$ and $\mathcal{P} \to \infty$, we can obtain that the stabilizability condition over an erasure channel is $\lambda_1^2 < \frac{1}{\epsilon}$, which degenerates to the results in Elia [2005] and Gupta et al. [2007].

When all the strictly unstable eigenvalues have the same magnitude, we can show that the sufficient condition (2.30) coincides with the necessary condition (2.5), as shown in the following corollary.

Corollary 2.1 *Suppose* $|\lambda_1| = \cdots = |\lambda_{d_u}| = \tilde{\lambda} > 1$ *and* $|\lambda_{d_u+1}| = \cdots = |\lambda_d| = 1$ *with* $1 \leq d_u \leq d$. *There exist coding and controlling strategies* $\{\mathscr{E}_t(\cdot)\}_{t\geq 0}, \{\mathscr{D}_t(\cdot)\}_{t\geq 0}$, *such that the system* (2.1) *can be mean-square stabilized over the channel* (2.2) *if and only if*

$$\ln\tilde{\lambda} < -\frac{1}{2}\ln\mathbb{E}\left\{\left(\frac{\sigma_\omega^2}{\sigma_\omega^2+\gamma_t^2\mathcal{P}}\right)^{\frac{1}{\nu_1+\cdots+\nu_{d_u}}}\right\}.$$

Remark 2.12 The results derived in this chapter for the power-constrained fading channel (2.2) can be easily extended to the following channel model:

$$r_t = \gamma_t(s_t + \omega_t), \tag{2.39}$$

which is suitable for modeling the digital erasure channel with $\{\omega_t\}$ denoting the quantization error and $\{\gamma_t\}$ representing the erasure process. If $\gamma_t = 0$, the communication channel cannot transmit any information. Otherwise, we can always multiply the received signal r_t by $1/\gamma_t$ at the decoder side, and thus the resulting channel is equivalent to an awgn channel. From this perspective, channel (2.39) is essentially the power-constrained lossy channel studied in Xu et al. [2016]. Thus, the results derived in Xu et al. [2016] apply directly to the channel (2.39).

2.5 Numerical Illustrations

2.5.1 Scalar Systems

The authors in Xiao and Xie [2011] derive the necessary and sufficient condition for mean-square stabilization of scalar lti systems over a power-constrained fading channel with linear encoders/decoders as $\frac{1}{2}\ln\left(1+\frac{\mu_\gamma^2\mathcal{P}}{\sigma_\gamma^2\mathcal{P}+\sigma_\omega^2}\right) > \ln|\lambda|$ with μ_γ and σ_γ^2 being the mean and variance of γ_t. We can similarly define the mean-square capacity of the power-constrained fading channel achieved with linear encoders/decoders as $C_m = \frac{1}{2}\ln\left(1+\frac{\mu_\gamma^2\mathcal{P}}{\sigma_\gamma^2\mathcal{P}+\sigma_\omega^2}\right)$. Assume that the fading follows the Bernoulli distribution, i.e. $\gamma_t \sim \text{Bernoulli}(\epsilon)$, and let $\mathcal{P} = 1$ and $\sigma_\omega^2 = 1$, the channel capacities in relation to the erasure probability are plotted in Figure 2.4. It is clear that $C_{\text{Shannon}} \geq C_a \geq C_m$ at any erasure probability ϵ. This result is obvious since we have proved that the Shannon capacity is no smaller than the anytime capacity. Besides, we have more freedom in designing the causal encoder/decoder pair compared with the linear encoder/decoder pair,

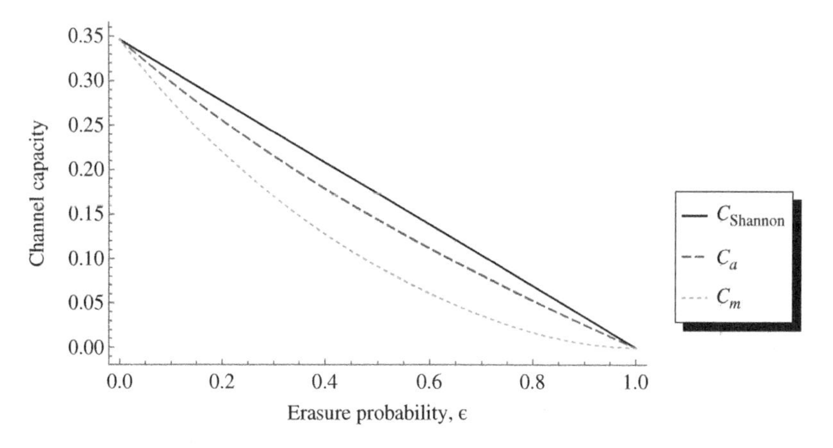

Figure 2.4 Comparison of different channel capacities for scalar systems. Source: L. Xu et al. 2017/with permission of IEEE.

thus allowing to achieve a higher capacity. The three kinds of capacity degenerate to the same value when $\epsilon = 0$ and $\epsilon = 1$, which represent the awgn channel case and the disconnected case, respectively. This fact is trivial for the disconnected case and is consistent with the awgn channel case in Sahai and Mitter [2006], Braslavsky et al. [2007], and Freudenberg et al. [2010], in which the authors show that the anytime capacity is equal to the Shannon capacity for awgn channels and causal encoder/decoder pair cannot provide any benefits in increasing the channel capacity.

2.5.2 Vector Systems

Consider a two-dimensional system (2.4) with $J = \begin{bmatrix} \lambda_1 & 0 \\ 0 & \lambda_2 \end{bmatrix}$, and the fading in (2.2) follows the Rayleigh distribution with probability density function $f(z;\sigma) = \frac{z}{\sigma^2}e^{-\frac{z^2}{2\sigma^2}}, z \geq 0$. Let $\mathcal{P} = 1, \sigma_\omega^2 = 1, \sigma = 2$, then the necessary stabilizability region, the sufficient stabilizability regions achieved with the optimal scheduler in Algorithm 2.1, the tdma scheduler in Algorithm 2.2, the adaptive tdma scheduler in Algorithm 2.3 and linear encoders/decoders in Xiao and Xie [2011], in terms of $(\ln|\lambda_1|, \ln|\lambda_2|)$ are plotted in Figure 2.5. We can observe that the region of $(\ln|\lambda_1|, \ln|\lambda_2|)$ that can be stabilized with the designed causal encoders/decoders

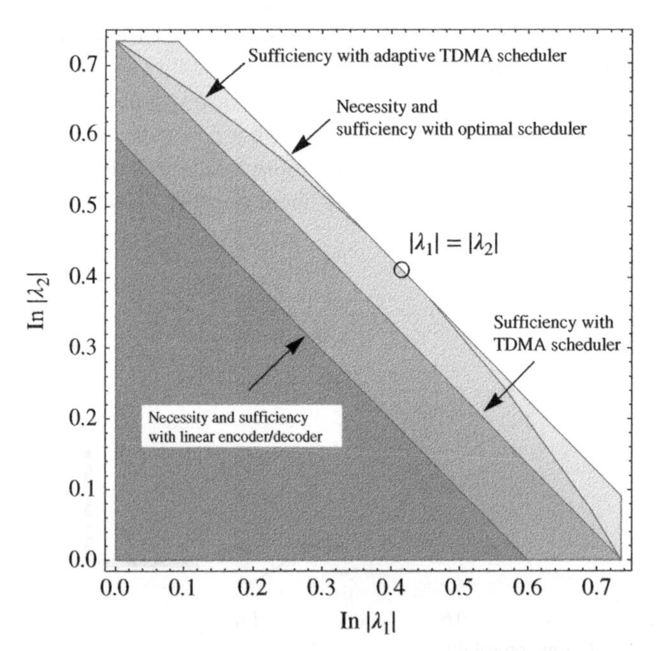

Figure 2.5 Comparison of stabilizability regions for two-dimensional systems. Source: L. Xu et al. 2017/with permission of IEEE.

is much larger than that by linear encoders/decoders in Xiao and Xie [2011]. Thus, by extending encoder/decoders from linear settings to causal requirements, we can tolerate more unstable systems. It is clear from the figure that the optimal scheduler proposed in Algorithm 2.1 covers the whole necessary stabilizability region. Besides, as noted in Remark 2.10, the adaptive tdma scheduler achieves a larger stabilizability region than that of the conventional tdma scheduler. Moreover, the adaptive tdma scheduler is optimal at three points, i.e. $|\lambda_1| = |\lambda_2|$, $|\lambda_1| = 1$ and $|\lambda_2| = 1$. This is consistent with Corollary 2.1.

2.6 Conclusions

This chapter has characterized the requirement for a power-constrained fading channel to allow the existence of coding and controlling strategies that can mean-square stabilize a discrete-time lti system. Fundamental limitations have been provided in terms of the system dynamics and channel parameters. Optimal communication designs have been provided for scalar systems and two-dimensional systems. For high-dimensional systems, tdma and adaptive tdma communication schemes have also been provided, which are shown to be optimal under certain situations. Numerical examples are provided to illustrate the derived results.

Bibliography

R. B. Ash and C. A. Doléans-Dade. *Probability and measure theory.* Harcourt/Academic Press, San Diego, CA, 2000.

J. H. Braslavsky, R. H. Middleton, and J. S. Freudenberg. Feedback stabilization over signal-to-noise ratio constrained channels. *IEEE Transactions on Automatic Control*, 52(8):1391–1403, 2007.

C. D. Charalambous and A. Farhadi. LQG optimality and separation principle for general discrete time partially observed stochastic systems over finite capacity communication channels. *Automatica*, 44(12):3181–3188, 2008.

G. Como, B. Bernhardsson, and A. Rantzer. *Information and control in networks.* Springer International Publishing, Cham, 2014.

T. M. Cover and J. A. Thomas. *Elements of information theory.* Wiley-Interscience, Hoboken, NJ, 2006.

S. Dey, A. S. Leong, and J. S. Evans. Kalman filtering with faded measurements. *Automatica*, 45(10):2223–2233, 2009.

N. Elia. Remote stabilization over fading channels. *Systems & Control Letters*, 54(3):237–249, 2005.

J. S. Freudenberg, R. H. Middleton, and V. Solo. Stabilization and disturbance attenuation over a Gaussian communication channel. *IEEE Transactions on Automatic Control*, 55(3):795–799, 2010.

A. Goldsmith. *Wireless communications.* Cambridge University Press, Cambridge, 2005.

V. Gupta, B. Hassibi, and R. M. Murray. Optimal LQG control across packet-dropping links. *Systems & Control Letters*, 56(6):439–446, 2007.

U. Kumar, J. Liu, V. Gupta, and J. N. Laneman. Stabilizability across a Gaussian product channel: Necessary and sufficient conditions. *IEEE Transactions on Automatic Control*, 59(9):2530–2535, 2014.

S. Lupashin, M. Hehn, M. W. Mueller, A. P. Schoellig, M. Sherback, and R. D'Andrea. A platform for aerial robotics research and demonstration: The flying machine arena. *Mechatronics*, 24(1):41–54, 2014.

N. C. Martins and M. A. Dahleh. Feedback control in the presence of noisy channels: Bode-like fundamental limitations of performance. *IEEE Transactions on Automatic Control*, 53(7):1604–1615, 2008.

P. Minero, M. Franceschetti, S. Dey, and G. N. Nair. Data rate theorem for stabilization over time-varying feedback channels. *IEEE Transactions on Automatic Control*, 54(2):243–255, 2009.

P. Minero, L. Coviello, and M. Franceschetti. Stabilization over Markov feedback channels: The general case. *IEEE Transactions on Automatic Control*, 58(2):349–362, 2013.

G. N. Nair and R. J. Evans. Stabilizability of stochastic linear systems with finite feedback data rates. *SIAM Journal on Control and Optimization*, 43(2):413–436, 2004.

A. Sahai and S. Mitter. The necessity and sufficiency of anytime capacity for stabilization of a linear system over a noisy communication link - Part I: Scalar systems. *IEEE Transactions on Information Theory*, 52(8):3369–3395, 2006.

J. Schalkwijk and T. Kailath. A coding scheme for additive noise channels with feedback-I: No bandwidth constraint. *IEEE Transactions on Information Theory*, 12(2):172–182, 1966.

E. I. Silva and S. A. Pulgar. Performance limitations for single-input LTI plants controlled over SNR constrained channels with feedback. *Automatica*, 49(2):540–547, 2013.

E. I. Silva, G. C. Goodwin, and D. E. Quevedo. Control system design subject to SNR constraints. *Automatica*, 46(2):428–436, 2010.

W. Su, T. Qi, J. Chen, and M. Fu. Optimal tracking design of an MIMO linear system with quantization effects. In *Proceedings of the 18th IFAC World Congress*, volume 18, pages 3268–3273, 2011.

S. Tatikonda and S. Mitter. Control over noisy channels. *IEEE Transactions on Automatic Control*, 49(7):1196–1201, 2004.

D. Tse and P. Viswanath. *Fundamentals of wireless communication*. Cambridge University Press, Cambridge, 2005.

N. Xiao and L. Xie. Analysis and design of discrete-time networked systems over fading channels. In *Proceedings of the 30th Chinese Control Conference*, pages 6562–6567, 2011.

L. Xu, Y. Mo, and L. Xie. Mean square stabilization of vector LTI systems over power constrained lossy channels. In *Proceedings of the 2016 American Control Conference*, pages 7129–7134, 2016.

K. You and L. Xie. Minimum data rate for mean square stabilizability of linear systems with Markovian packet losses. *IEEE Transactions on Automatic Control*, 56(4):772–785, 2011.

S. Yuksel and T. Basar. *Stochastic networked control systems: stabilization and optimization under information constraints*. Springer New York, New York: Imprint: Birkhauser, 2013.

A. A. Zaidi, T. J. Oechtering, S. Yuksel, and M. Skoglund. Stabilization of linear systems over Gaussian networks. *IEEE Transactions on Automatic Control*, 59(9):2369–2384, 2014.

A. A. Zaidi, S. Yüksel, T. J. Oechtering, and M. Skoglund. On the tightness of linear policies for stabilization of linear systems over Gaussian networks. *Systems & Control Letters*, 88:32–38, 2016.

3

Stabilization over Gaussian Finite-State Markov Channels

3.1 Introduction

The case with i.i.d. channel fading has been studied in Chapter 2. However, the i.i.d. assumption fails to capture channel correlations. Since Markov models are simple and effective to capture temporal correlations of channel conditions [Goldsmith, 2005, Goldsmith and Varaiya, 1996, Viswanathan, 1999], we are interested in the stabilization problem of discrete-time linear time-invariant (LTI) systems controlled over Gaussian finite-state Markov channels [Liu et al., 2015], where the channel fading is modeled by a time-homogeneous Markov process. Due to the existence of correlations of channel conditions over time, the methods used to deal with the i.i.d. channel fading in Chapter 2 cannot be applied directly to the Markov channel fading case. Besides, Chapter 2 only considers the state feedback case and the plant under investigation is free of process and measurement noises. The output feedback case and how plant noises affect the stabilizability of the networked control system have not been studied. In this chapter, we propose observer/estimator designs and extend the channel resource allocation schemes in Chapter 2 to the Gaussian Markov channel case and derive necessary and sufficient stabilization conditions by utilizing the stability of a Markov jump linear system (MJLS) and the i.i.d. property of the sojourn time of the Markov chain [Xie and Xie, 2009].

This chapter is organized as follows: the problem formulation is given in Section 3.2. The existence of fundamental limitations for stabilization is demonstrated in Section 3.3. Sufficient stabilization conditions for Gaussian finite-state Markov channels and power-constrained Markov lossy channels are provided in Sections 3.4 and 3.5, respectively. This chapter ends with some concluding remarks in Section 3.6.

Control over Communication Networks: Modeling, Analysis, and Design of Networked Control Systems and Multi-Agent Systems over Imperfect Communication Channels, First Edition.
Jianying Zheng, Liang Xu, Qinglei Hu, and Lihua Xie.

3.2 Problem Formulation

This chapter studies the following discrete-time linear system:

$$\begin{aligned} x_{t+1} &= Ax_t + Bu_t + v_t, \\ y_t &= Cx_t + w_t, \end{aligned} \tag{3.1}$$

where $x_t \in \mathbb{R}^n$ is the system state; $y_t \in \mathbb{R}^p$ is the system output; $u_t \in \mathbb{R}$ is the control input; v_t, w_t are the process noise and measurement noise, respectively; (A, B) is stabilizable; (C, A) is observable; $\{v_t\}_{t\geq 0}$ and $\{w_t\}_{t\geq 0}$ are i.i.d. and with zero mean and bounded covariance matrices and are independent of the initial state x_0, which follows a zero mean Gaussian distribution with a bounded covariance matrix. Without loss of generality, we make the following assumption as in Zaidi et al. [2014a] and Minero et al. [2009]:

Assumption 3.1 All the eigenvalues of A are either on or outside the unit circle.

This chapter considers a networked control setting, where y_t is observed and encoded with the law $\mathscr{E}_t(\cdot)$ and transmitted to the controller through a Gaussian Markov channel to generate the control signal u_t with the law $\mathscr{D}_t(\cdot)$. The Markov channel corrupted with Gaussian noises is modeled as

$$r_t = \gamma_t s_t + \omega_t, \tag{3.2}$$

where s_t denotes the channel input satisfying an average power constraint, i.e. $\mathbb{E}\left\{s_t^2\right\} \leq P$; r_t is the channel output; γ_t is the channel fading which represents the variation of received signal power over time and ω_t is an additive white Gaussian noise (AWGN) with zero-mean and bounded variance σ_ω^2. Different Markov models can be assumed for γ_t. In this chapter, we are interested in two kinds of Gaussian Markov channels: the Gaussian finite-state Markov channel and the power-constrained Markov lossy channel.

Gaussian finite-state Markov channels: The channel state $\{\gamma_t\}_{t\geq 0}$ is modeled as a time-homogeneous ergodic Markov process. Specifically, γ_t takes values in a finite set of distinct nonnegative values $\{\mathfrak{r}_1, \mathfrak{r}_2, \dots, \mathfrak{r}_\mathfrak{l}\}$, which represents different fading levels [Liu et al., 2015]. The Markov transition probability matrix Q is defined by $Q = [\mathfrak{q}_{ij}]$ with

$$\mathfrak{q}_{ij} = \Pr\{\gamma_{t+1} = \mathfrak{r}_j | \gamma_t = \mathfrak{r}_i\}. \tag{3.3}$$

Power-constrained Markov lossy channels: The channel state $\{\gamma_t\}_{t\geq 0}$ is modeled as a Markov lossy process. That is, γ_t only switches between two states: the state $\mathfrak{r}_1 = 0$ and the state $\mathfrak{r}_2 = 1$, where $\mathfrak{r}_1 = 0$ indicates the appearance of channel fading and the transmission fails, and $\mathfrak{r}_2 = 1$ means that the channel is free of fading and the transmission is successful. Therefore, the Markov process has the following transition probability matrix:

$$Q = \begin{bmatrix} 1-\mathfrak{q} & \mathfrak{q} \\ \mathfrak{p} & 1-\mathfrak{p} \end{bmatrix}, \tag{3.4}$$

where $\mathfrak{p}$ represents the failure rate and $\mathfrak{q}$ denotes the recovery rate. To avoid any trivial case, $\mathfrak{p}$ and $\mathfrak{q}$ are assumed to be strictly positive and less than 1, i.e. $0 < \mathfrak{p}, \mathfrak{q} < 1$, so that the Markov process is ergodic. The power-constrained Markov lossy channel is one special kind of Gaussian finite-state Markov channels and has several unique properties that allow to derive refined results compared to Gaussian finite-state Markov channels.

For both kinds of channels, we assume that $\{\omega_t\}_{t\geq 0}$ is i.i.d.; x_0, $\{v_t\}_{t\geq 0}$, $\{w_t\}_{t\geq 0}$, $\{\gamma_t\}_{t\geq 0}$, and $\{\omega_t\}_{t\geq 0}$ are independent; the channel state information is known at the receiver side, and the channel output and the channel state are fed back to the transmitter through a noiseless feedback channel with one-step delay as in Chapter 2. The feedback configuration and the information structure of the sensor and controller are depicted in Figure 3.1.

Throughout the chapter, a stochastic system with state x_t is mean-square stable if $\sup_t \mathbb{E}\left\{x_t'x_t\right\} < \infty$. We try to characterize the requirements on Gaussian finite-state Markov channels and power-constrained Markov lossy channels such that there exist coding and controlling strategies $\{\mathscr{E}_t(\cdot)\}_{t\geq 0}$, $\{\mathscr{D}_t(\cdot)\}_{t\geq 0}$ which can stabilize the LTI dynamics (3.1). In the following, we present several preliminary results that would be used in the subsequent analysis.

3.2.1 Stability of Markov Jump Linear Systems

Denote the instantaneous channel capacity as $c_t = \frac{1}{2}\ln\left(1 + \frac{\gamma_t^2 P}{\sigma_\omega^2}\right)$. Since $\{\gamma_t\}_{t\geq 0}$ is Markovian, so is $\{c_t\}_{t\geq 0}$ and c_t takes values in a finite set $\{\mathfrak{c}_1, \ldots, \mathfrak{c}_l\}$ with

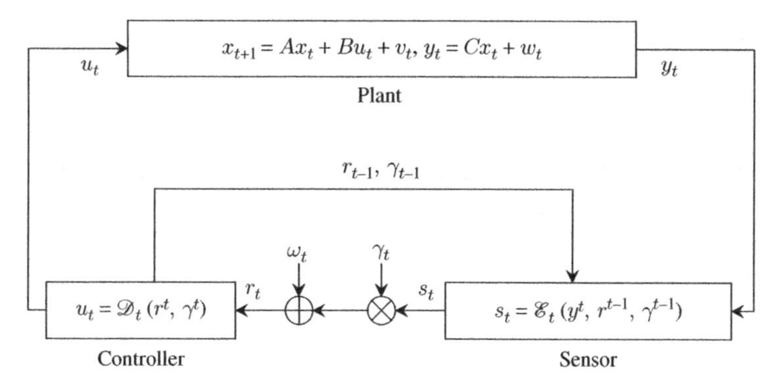

Figure 3.1 Networked control over Gaussian Markov channels. Source: L. Xu et al. 2018/with permission of IEEE.

$\mathfrak{c}_i = \frac{1}{2}\ln\left(1 + \frac{\mathfrak{r}_i^2 p}{\sigma_\omega^2}\right)$ and is with the same Markov transition probability (3.3). Consider the MJLS defined by

$$z_{t+1} = \lambda^2 e^{-\frac{2}{o}c_t} z_t + a, \tag{3.5}$$

where $z_t \in \mathbb{R}$ with $z_0 < \infty$; $\lambda \in \mathbb{R}$; $o \in \mathbb{N}^+$; $a \geq 0$ and $\{c_t\}_{t\geq 0}$ is the Markov process described above. Let $H_o = Q' D_o$ with $D_o = \text{diag}(e^{-\frac{2}{o}\mathfrak{c}_1}, \ldots, e^{-\frac{2}{o}\mathfrak{c}_l})$. Similar to Lemma 1 in Minero et al. [2013] and Coviello et al. [2011], we have the following necessary and sufficient condition characterizing the first moment stability of (3.5).

Lemma 3.1 *The first moment of the system (3.5) is stable, i.e.* $\sup_t \mathbb{E}\left\{|z_t|\right\} < \infty$, *if and only if*

$$\lambda^2 < \frac{1}{\rho(H_o)}.$$

Remark 3.1 In this chapter, we are interested in the mean-square stability of linear systems. Since the mean-square value of the linear system state conditioned on the fading process evolves as a MJLS [Xie and Xie, 2009], to study the mean-square stability of the original system, we only need to study the first moment stability of the corresponding MJLS.

3.2.2 Sojourn Times for Markov Lossy Process

Associated with the Markov lossy process $\{\gamma_t\}_{t\geq 0}$, a stochastic time sequence $\{T_k\}_{k\geq 0}$ is introduced to denote the time at which the transmission is successful. Without loss of generality, let $\gamma_0 = \mathfrak{r}_2$ [You and Xie, 2011]. Then $T_0 = 0$ and T_k, $k \geq 1$ is precisely defined by

$$\begin{aligned} T_1 &= \inf\{k : k \geq 1, \gamma_k = 1\}, \\ T_2 &= \inf\{k : k \geq T_1, \gamma_k = 1\}, \\ &\vdots \quad \vdots \\ T_k &= \inf\{k : k \geq T_{k-1}, \gamma_k = 1\}. \end{aligned} \tag{3.6}$$

By the ergodic property of the Markov process $\{\gamma_k\}_{k\geq 0}$, $T_k, \forall k \in \mathbb{N}$ is finite almost everywhere (abbreviated as a.e.). Thus, the integer valued sojourn time $\{T_k^*\}_{k>0}$, which denotes the time duration between two successive successful transmissions, is well-defined a.e., where

$$T_k^* = T_k - T_{k-1} > 0. \tag{3.7}$$

Moreover, we have the following characterization of the probability distribution of sojourn times $\{T_k^*\}_{k>0}$.

Lemma 3.2 *(Xie and Xie [2009]) The sojourn times $\{T_k^*\}_{k>0}$ are i.i.d. Furthermore, the distribution of T_k^* is explicitly expressed as*

$$\Pr(T_k^* = i) = \begin{cases} 1 - \mathfrak{p} & i = 1, \\ \mathfrak{p}\mathfrak{q}(1-\mathfrak{q})^{i-2} & i > 1. \end{cases}$$

3.3 Fundamental Limitation

Let $\lambda_1, \dots, \lambda_d$ denote the distinct unstable eigenvalues (if λ_i is complex, its conjugate is excluded from this list) of A with $|\lambda_1| \geq |\lambda_2| \geq \cdots \geq |\lambda_d|$. Let m_i represent the algebraic multiplicity of λ_i. The real Jordan canonical form J of A then has a form that $J = \text{diag}(J_1, \dots, J_d) \in \mathbb{R}^{n\times n}$ [Nair and Evans, 2004], where $J_i \in \mathbb{R}^{\nu_i \times \nu_i}$ and $|\det J_i| = |\lambda_i|^{\nu_i}$, with $\nu_i = m_i$ if $\lambda_i \in \mathbb{R}$, and $\nu_i = 2m_i$ otherwise. It is clear that the mean-square stability of (3.1) is equivalent to the mean-square stability of

$$x_{t+1} = Jx_t + OBu_t + Ov_t, \tag{3.8}$$

$$y_t = CO^{-1}x_t + w_t, \tag{3.9}$$

for some invertible matrix O.

The following theorem characterizes a fundamental limitation for mean-square stabilization over Gaussian finite-state Markov channels. The necessity is obtained via an information theoretic argument as in Section 2.3, but with differences due to the application of output feedback and the existence of process and measurement noises and the correlated channel fading.

Theorem 3.1 *There exist coding strategy $\{\mathscr{E}_t(\cdot)\}_{t\geq 0}$ and controlling strategy $\{\mathscr{D}_t(\cdot)\}_{t\geq 0}$, such that the system (3.1) can be mean-square stabilized over the Gaussian finite-state Markov channel only if $[|\lambda_1|, \dots, |\lambda_d|]' \in \mathbb{R}^d$ satisfy*

$$\left(\prod_{i=1}^{d} |\lambda_i|^{\varrho_i o_i}\right)^{\frac{2}{o}} < \frac{1}{\rho(H_o)}, \tag{3.10}$$

for all $o_i \in \{0, \dots, m_i\}$, $i = 1, \dots, d$ with $o = \sum_{i=1}^{d} \varrho_i o_i$, where $\varrho_i = 1$ if $\lambda_i \in \mathbb{R}$ and $\varrho_i = 2$ otherwise.

Proof: We use uppercase letters $\mathcal{X}, \mathcal{R}, \Gamma$ to denote random variables of the system state, the channel output, and the channel fading, respectively. We use the lowercase letters x, r, γ to denote their realizations. Following a similar line of arguments as in the proof of Lemma 2.1, we can show that

$$\mathcal{N}_{\gamma^t}(\mathcal{X}_{t+1}|\mathcal{R}^t) \geq (\det A)^{\frac{2}{n}} e^{-\frac{2}{n}c_t} \mathcal{N}_{\gamma^{t-1}}(\mathcal{X}_t|\mathcal{R}^{t-1}). \tag{3.11}$$

In view of Proposition II.1 in Freudenberg et al. [2010], a necessary condition to ensure the mean-square stability of $\mathcal{X}_t$ is that the first moment of $\mathcal{N}_{\gamma^t}(\mathcal{X}_{t+1}|\mathcal{R}^t)$ should converge to zero asymptotically. Thus, the MJLS $z_{k+1} = (\det A)^{\frac{2}{n}} e^{-\frac{2}{n}c_t} z_k$ should be stable in the first moment. Following Lemma 3.1, a necessary condition to ensure the mean-square stability can be obtained as

$$(\det A)^{\frac{2}{n}} < \frac{1}{\rho(H_n)}. \tag{3.12}$$

Each block J_i has an invariant real subspace $\mathcal{A}_{o_i}$ of dimension $\varrho_i o_i$, for any $o_i \in \{0, \dots, m_i\}$. Consider the subspace $\mathcal{A}$ formed by taking the product of $\mathcal{A}_{o_i}$, $i = 1, \dots, d$. The total dimension of $\mathcal{A}$ is $\sum_{i=1}^{d} \varrho_i o_i$ and the real Jordan form for the dynamics in the subspace $\mathcal{A}$ is J^{ν} with $|\det J^{\nu}| = \prod_{i=1}^{d} |\lambda_i|^{\varrho_i o_i}$. Since (3.1) is mean-square stabilizable, the dynamics in the subspace $\mathcal{A}$ is also mean-square stabilizable. Following a similar line of arguments as in the derivation of (3.12), the fundamental limitation (3.10) can be obtained. □

Let $\delta = \frac{\sigma_\omega^2}{P+\sigma_\omega^2}$. We can derive the necessity for control over power-constrained Markov lossy channels from Theorem 3.1 directly. First, the following lemma is needed.

Lemma 3.3 *Let Q be defined in (3.4); $D = \begin{bmatrix} 1 & 0 \\ 0 & \delta \end{bmatrix}$ with $0 < \mathfrak{q}, \mathfrak{p}, \delta < 1$; $\lambda \in \mathbb{R}$, $|\lambda| \geq 1$ and T_k^* be defined in (3.7). The following statements are equivalent,*

1. $\lambda^2 \rho(Q'D) < 1$,
2. $\mathbb{E}\left\{\lambda^{2T_k^*}\right\} \delta < 1$,
3. $$1 - \lambda^2(1-\mathfrak{q}) > 0, \tag{3.13}$$
$$\lambda^2 \delta \left[1 + \frac{\mathfrak{p}(\lambda^2 - 1)}{1 - \lambda^2(1-\mathfrak{q})}\right] < 1. \tag{3.14}$$

Proof: (2)↔(3): In view of the probability distribution of T_k^* in Lemma 3.2, we have

$$\begin{aligned}
\mathbb{E}\left\{\lambda^{2T_k^*}\right\} &= \sum_{i=1}^{\infty} \Pr(T_k^* = i)\lambda^{2i} \\
&= \Pr(T_k^* = 1)\lambda^2 + \sum_{i=2}^{\infty} \Pr(T_k^* = i)\lambda^{2i} \\
&= (1-\mathfrak{p})\lambda^2 + \sum_{i=2}^{\infty} \mathfrak{p}\mathfrak{q}(1-\mathfrak{q})^{i-2}\lambda^{2i}.
\end{aligned}$$

To guarantee the boundedness of $\mathbb{E}\left\{\lambda^{2T_k^*}\right\}$, we should have $\lambda^2(1-\mathfrak{q})<1$. Then it follows

$$\mathbb{E}\left\{\lambda^{2T_k^*}\right\} = (1-\mathfrak{p})\lambda^2 + \frac{\mathfrak{p}\mathfrak{q}}{(1-\mathfrak{q})^2}\frac{(1-\mathfrak{q})^2\lambda^4}{1-\lambda^2(1-\mathfrak{q})}$$
$$= \lambda^2\left[1+\frac{\mathfrak{p}(\lambda^2-1)}{1-\lambda^2(1-\mathfrak{q})}\right].$$

Summarizing the above results, we have

$$\mathbb{E}\left\{\lambda^{2T_k^*}\right\} = \begin{cases} \infty, & \text{if } \lambda^2(1-\mathfrak{q})>1, \\ \lambda^2\left[1+\frac{\mathfrak{p}(\lambda^2-1)}{1-\lambda^2(1-\mathfrak{q})}\right], & \text{if } \lambda^2(1-\mathfrak{q})<1. \end{cases}$$

Then the equivalence of (2) and (3) is straightforward from the expression of $\mathbb{E}\left\{\lambda^{2T_k^*}\right\}$.

(1)→(3): Let

$$H = Q'D = \begin{bmatrix} 1-\mathfrak{q} & \mathfrak{p}\delta \\ \mathfrak{q} & (1-\mathfrak{p})\delta \end{bmatrix}.$$

Since H is a nonnegative matrix, in view of Corollary 8.1.20 in Horn and Johnson [1985], $1-\mathfrak{q}\le\rho(H)<\frac{1}{\lambda^2}$, which is (3.13). Suppose the two eigenvalues of H are ζ_1 and ζ_2, then $\zeta_1+\zeta_2=\mathrm{tr}(H)$, $\zeta_1\zeta_2=\det(H)$ with $\mathrm{tr}(H)=(1-\mathfrak{q})+(1-\mathfrak{p})\delta$ and $\det(H)=(1-\mathfrak{p}-\mathfrak{q})\delta$. Since

$$\mathrm{tr}(H)^2-4\det(H) = ((1-\mathfrak{q})+(1-\mathfrak{p})\delta)^2-4(1-\mathfrak{p}-\mathfrak{q})\delta$$
$$= ((1-\mathfrak{q})-(1-\mathfrak{p})\delta)^2+4\mathfrak{p}\mathfrak{q}\delta>0,$$

we know that the spectral radius of H is

$$\rho(H) = \frac{\mathrm{tr}(H)+\sqrt{\mathrm{tr}(H)^2-4\det(H)}}{2}.$$

Since $\lambda^2\rho(H)<1$, we have that $\lambda^2\sqrt{\mathrm{tr}(H)^2-4\det(H)}<2-\lambda^2\,\mathrm{tr}(H)$. Taking square of both sides, we obtain

$$\lambda^4\det(H)-\lambda^2\mathrm{tr}(H)+1>0.$$

Substituting the expression of $\mathrm{tr}(H)$ and $\det(H)$ into the above inequality, we have

$$\lambda^4(1-\mathfrak{q}-\mathfrak{p})\delta-\lambda^2[(1-\mathfrak{q})+(1-\mathfrak{p})\delta]+1>0,$$

which implies

$$\lambda^2\delta[(1-\mathfrak{p})-\lambda^2(1-\mathfrak{p}-\mathfrak{q})]<1-\lambda^2(1-\mathfrak{q}).$$

Dividing both sides by $1-\lambda^2(1-\mathfrak{q})$, we can obtain (3.14).

(3)→(1): We first note that $\lambda^2\delta < 1$ from (3.14). In view of (3.13), we further have

$$2 - \lambda^2\mathrm{tr}(H) = 1 - \lambda^2(1 - \mathfrak{q}) + 1 - \lambda^2\delta(1 - \mathfrak{p}) > 0.$$

Then (3)→(1) can be proved by reversing the proof of (1)→(3). The proof is completed. □

The fundamental limitation for control over power-constrained Markov lossy channels is stated below.

Theorem 3.2 *There exist coding and controlling strategies $\{\mathscr{E}_t(\cdot)\}_{t\geq 0}$, $\{\mathscr{D}_t(\cdot)\}_{t\geq 0}$, such that the system* (3.1) *can be mean-square stabilized over the power- constrained Markov lossy channel only if* $[|\lambda_1|, \dots, |\lambda_d|]' \in \mathbb{R}^d$ *satisfy*

$$1 - \left(\prod_{i=1}^{d} |\lambda_i|^{\varrho_i o_i}\right)^{\frac{2}{o}} (1 - \mathfrak{q}) > 0, \tag{3.15}$$

$$\delta^{\frac{1}{o}} \left(\prod_{i=1}^{d} |\lambda_i|^{\varrho_i o_i}\right)^{\frac{2}{o}} \left[1 + \frac{\mathfrak{p}\left(\left(\prod_{i=1}^{d} |\lambda_i|^{\varrho_i o_i}\right)^{\frac{2}{o}} - 1\right)}{1 - (1 - \mathfrak{q})\left(\prod_{i=1}^{d} |\lambda_i|^{\varrho_i o_i}\right)^{\frac{2}{o}}}\right] < 1, \tag{3.16}$$

for all $o_i \in \{0, \dots, m_i\}$, $i = 1, \dots, d$ *with* $o = \sum_{i=1}^{d} \varrho_i o_i$.

Proof: Since

$$H_o = Q' D_o = \begin{bmatrix} 1 - \mathfrak{q} & \mathfrak{p} \\ \mathfrak{q} & 1 - \mathfrak{p} \end{bmatrix} \begin{bmatrix} 1 & 0 \\ 0 & \delta^{\frac{1}{o}} \end{bmatrix},$$

for power-constrained Markov lossy channels, in view of Theorem 3.1 and Lemma 3.3, the necessity can be obtained. □

In the subsequent analysis, we will show that the necessary conditions in Theorems 3.1 and 3.2 are also sufficient for scalar systems and certain high-dimensional systems.

3.4 Stabilization over Finite-State Markov Channels

In this section, we provide a sufficient stabilization condition for control over Gaussian finite-state Markov channels via the construction of observer, estimator, controller, channel encoder, decoder, and scheduler. The observer/estimator/controller is reproduced from Minero et al. [2013] and Nair and Evans [2004], which mimics the optimal estimation and control scheme in linear quadratic

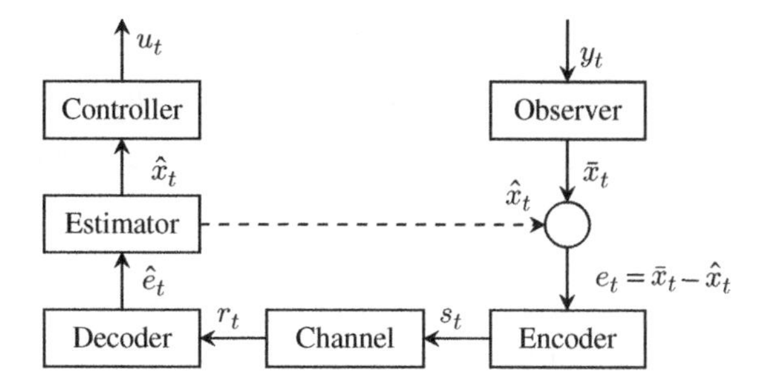

Figure 3.2 Communication structure. Source: L. Xu et al. 2018/with permission of IEEE.

Gaussian (LQG) control [Tatikonda et al., 2004]. The channel encoder/decoder/scheduler design is borrowed from Chapter 2, which adopts a time division multiple access (TDMA) scheme to transmit multiple sources over a scalar channel.

3.4.1 Communication Structure

The entire communication scheme is shown in Figure 3.2. The observer and estimator can be regarded as the source encoder and decoder, which take the measurement signal y_t to estimate the system state $\hat{x}_t$. The channel encoder and decoder are designed to reliably transmit source signals over the uncertain channel. Since the observer/encoder is aware of the one-step delayed channel fading and channel output via the feedback link, it can thus simulate the decoder/estimator/controller to obtain the estimated state $\hat{x}_t$ and the control input u_t.

3.4.2 Observer/Estimator/Controller Design

The Luenberger observer is designed as

$$\bar{x}_{t+1} = A\bar{x}_t + Bu_t - L(y_t - C\bar{x}_t), \tag{3.17}$$

where $\bar{x}_0 = 0$ and L is selected such that $A + LC$ is Hurwitz. The estimator generates the estimate $\hat{x}_t$ with

$$\hat{x}_{t+1} = A\hat{x}_t + A\hat{e}_t + Bu_t, \tag{3.18}$$

where $\hat{x}_0 = 0$ and $\hat{e}_t$ is the output of the channel decoder. The controller is given by

$$u_t = K\hat{x}_t, \tag{3.19}$$

where K is selected such that $A + BK$ is Hurwitz. With the above observer, estimator, and controller design, we have the following result.

Lemma 3.4 *If there exists a pair of channel encoder and decoder, such that* $\sup_t \mathbb{E}\left\{\|e_t\|^2\right\} < \infty$ *with* $e_t = \bar{x}_t - \hat{x}_t$*, the system* (3.1) *is mean-square stabilizable over the Gaussian finite-state Markov channel with the designed communication structure.*

Proof: In view of (3.1) and (3.17), we have

$$x_{t+1} - \bar{x}_{t+1} = (A + LC)(x_t - \bar{x}_t) + v_t + Lw_t. \tag{3.20}$$

Since L is selected such that $A + LC$ is stable, we have

$$\sup_t \mathbb{E}\left\{\|x_t - \bar{x}_t\|^2\right\} < \infty.$$

From the observer dynamics (3.17) and the controller (3.19), we have

$$\begin{aligned}\bar{x}_{t+1} &= (A + BK)\bar{x}_t - BK(\bar{x}_t - \hat{x}_t) - L(y_t - C\bar{x}_t)\\ &= (A + BK)\bar{x}_t - BK(\bar{x}_t - \hat{x}_t) - LC(x_t - \bar{x}_t) - Lw_t.\end{aligned}$$

Since $A + BK$ is Hurwitz and $\sup_t \mathbb{E}\left\{\|x_t - \bar{x}_t\|^2\right\} < \infty$, if $\sup_t \mathbb{E}\left\{\|e_t\|^2\right\} < \infty$, we have $\sup_t \mathbb{E}\left\{\|\bar{x}_t\|^2\right\} < \infty$. Therefore, we have

$$\begin{aligned}\sup_t \mathbb{E}\left\{\|x_t\|^2\right\} &= \sup_t \mathbb{E}\left\{\|x_t - \bar{x}_t + \bar{x}_t\|^2\right\}\\ &\le \sup_t \mathbb{E}\left\{\|x_t - \bar{x}_t\|^2\right\} + \sup_t \mathbb{E}\left\{\|\bar{x}_t\|^2\right\} < \infty,\end{aligned}$$

which implies that the original system (3.1) is mean-square stable. The proof is completed. □

In view of the above lemma, we are now to design the channel encoder/decoder to ensure that $\sup_t \mathbb{E}\left\{\|e_t\|^2\right\} < \infty$. The dynamics for e_t is

$$e_{t+1} = A(e_t - \hat{e}_t) + \Phi_t, \tag{3.21}$$

where $\Phi_t = -LC(x_t - \bar{x}_t) - Lw_t$.

From (3.20), we have that

$$x_t - \bar{x}_t = (A + LC)^t x_0 + \sum_{i=0}^{t-1} (A + LC)^{t-1-i}(v_i + Lw_i).$$

Since x_0, $\{v_t\}_{t\ge0}$, $\{w_t\}_{t\ge0}$ are independent and with zero mean and bounded variance, $x_t - \bar{x}_t$ and thus Φ_t are with zero mean and bounded variance.

Remark 3.2 Assumption 3.1 can be justified from Lemma 3.4. Assume $A = \text{diag}(J_u, J_s)$, where J_u contains eigenvalues that are either on or outside the unit circle and J_s contains eigenvalues that are within the unit circle. Decompose the dynamics (3.21) into stable part and unstable part according to A as

$$e_{u,t+1} = J_u e_{u,t} + \Phi_{u,t} - J^u \hat{e}_{u,t}, \tag{3.22}$$

$$e_{s,t+1} = J_s e_{s,t} + \Phi_{s,t} - J^s \hat{e}_{s,t}, \tag{3.23}$$

where $e_{u,t}, e_{s,t}, \Phi_{u,t}, \Phi_{s,t}, \hat{e}_{u,t}, \hat{e}_{s,t}$ are the corresponding partitions of e_t, Φ_t, and $\hat{e}_t$. Since Φ_t is mean-square bounded, if we let $\hat{e}_{s,t} = 0$ at the decoder side, (3.23) is mean-square stable. Thus, we do not need to consider the transmission of the information corresponding to stable eigenvalues. Therefore, we can ignore the eigenvalues that are in the unit circle without loss of generality.

3.4.3 Encoder/Decoder/Scheduler Design

To transmit the n-dimensional vector e_t through the scalar channel, the TDMA strategy is used. There are n encoder/decoder pairs to transmit the n sources $\{e_{1,t}, \dots, e_{n,t}\}$ with $e_{i,t}$ being the ith value of e_t and a scheduler to multiplex the channel use. Suppose at time t, the ith encoder/decoder pair is scheduled to use the channel. The encoder i first generates a symbol $s_{i,t}$, which is a scaled version of $e_{i,t}$ to satisfy the channel input power constraint and transmits it to the decoder through the communication channel. The decoder i then forms the minimal mean-square error estimate $\hat{e}_{i,t}$ based on the channel output $r_{i,t}$. The estimator maintains an array $\hat{e}_t = [\hat{e}_{1,t}, \dots, \hat{e}_{n,t}]'$ that represents the estimate of e_t, which is set to 0 at $t = 0$. When the information about $e_{i,t}$ is transmitted, only $\hat{e}_{i,t}$ is updated at the estimator side. The channel encoder/decoder/scheduler structure is illustrated in Figure 3.3.

If at time t, the encoder i is scheduled to use the channel, then the encoder generates

$$s_{i,0} = 0, \quad s_{i,t} = \sqrt{\frac{P}{\sigma^2_{e_{i,t}}}} e_{i,t}, \; t \geq 1, \tag{3.24}$$

where $\sigma^2_{e_{i,t}}$ represents the variance of $e_{i,t}$. The decoder i satisfies

$$\hat{e}_{i,t} = \frac{\mathbb{E}_{\gamma_t}\{r_{i,t}e_{i,t}\}}{\mathbb{E}_{\gamma_t}\{r^2_{i,t}\}} r_{i,t}. \tag{3.25}$$

It is clear from (3.21) and the designed communication scheme that $\mathbb{E}\left\{e_t\right\} = 0$ and $\mathbb{E}\left\{\hat{e}_t\right\} = 0$.

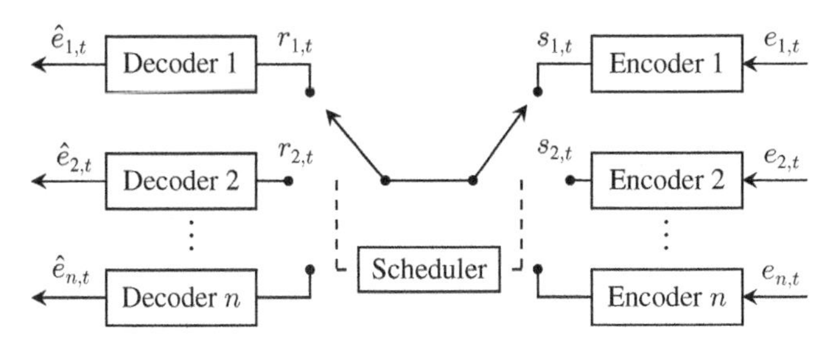

Figure 3.3 Channel encoder/decoder/scheduler structure. Source: L. Xu et al. 2018/with permission of IEEE.

The scheduling Algorithm 3.1 is designed, which adopts a TDMA strategy and allocates a fixed transmission period to each encoder/decoder pair, where $\tau_i, i = 1, \dots, n$ are scheduler parameters to be specified later. We assume that both the encoder and the decoder know the scheduling algorithm. Since the switching among transmissions is only determined by time, we do not need to consider the coordination among encoders and decoders.

3.4.4 Sufficient Stabilizability Conditions

Theorem 3.3 *If*

$$\prod_{i=1}^{d} |\lambda_i|^{2v_i} < \frac{1}{\rho(H_1)}, \tag{3.26}$$

there exist τ_i, $i = 1, \dots, n$, such that the system (3.1) *can be mean-square stabilized over the Gaussian finite-state Markov channel with the proposed TDMA communication scheme.*

Algorithm 3.1: TDMA Scheduler for Gaussian Finite-state Markov Channels

---

In the k-th round of transmissions

- The first encoder/decoder pair is scheduled to use the channel for τ_1 times.
- ...
- The jth encoder/decoder pair is scheduled to use the channel for τ_j times.
- ...
- The nth encoder/decoder pair is scheduled to use the channel for τ_n times.
- Repeat this process.

In view of Lemma 3.4, if (3.21) is mean-square stable, the system (3.1) can be mean-square stabilized over the Gaussian finite-state Markov channel. Thus, the key in proving Theorem 3.3 is to show that there exist τ_is such that (3.21) is mean-square stable. Moreover, with the designed TDMA communication scheme, we can show that each subsystem in (3.21) is described by a MJLS. If (3.26) holds, we have that $|\lambda_i|^{\frac{2\sum_j \ln|\lambda_j|}{\ln|\lambda_i|}} \rho(H_1) < 1$, for $i = 1, \dots, n$ (for the case that $\lambda_1, \dots, \lambda_d$ are real and $m_i = v_i = 1$). If τ_i is selected such that $\frac{\tau_i}{\sum_j \tau_j} = \frac{\ln|\lambda_i|}{\sum_j \ln|\lambda_j|}$, the MJLS is stable, which further implies the mean-square stability of (3.21). Then the original system is mean-square stable. The detailed proof is provided as below.

Proof: Without loss of generality, we assume that $\lambda_1, \dots, \lambda_d$ are real and $m_i = v_i = 1$. For other cases, the theorem can be proved by combining the following analysis with a similar line of arguments used in Zaidi et al. [2014b].

In the first step, we shall derive the dynamics for the mean-square value of $e_{i,t}$. From (3.21), we obtain

$$e_{i,t+1} = \lambda_i(e_{i,t} - \hat{e}_{i,t}) + \Phi_{i,t}. \tag{3.27}$$

Analogous to the analysis in Xu et al. [2016], we can show that with the encoder (3.24) and the decoder (3.25),

$$\mathbb{E}_{\gamma^{t+1}}\{e_{i,t+1}^2\} = \lambda_i^2 e^{-2c_i}\mathbb{E}_{\gamma^t}\{e_{i,t}^2\} + \mathbb{E}\left\{\Phi_{i,t}^2\right\}, \tag{3.28}$$

if the ith encoder/decoder pair is scheduled to use the channel at time t. Let $\tau = \sum_{i=1}^n \tau_i$ and $\eta_{i,k\tau} = \mathbb{E}\{e_{1,k\tau}^2 | \gamma_{k\tau} = \mathfrak{r}_i\}\Pr(\gamma_{k\tau} = \mathfrak{r}_i)$. Since from time $k\tau + 1$ to $k\tau + \tau_1$, the first encoder/decoder pair is scheduled to use the channel from Algorithm 3.1, we have that

$$\begin{aligned}
\eta_{j,k\tau+1} &= \mathbb{E}\left\{e_{1,k\tau+1}^2 | \gamma_{k\tau+1} = \mathfrak{r}_j\right\} \Pr(\gamma_{k\tau+1} = \mathfrak{r}_j) \\
&= \sum_{i=1}^{l} \Pr(\gamma_{k\tau} = \mathfrak{r}_i | \gamma_{k\tau+1} = \mathfrak{r}_j)\Pr(\gamma_{k\tau+1} = \mathfrak{r}_j) \\
&\quad \times \mathbb{E}\left\{e_{1,k\tau+1}^2 | \gamma_{k\tau+1} = \mathfrak{r}_j, \gamma_{k\tau} = \mathfrak{r}_i\right\} \\
&\overset{(a)}{=} \sum_{i=1}^{l} \Pr(\gamma_{k\tau+1} = \mathfrak{r}_j | \gamma_{k\tau} = \mathfrak{r}_i)\Pr(\gamma_{k\tau} = \mathfrak{r}_i) \\
&\quad \times \mathbb{E}\left\{e_{1,k\tau+1}^2 | \gamma_{k\tau+1} = \mathfrak{r}_j, \gamma_{k\tau} = \mathfrak{r}_i\right\} \\
&\overset{(b)}{=} \sum_{i=1}^{l} \mathfrak{q}_{ij} \Pr(\gamma_{k\tau} = \mathfrak{r}_i)\mathbb{E}\left\{e_{1,k\tau+1}^2 | \gamma_{k\tau} = \mathfrak{r}_i\right\} \\
&\overset{(c)}{\leq} \sum_{i=1}^{l} \mathfrak{q}_{ij} \Pr(\gamma_{k\tau} = \mathfrak{r}_i)\frac{\lambda_1^2}{e^{2c_i}}\mathbb{E}\left\{e_{1,k\tau}^2 | \gamma_{k\tau} = \mathfrak{r}_i\right\} + \mathbb{E}\left\{\Phi_{1,k\tau}^2\right\} \\
&= \sum_{i=1}^{l} \frac{\lambda_1^2}{e^{2c_i}}\mathfrak{q}_{ij}\eta_{i,k\tau} + \mathbb{E}\left\{\Phi_{1,k\tau}^2\right\},
\end{aligned}$$

where (a) follows from the Bayes law; (b) is due to the fact that $e_{1,k\tau+1}$ is independent of $\gamma_{k\tau+1}$; and (c) arises from (3.28). Let $\eta_{k\tau} = [\eta_{1,k\tau}, \eta_{2,k\tau}, \dots, \eta_{l,k\tau}]'$, then we have $\eta_{k\tau+1} \leq \lambda_1^2 Q' D_1 \eta_{k\tau} + \mathbf{1}\mathbb{E}\left\{\Phi_{1,k\tau}^2\right\}$, where $\mathbf{1}$ is a vector with all elements being one. With similar derivations, we have that

$$\eta_{k\tau+\tau_1} \leq \lambda_1^{2\tau_1} H_1^{\tau_1}\eta_{k\tau} + \sum_{i=0}^{\tau_1-1}(\lambda_1^2 H_1)^{\tau_1-1-i}\mathbf{1}\mathbb{E}\left\{\Phi_{1,k\tau+i}^2\right\}. \tag{3.29}$$

Since from the time $k\tau + \tau_1 + 1$ to $(k+1)\tau$, there are no scheduled transmissions for the first encoder/decoder pair, similar to the derivation of (3.29), we have

$$\eta_{(k+1)\tau} \le \lambda_1^{2(\tau-\tau_1)}(Q')^{\tau-\tau_1}\eta_{k\tau+\tau_1} + \sum_{i=0}^{\tau-\tau_1-1}(\lambda^2 Q')^{\tau-\tau_1-1-i}\mathbf{1}\mathbb{E}\left\{\Phi^2_{1,k\tau+\tau_1+i}\right\}. \quad (3.30)$$

Combining (3.29) and (3.30), we have that

$$\eta_{(k+1)\tau} \le \lambda_1^{2\tau}(Q')^{\tau-\tau_1}H_1^{\tau_1}\eta_{k\tau} + \Psi_{k\tau}, \quad (3.31)$$

where

$$\Psi_{k\tau} = \lambda_1^{2(\tau-\tau_1)}(Q')^{\tau-\tau_1}\sum_{i=0}^{\tau_1-1}(\lambda_1^2 H_1)^{\tau_1-1-i}\mathbf{1}\mathbb{E}\left\{\Phi^2_{1,k\tau+i}\right\}$$
$$+ \sum_{i=0}^{\tau-\tau_1-1}(\lambda^2 Q')^{\tau-\tau_1-1-i}\mathbf{1}\mathbb{E}\left\{\Phi^2_{1,k\tau+\tau_1+i}\right\}$$

and $\Psi_{k\tau}$ is bounded.

In the second step, we will show that if the sufficient condition (3.26) is satisfied, there exist τ_is such that (3.31) is mean-square stable.

If (3.26) holds, we have $\ln \rho(H_1) + 2\sum_j \ln|\lambda_j| < 0$. Therefore, there exists $\varsigma > 0$, such that $\ln \rho(H_1) + 2\sum_j \ln|\lambda_j| + \varsigma = 0$, which also implies $2\ln|\lambda_i| + \alpha_i \ln \rho(H_1) = -\frac{\varsigma}{n} < 0$, with $\alpha_i = \frac{2\ln|\lambda_i| + \frac{\varsigma}{n}}{2\sum_j \ln|\lambda_j| + \varsigma} > 0$ and $\sum_i \alpha_i = 1$. Thus, we have $\lambda_i^2 \rho(H_1)^{\alpha_i} < 1$ for all $i = 1, \ldots, n$. Let $\iota = \min_i(2\log_{\rho(H_1)}|\lambda_i| + \alpha_i) > 0$. Since for every $\alpha_i \in \mathbb{R}$, there exists a rational sequence $\{\beta_{i,k}\}_{k\ge0}$, such that $\lim_{k\to\infty}\beta_{i,k} = \alpha_i$, we have $\lim_{k\to\infty}\frac{\beta_{i,k}}{\sum_j \beta_{j,k}} = \frac{\alpha_i}{\sum_j \alpha_j} = \alpha_i$. Therefore, for the given ι, there exists $M \in \mathbb{N}^+$, such that $\left|\frac{\beta_{i,M}}{\sum_j \beta_{j,M}} - \alpha_i\right| < \iota$. Let $\vartheta_i = \frac{\sum_j \beta_{j,M}}{\beta_{i,M}}$. Then $\vartheta_i^{-1} > \alpha_i - \iota \ge -2\log_{\rho(H_1)}|\lambda_i|$. Thus, we have $\lambda_i^{2\vartheta_i}\rho(H_1) < 1$.

In view of Lemma 5.6.10 in Horn and Johnson [1985], there exists a norm $\|\cdot\|$ such that $\kappa_i := \|\lambda_i^{2\vartheta_i}H_1\| < 1$. From the equivalence of norms, we have that $\|\cdot\| \le \epsilon\|\cdot\|_1$ for some $\epsilon > 1$. Then $\tau_i \in \mathbb{N}^+$ is selected to satisfy that $\tau_i > -\log_{\kappa_i}\epsilon$ and $\beta_{i,M} = \frac{\tau_i}{\overline{\tau}}$ for all $i = 1, \ldots, n$ and for some $\overline{\tau}$. The existence of such τ_is can always be guaranteed by first writing rational numbers $\beta_{i,M}$s into fractions and then reducing fractions to a common denominator and finally scaling the numerators and denominators simultaneously to obtain a sufficiently large numerator τ_i which satisfies $\tau_i > -\log_{\kappa_i}\epsilon$.

Then we have from (3.31) that

$$\begin{aligned}
\|\eta_{(k+1)\tau}\| &\le \|(Q')^{\tau-\tau_1}\|\|\lambda_1^{2\tau}H_1^{\tau_1}\|\|\eta_{k\tau}\| + \|\Psi_{k\tau}\| \\
&\le \kappa_1^{\tau_1}\|(Q')^{\tau-\tau_1}\|\|\eta_{k\tau}\| + \|\Psi_{k\tau}\| \\
&\le \kappa_1^{\tau_1}\epsilon\|(Q')^{\tau-\tau_1}\|_1\|\eta_{k\tau}\| + \|\Psi_{k\tau}\| \\
&\le \kappa_1^{\tau_1}\epsilon\|\eta_{k\tau}\| + \|\Psi_{k\tau}\|.
\end{aligned}$$

Since $\kappa_1^{\tau_1}\epsilon < 1$, we know that $\|\eta_{k\tau}\|$ is mean-square bounded. Since $\mathbb{E}\left\{e_{1,k\tau}^2\right\} = \sum_i^l \eta_{i,k\tau}$, we further have that $\mathbb{E}\left\{e_{1,k\tau}^2\right\}$ is mean-square bounded.

Similarly, we can also prove that $\sup_k \mathbb{E}\left\{e_{i,k\tau}^2\right\} < \infty$ for all $i = 2, \dots, n$. Therefore, e_t is mean-square bounded. In view of Lemma 3.4, the closed-loop system is mean-square stable. The proof is completed. □

Remark 3.3 Suppose $\mathfrak{q}_{ij} = \mathfrak{q}_j$ for $i,j = 1, \dots, \mathfrak{l}$, then the Gaussian finite-state Markov channel degenerates to the power-constrained fading channel with finite i.i.d. channel states. The stabilization condition in Theorem 3.3 becomes

$$\prod_{i=1}^{d} |\lambda_i|^{2\nu_i} \left(\sum_{i=1}^{\mathfrak{l}} \mathfrak{q}_i \frac{\sigma_\omega^2}{\sigma_\omega^2 + \mathfrak{r}_i^2 \mathcal{P}} \right) < 1,$$

which coincides with Theorem 2.4.

The sufficient condition is also necessary for scalar systems as shown in the following corollary.

Corollary 3.1 *Suppose $A = \lambda_1$ with $\lambda_1 \in \mathbb{R}$ and $|\lambda_1| \geq 1$. There exist coding and controlling strategies $\{\mathscr{E}_t(\cdot)\}_{t\geq 0}, \{\mathscr{D}_t(\cdot)\}_{t\geq 0}$, such that the system (3.1) can be mean-square stabilized over the Gaussian finite-state Markov channel if and only if*

$$\lambda_1^2 < \frac{1}{\rho(H_1)}.$$

Generally, there exists a gap between the necessity (3.10) and the sufficiency (3.26) for high- dimensional systems. In Section 3.5, we will study power-constrained Markov lossy channels and derive improved results.

3.5 Stabilization over Markov Lossy Channels

In this section, by utilizing the properties of the Markov lossy process, we propose communication scheduling algorithms for power-constrained Markov lossy channels and show that they can achieve larger stabilizability regions than that with the TDMA scheduler. We first start with two-dimensional systems.

3.5.1 Two-Dimensional Systems

The necessary and sufficient condition to ensure the mean-square stabilizability for two-dimensional systems controlled over power-constrained Markov lossy channels is stated in the following theorem.

Theorem 3.4 *Suppose* $n = 2$. *There exist coding and controlling strategies* $\{\mathcal{E}_t(\cdot)\}_{t\geq 0}$, $\{\mathcal{D}_t(\cdot)\}_{t\geq 0}$, *such that the system* (3.1) *can be mean-square stabilized over the power- constrained Markov lossy channel if and only if* (3.15) *and* (3.16) *hold.*

For the case of two-dimensional systems with eigenvalues of equal magnitude, the communication scheme designed in Section 3.5.2 is shown to be optimal (in the sense that it achieves the largest stabilizability region indicated by the necessary condition in Theorem 3.2); see Corollary 3.2. In this subsection, we only provide the optimal communication scheme for two-dimensional systems with eigenvalues having different magnitudes, i.e. $A = \begin{bmatrix} \lambda_1 & 0 \\ 0 & \lambda_2 \end{bmatrix}$ with $\lambda_1, \lambda_2 \in \mathbb{R}$ and $|\lambda_1| > |\lambda_2| \geq 1$. In view of Theorem 3.4, we only need to prove that the following conditions are sufficient:

$$(1 - \mathfrak{q})\lambda_1^2 < 1, \tag{3.32}$$

$$\delta\lambda_1^2 \left[1 + \frac{\mathfrak{p}(\lambda_1^2 - 1)}{1 - (1 - \mathfrak{q})\lambda_1^2}\right] < 1, \tag{3.33}$$

$$\delta^{\frac{1}{2}}|\lambda_1\lambda_2| \left[1 + \frac{\mathfrak{p}(|\lambda_1\lambda_2| - 1)}{1 - (1 - \mathfrak{q})|\lambda_1\lambda_2|}\right] < 1. \tag{3.34}$$

Remark 3.4 A small δ, a large $\mathfrak{q}$, and a small $\mathfrak{p}$ are always preferred, which correspond to a more reliable channel, and thus can tolerate more unstable systems. This is confirmed from (3.32), (3.33), and (3.34).

3.5.1.1 Optimal Scheduler Design

The communication structure is designed similarly as in Section 3.4 with the same observer/estimator/controller design and the channel encoder/decoder design.

The scheduling Algorithm 3.2 is then proposed, where $\phi = 2\frac{\ln|\lambda_1| - \ln|\lambda_2|}{\ln\delta}$ and τ_1 is the scheduler parameter to be specified later. Since the switching among transmissions in Algorithm 3.2 relies on the channel state information, which is known to the decoder and the encoder via the channel feedback, we do not need to consider the coordination among encoders and decoders. Algorithm 3.2 is based on the optimal scheduling algorithm for control over power-constrained lossy channels in Section 2.4.2, where it is shown that such allocation of channel resources is optimal for the stabilization of two-dimensional systems with i.i.d. channel states. Even though the channel state $\{\gamma_t\}_{t\geq 0}$ is correlated over time for the power- constrained Markov lossy channel, the sojourn time $\{T_k^*\}_{k>0}$ is i.i.d. We may study the channel from the perspective of the i.i.d. sojourn time sequence and expect that Algorithm 3.2 is optimal as well.

The right-hand side of (3.36) is the requirement on the minimal successful transmission numbers $\tau_{2,k}$. Even if transmissions fail consecutively, since $\tau_1 + (T_k^1 + t)\phi$

Algorithm 3.2: Chasing and Optimal Stopping Scheduler for Power-Constrained Markov Lossy Channels

In the k-th round,

- The first encoder/decoder pair is scheduled to use the channel until the transmissions succeed for τ_1 times. Denote the time period to achieve this object as T_k^1.
- – If

$$T_k^1 < -\frac{\tau_1}{\phi}, \tag{3.35}$$

  the second encoder/decoder pair is scheduled to use the channel until the transmissions succeed for $\tau_{2,k}$times with

$$\tau_{2,k} > \tau_1 + (T_k^1 + T_k^2)\phi, \tag{3.36}$$

  where T_k^2 denotes the minimal period of achieving this object.
  – Otherwise, set $T_k^2 = 0$ and do not conduct any transmissions.
- Repeat.

is diminishing with time t, the stopping condition (3.36) would be satisfied eventually, which means that T_k^2 is bounded. To make notions clear, we plot the scheduled transmissions and the first round transmission in Figure 3.4 and Figure 3.5, respectively, where the definitions of T_k, T_k^*, T_k^1, T_k^2, and the new symbols $\overline{T}_k, \check{T}_k$ are summarized in Table 3.1. It is clear from Algorithm 3.2 that $\overline{T}_i$ and $\overline{T}_j$ are i.i.d.; T_i^2 is independent of T_j^2 for any $i \neq j$. Besides, we have $T_1^1 = T_1^* + \cdots + T_{\tau_1}^*$, $T_1^2 = T_{\tau_1+1}^* + \cdots + T_{\tau_1+\tau_{2,1}}^*$.

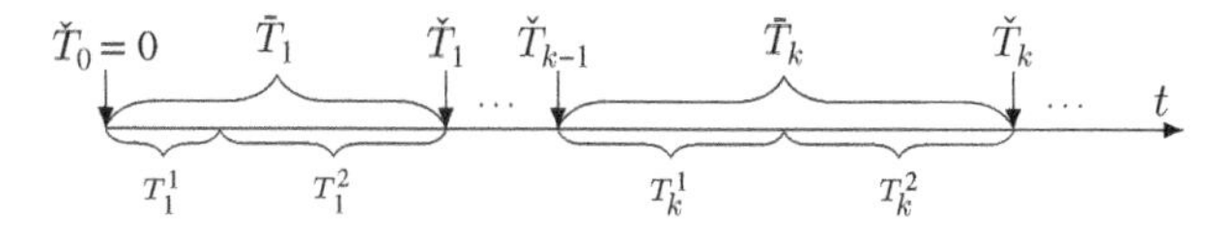

Figure 3.4 Scheduled transmissions with Algorithm 3.2.

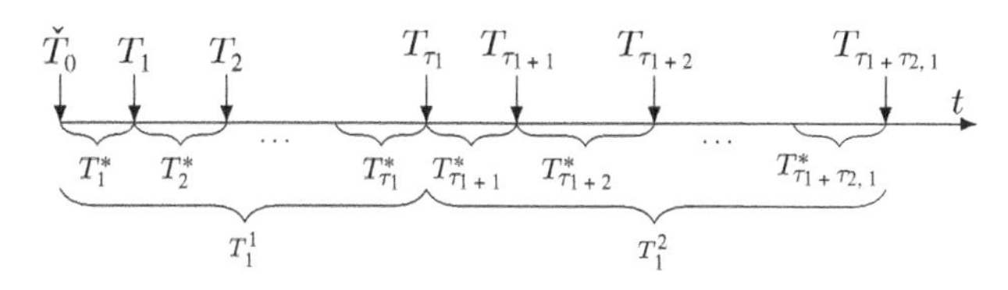

Figure 3.5 The first round transmission with Algorithm 3.2. Source: L. Xu et al., 2018/with permission of IEEE.

Table 3.1 Lists of transmission-related definitions.

$T_k, k \geq 0$	**The time when the transmission is successful as defined in (3.6)**
$T_k^*, k \geq 1$	Time duration between two successive successful transmissions as defined in (3.7)
$T_k^1, k \geq 1$	The period to transmit the first encoder/decoder pair in the k-th round transmission
$T_k^2, k \geq 1$	The period to transmit the second encoder/decoder pair in the k-th round transmission
$\overline{T}_k, k \geq 1$	The total time to complete the k-th round of transmissions, i.e. $\overline{T}_k = T_k^1 + T_k^2$
$\check{T}_k, k \geq 0$	The time when k rounds of transmissions are completed, i.e. $\check{T}_k = \sum_{j=1}^{k} \overline{T}_j$

Source: L. Xu et al., 2018/with permission of IEEE.

3.5.1.2 Scheduler Parameter Selection

If (3.32) holds, we have

$$\mathbb{E}\left\{\lambda_1^{2T_1^*}\right\} = \lambda_1^2 \left[1 + \frac{\mathfrak{p}(\lambda_1^2 - 1)}{1 - (1 - \mathfrak{q})\lambda_1^2}\right] > 1.$$

Since $(1 - \mathfrak{q})|\lambda_1 \lambda_2| < 1$ from (3.32), if (3.34) holds, we have

$$\delta^{\frac{1}{2}} \mathbb{E}\left\{|\lambda_1 \lambda_2|^{T_1^*}\right\} = \delta^{\frac{1}{2}} |\lambda_1 \lambda_2| \left[1 + \frac{\mathfrak{p}(|\lambda_1 \lambda_2| - 1)}{1 - (1 - \mathfrak{q})|\lambda_1 \lambda_2|}\right] < 1.$$

Since $\mathbb{E}\{e^{\theta + bT_1^*}\}$ with $b = 2 \ln |\lambda_1| - \phi\theta$ is increasing in θ, when $\theta = 0$,

$$\mathbb{E}\{e^{\theta + bT_1^*}\} = \mathbb{E}\{\lambda_1^{2T_1^*}\} > 1,$$

and when $\theta = \frac{1}{2} \ln \delta$,

$$\mathbb{E}\{e^{\theta + bT_1^*}\} = \delta^{\frac{1}{2}} \mathbb{E}\{|\lambda_1 \lambda_2|^{T_1^*}\} < 1,$$

we know that there exists θ^* with $\frac{1}{2} \ln \delta < \theta^* < 0$, such that

$$\mathbb{E}\left\{e^{\theta^* + bT_1^*}\right\} = 1. \tag{3.37}$$

The scheduler parameter τ_1 is then selected to satisfy

$$\tau_1 > \max \left\{ \frac{-2 \ln 2 + \theta^*(1 - \phi)}{\ln \delta - 2\theta^*}, \frac{-\ln 2}{\ln \left(\delta \lambda_1^2 \left[1 + \frac{\mathfrak{p}(\lambda_1^2 - 1)}{1 - (1 - \mathfrak{q})\lambda_1^2}\right]\right)} \right\}. \tag{3.38}$$

3.5.1.3 Sufficiency Proof of Theorem 3.4

Denote the event $T_1^1 < -\frac{\tau_1}{\phi}$ as ξ. Let $\mathcal{B}_t = \sum_{i=T_1^1+1}^{t} \gamma_i$ and $Y_t = e^{\theta^* \mathcal{B}_t + bt}$, $t \geq T_1^1 + 1$. For $k > \tau_1$, we have

$$\mathbb{E}_\xi \left\{ Y_{T_k} | Y_{T_{k-1}}, \ldots, Y_{T_{\tau_1}} \right\} = Y_{T_{k-1}} \mathbb{E}_\xi \left\{ e^{\theta^* \sum_{i=T_{k-1}+1}^{T_k} \gamma_i + bT_k^*} \right\}$$
$$= Y_{T_{k-1}} \mathbb{E}_\xi \left\{ e^{\theta^* + bT_k^*} \right\} \overset{(a)}{=} Y_{T_{k-1}},$$

where (a) follows from the fact that $\{T_k^*\}_{k>0}$ are i.i.d. and (3.37). Thus, Y_{T_k} is a martingale in $k > \tau_1$. Then in view of the optional stopping theorem [Ash and Doléans-Dade, 2000], we have $\mathbb{E}\{Y_{T_{\tau_1+\tau_{2,1}}}\} = \mathbb{E}\{Y_{T_{\tau_1+1}}\} = 1$. However, by our stopping condition (3.36), we know that $\mathcal{B}_{T_{\tau_1+\tau_{2,1}}} = \tau_{2,1} = \tau_1 + (T_1^1 + T_1^2)\phi + c$ for some $c \geq 0$. Therefore, $\mathbb{E}_\xi \left\{ Y_{T_{\tau_1+\tau_{2,1}}} \right\} = \mathbb{E}_\xi \left\{ e^{\theta^* \tau_1 + \theta^* \phi (T_1^1 + T_1^2) + \theta^* c + bT_1^2} \right\} = 1$, which implies that

$$\mathbb{E}_\xi \left\{ e^{(\theta^* \phi + b) T_1^2} \right\} = \mathbb{E}_\xi \left\{ \lambda_1^{2T_1^2} \right\} = e^{-\theta^* \tau_1 - \theta^* \phi T_1^1 - \theta^* c}. \tag{3.39}$$

We then show that c is bounded. If at time $\check{T}_1$, the transmission is successful, then $\tau_{2,1} - 1 \leq \tau_1 + (\check{T}_1 - 1)\phi$ since the stopping condition (3.36) is not satisfied at time $\check{T}_1 - 1$. In the consideration that $c = \tau_{2,1} - \tau_1 - \check{T}_1\phi$, we have an upper bound for c as $c \leq 1 - \phi$. Similarly, if at time $\check{T}_1$, the transmission fails, then $\tau_{2,1} \leq \tau_1 + (\check{T}_1 - 1)\phi$. An upper bound for c is therefore given by $c \leq -\phi$. Thus, in general, an upper bound for c can be given by $c \leq 1 - \phi$.

Since $\theta^* > \frac{1}{2} \ln \delta$, $1 - \frac{\theta^*}{\ln \delta} > 0$, which means $2 \ln |\lambda_1| - 2 \ln |\lambda_2| - \theta^* \phi > 0$. When $T_1^1 \geq -\frac{\tau_1}{\phi}$, we further have

$$T_1^1 (2 \ln |\lambda_1| - 2 \ln |\lambda_2| - \theta^* \phi) > \theta^* \tau_1 - \tau_1 \ln \delta + \theta^* c - \ln 2.$$

With some manipulations, we can show that when $T_1^1 \geq -\frac{\tau_1}{\phi}$,

$$\Omega := \lambda_2^{2T_1^1} - 2\lambda_1^{2T_1^1} e^{-\theta^* \tau_1 - \theta^* \phi T_1^1 - \theta^* c} \delta^{\tau_1} < 0.$$

Denote the event that $T_1^1 \geq -\frac{\tau_1}{\phi}$ as ψ. In view of the conditional expectation, we have

$$\mathbb{E}\left\{ \lambda_1^{2\overline{T}_1} \delta^{\tau_1} + \lambda_2^{2\overline{T}_1} \delta^{\tau_{2,1}} \right\}$$
$$= \mathbb{E}\left\{ \mathbb{E}_\xi \left\{ \lambda_1^{2\overline{T}_1} \delta^{\tau_1} + \lambda_2^{2\overline{T}_1} \delta^{\tau_{2,1}} \right\} \right\} + \mathbb{E}\left\{ \mathbb{E}_\psi \left\{ \lambda_1^{2\overline{T}_1} \delta^{\tau_1} + \lambda_2^{2\overline{T}_1} \delta^{\tau_{2,1}} \right\} \right\}$$
$$\overset{(a)}{\leq} \mathbb{E}\left\{ \mathbb{E}_\xi \left\{ 2\lambda_1^{2\overline{T}_1} \delta^{\tau_1} \right\} \right\} + \mathbb{E}\left\{ \mathbb{E}_\psi \left\{ \lambda_1^{2T_1^1} \delta^{\tau_1} + \lambda_2^{2T_1^1} \right\} \right\}$$
$$\overset{(b)}{=} \mathbb{E}\left\{ \mathbb{E}_\xi \left\{ 2\lambda_1^{2T_1^1} e^{-\theta^* \tau_1 - \theta^* \phi T_1^1 - \theta^* c} \delta^{\tau_1} \right\} \right\} + \mathbb{E}\left\{ \mathbb{E}_\psi \left\{ \lambda_1^{2T_1^1} \delta^{\tau_1} + \lambda_2^{2T_1^1} \right\} \right\}$$

$$\leq \mathbb{E}\left\{2\lambda_1^{2T_1^1} e^{-\theta^*\tau_1-\theta^*\phi T_1^1-\theta^* c}\delta^{\tau_1}\right\} + \mathbb{E}\left\{\mathbb{E}_\psi\left\{\lambda_1^{2T_1^1}\delta^{\tau_1} + \Omega\right\}\right\}$$

$$\stackrel{(c)}{\leq} \mathbb{E}\left\{2\lambda_1^{2T_1^1} e^{-\theta^*\tau_1-\theta^*\phi T_1^1-\theta^* c}\delta^{\tau_1}\right\} + \mathbb{E}\left\{\lambda_1^{2T_1^1}\delta^{\tau_1}\right\}, \tag{3.40}$$

where (*a*) follows from (3.35) and (3.36); (*b*) follows from (3.39) and (*c*) is due to the fact that when $T_1^1 \geq -\frac{\tau_1}{\phi}$, $\Omega < 0$. Since

$$\begin{aligned}
\mathbb{E}\left\{2\lambda_1^{2T_1^1} e^{-\theta^*\tau_1-\theta^*\phi T_1^1-\theta^* c}\delta^{\tau_1}\right\} &\stackrel{(a)}{\leq} 2\delta^{\tau_1} e^{-\theta^*\tau_1-\theta^*(1-\phi)}\mathbb{E}\left\{e^{bT_1^1}\right\} \\
&= 2\delta^{\tau_1} e^{-\theta^*\tau_1-\theta^*(1-\phi)}\mathbb{E}\{e^{bT_1^*}\}^{\tau_1} \\
&\stackrel{(b)}{=} 2e^{-\theta^*(1-\phi)}(\delta e^{-2\theta^*})^{\tau_1},
\end{aligned}$$

where (*a*) follows from the fact that $c \leq 1-\phi$; (*b*) follows from (3.37), we have

$$\begin{aligned}
&\mathbb{E}\left\{\lambda_1^{2\overline{T}_1}\delta^{\tau_1} + \lambda_2^{2\overline{T}_1}\delta^{\tau_{2,1}}\right\} \\
&\quad\leq 2e^{-\theta^*(1-\phi)}(\delta e^{-2\theta^*})^{\tau_1} + \mathbb{E}\left\{\lambda_1^{2T_1^*}\delta\right\}^{\tau_1} \\
&\quad= 2e^{-\theta^*(1-\phi)}(\delta e^{-2\theta^*})^{\tau_1} + \left(\delta\lambda_1^2\left[1 + \frac{p(\lambda_1^2-1)}{1-(1-q)\lambda_1^2}\right]\right)^{\tau_1}.
\end{aligned}$$

Since $\delta e^{-2\theta^*} < 1$, if (3.33) holds and τ_1 is selected to satisfy (3.38), we have $\mathbb{E}\{\lambda_1^{2\overline{T}_1}\delta^{\tau_1} + \lambda_2^{2\overline{T}_1}\delta^{\tau_{2,1}}\} < 1$, which further ensures

$$\mathbb{E}\left\{\lambda_1^{2\overline{T}_1}\delta^{\tau_1}\right\} < 1, \quad \mathbb{E}\left\{\lambda_2^{2\overline{T}_1}\delta^{\tau_{2,1}}\right\} < 1. \tag{3.41}$$

Next, we will show that the randomly sampled sequence $\mathbb{E}\left\{e_{1,\check{T}_k}^2\right\}, k \geq 0$ is mean-square bounded. Conditioned on the sequence $\{\gamma_{\check{T}_{k-1}}, \gamma_{\check{T}_{k-1}+1}, \ldots, \gamma_{\check{T}_{k-1}+\overline{T}_k}\}$ and from (3.28), we have

$$\begin{aligned}
\mathbb{E}\left\{e_{1,\check{T}_k}^2\right\} &= \mathbb{E}\left\{e_{1,\check{T}_{k-1}+\overline{T}_k}^2\right\} \\
&= \prod_{j=0}^{\overline{T}_k-1}\lambda_1^2\delta^{\gamma_{\check{T}_{k-1}+j}}\mathbb{E}\left\{e_{1,\check{T}_{k-1}}^2\right\} + \sum_{i=0}^{\overline{T}_k-1}\prod_{j=i+1}^{\overline{T}_k-1}\lambda_1^2\delta^{\gamma_{\check{T}_{k-1}+j}}\mathbb{E}\left\{\Phi_{1,\check{T}_{k-1}+i}^2\right\} \\
&= \lambda_1^{2\overline{T}_k}\delta^{\tau_1}\mathbb{E}\left\{e_{1,\check{T}_{k-1}}^2\right\} + \sum_{i=0}^{\overline{T}_k-1}\prod_{j=i+1}^{\overline{T}_k-1}\lambda_1^2\delta^{\gamma_{\check{T}_{k-1}+j}}\mathbb{E}\left\{\Phi_{1,\check{T}_{k-1}+i}^2\right\} \\
&\stackrel{(a)}{\leq} \lambda_1^{2\overline{T}_k}\delta^{\tau_1}\mathbb{E}\left\{e_{1,\check{T}_{k-1}}^2\right\} + \sum_{i=0}^{\overline{T}_k-1}\lambda_1^{2(\overline{T}_k-i-1)}\mathbb{E}\left\{\Phi_{1,\check{T}_{k-1}+i}^2\right\} \\
&\leq \lambda_1^{2\overline{T}_k}\delta^{\tau_1}\mathbb{E}\left\{e_{1,\check{T}_{k-1}}^2\right\} + \sup_t \mathbb{E}\left\{\Phi_{1,t}^2\right\}\sum_{i=0}^{\overline{T}_k-1}\lambda_1^{2(\overline{T}_k-i-1)} \\
&\leq \lambda_1^{2\overline{T}_k}\delta^{\tau_1}\mathbb{E}\left\{e_{1,\check{T}_{k-1}}^2\right\} + \sup_t \mathbb{E}\left\{\Phi_{1,t}^2\right\}\frac{\lambda_1^{2(\overline{T}_k-1)} - \lambda_1^{-2}}{1-\lambda_1^{-2}},
\end{aligned}$$

where (a) follows from the fact that $\delta^{\gamma_k} \leq 1$ for any k. Thus, we have that

$$\mathbb{E}\left\{e^2_{1,\check{T}_k}\right\} \leq \mathbb{E}\left\{\lambda_1^{2\overline{T}_k}\delta^{\tau_1}\right\}\mathbb{E}\left\{e^2_{1,\check{T}_{k-1}}\right\} + \sup_t \mathbb{E}\left\{\Phi^2_{1,t}\right\}\mathbb{E}\left\{\frac{\lambda_1^{2(\overline{T}_k-1)} - \lambda_1^{-2}}{1-\lambda_1^{-2}}\right\}. \tag{3.42}$$

Since $\{\overline{T}_k\}_{k\geq 1}$ are i.i.d., we have

$$\mathbb{E}\{\lambda_1^{2\overline{T}_k}\delta^{\tau_1}\} < 1$$

and $\sup_t \mathbb{E}\{\Phi^2_{1,t}\}\mathbb{E}\left\{\frac{\lambda_1^{2(\overline{T}_k-1)}-\lambda_1^{-2}}{1-\lambda_1^{-2}}\right\}$ is bounded from (3.41), then the randomly sampled sequences $\mathbb{E}\left\{e^2_{1,\check{T}_k}\right\}, k \geq 0$ is bounded from (3.42). Similarly, we can also prove that $\mathbb{E}\left\{e^2_{2,\check{T}_k}\right\}, k \geq 0$ is bounded.

For any t, there must exist k such that $t \in [\check{T}_k, \check{T}_{k+1}]$. Thus, conditioned on the lossy process $\{\gamma_t\}_{t\geq 0}$, we obtain that for $i = 1, 2$

$$\begin{aligned}
\mathbb{E}\left\{e^2_{i,t}\right\} &= \prod_{j=\check{T}_k}^{t-1}\lambda_i^2\delta^{\gamma_j}\mathbb{E}\left\{e^2_{i,\check{T}_k}\right\} + \sum_{i=0}^{t-\check{T}_k-1}\prod_{j=\check{T}_k+i+1}^{t-1}\lambda_i^2\delta^{\gamma_j}\mathbb{E}\left\{\Phi^2_{i,\check{T}_k+i}\right\} \\
&\leq \lambda^{2(t-\check{T}_k-1)}\mathbb{E}\left\{e^2_{i,\check{T}_k}\right\} + \sup_t \mathbb{E}\left\{\Phi^2_{i,t}\right\}\frac{\lambda_i^{2(t-\check{T}_k-1)} - \lambda_i^{-2}}{1-\lambda_i^{-2}} \\
&\leq \lambda^{2(\check{T}_{k+1}-\check{T}_k-1)}\mathbb{E}\left\{e^2_{i,\check{T}_k}\right\} + \sup_t \mathbb{E}\left\{\Phi^2_{i,t}\right\}\frac{\lambda_i^{2(\check{T}_{k+1}-\check{T}_k-1)} - \lambda_i^{-2}}{1-\lambda_i^{-2}} \\
&\leq \lambda^{2(\overline{T}_k-1)}\mathbb{E}\left\{e^2_{i,\check{T}_k}\right\} + \sup_t \mathbb{E}\left\{\Phi^2_{i,t}\right\}\frac{\lambda_i^{2(\overline{T}_k-1)} - \lambda_i^{-2}}{1-\lambda_i^{-2}}.
\end{aligned}$$

Thus, we have

$$\mathbb{E}\left\{e^2_{i,t}\right\} \leq \mathbb{E}\left\{\lambda^{2(\overline{T}_k-1)}\right\}\mathbb{E}\left\{e^2_{i,\hat{T}_k}\right\} + \sup_t \mathbb{E}\left\{\Phi^2_{i,t}\right\}\mathbb{E}\left\{\frac{\lambda_i^{2(\overline{T}_k-1)} - \lambda_i^{-2}}{1-\lambda_i^{-2}}\right\}.$$

Since $\mathbb{E}\left\{\lambda_i^{2\overline{T}_k}\right\}$ and $\mathbb{E}\left\{e^2_{i,\hat{T}_k}\right\}$ are bounded, we know that $\mathbb{E}\left\{e^2_{i,t}\right\}$ is bounded. In view of Lemma 3.4, the sufficiency is proved.

3.5.2 High-Dimensional Systems

The key difficulty in stabilizing multidimensional systems over fading channels is to allocate channel resources among different subdynamics optimally. We can show that the desired optimal allocation is determined by the magnitudes of eigenvalues and the realization of the channel fading. To optimally schedule the current transmission, we need to know the future fading realizations, as shown in Section 2.4.3, which is not available due to the casualty constraint. For two-dimensional systems, we can adopt Algorithm 3.2 to overcome this problem, which first allocates a constant amount of channel resources to the first subdynamics and then optimally stops the transmissions for the second subdynamics. But this method is not applicable to three- or higher-dimensional systems since to optimally stop the transmissions for the second or subsequent subdynamics, we need the information of the channel fading realizations from the future transmissions for all the subdynamics, which is not possible due to the causal availability of the channel state information. In this subsection, an adaptive TDMA scheduling algorithm is proposed for high-dimensional systems, which is adaptive to the lossy process and outperforms the scheduling Algorithm 3.1 as shown later. The adaptive TDMA scheduler is stated in Algorithm 3.3, where $\tau_1, \ldots, \tau_n$ are scheduler parameters to be specified later.

Algorithm 3.3: Adaptive TDMA Scheduler for Power-Constrained Markov Lossy Channels

- The first encoder/decoder pair is scheduled to use the channel, until the transmissions succeed for τ_1 times.
- The second encoder/decoder pair is scheduled to use the channel, until the transmissions succeed for τ_2 times.
- ...
- The nth encoder/decoder pair is scheduled to use the channel, until the transmissions succeed for τ_n times.
- Repeat.

Let T_k^i denote the period for the ith encoder/decoder pair to achieve τ_i successful transmissions in the kth round and define $\overline{T}_k$, $\check{T}_k$ analogously as in Section 3.5.1. The scheduled transmissions with Algorithm 3.3 is depicted in Figure 3.6. It is clear that T_k^i is independent of T_k^j, and $\overline{T}_i$ and $\overline{T}_j$ are i.i.d. for any $i \neq j$. A sufficient stabilizability result with Algorithm 3.3 is stated in the following theorem.

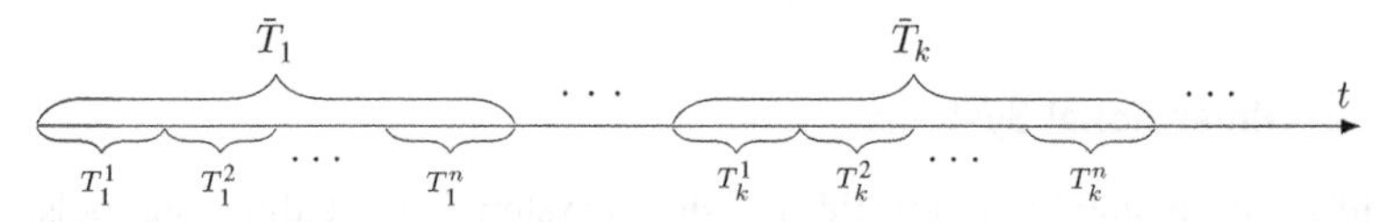

Figure 3.6 Transmissions with the adaptive TDMA scheduler.

Theorem 3.5 *There exist coding and controlling strategies* $\{\mathscr{E}_t(\cdot)\}_{t\geq 0}$, $\{\mathscr{D}_t(\cdot)\}_{t\geq 0}$, *such that the system* (3.1) *can be mean-square stabilized over the power- constrained Markov lossy channel, if there exist* α_i, $i = 1, \ldots, d$ *with* $0 < \alpha_i \leq 1$ *and* $\sum_{i=1}^{d} \alpha_i = 1$ *such that*

$$(1 - \mathfrak{q})|\lambda_1|^2 < 1, \tag{3.43}$$

$$\delta^{\frac{\alpha_1}{v_i}} |\lambda_i|^2 \left[1 + \frac{\mathfrak{p}(|\lambda_i|^2 - 1)}{1 - (1 - \mathfrak{q})|\lambda_i|^2}\right] < 1, \tag{3.44}$$

for all $i = 1, \ldots, d$.

Proof: Here we only consider the case that $\lambda_1, \ldots, \lambda_d$ are real and $m_i = v_1 = 1$. We can easily extend the analysis to other cases by combining the following analysis with similar arguments used in Zaidi et al. [2014b]. In view of Lemma 3.3, the sufficient condition in Theorem 3.5 is equivalent to the following condition:

$$\mathbb{E}\left\{\lambda_i^{2T_1^*}\right\}\delta^{\alpha_i} < 1, \quad i = 1, \ldots, n. \tag{3.45}$$

Let $\iota = \min_i \left(\log_\delta \mathbb{E}\left\{\lambda_i^{2T_1^*}\right\} + \alpha_i\right)$. For any α_i, there exists a rational sequence $\{\beta_{i,k}\}_{k\geq 0}$, such that $\lim_{k\to\infty}\beta_{i,k} = \alpha_i$. Then

$$\lim_{k\to\infty} \frac{\beta_{i,k}}{\sum_j \beta_{j,k}} = \frac{\alpha_i}{\sum_j \alpha_j} = \alpha_i.$$

Therefore, for the given ι, there exists $M \in \mathbb{N}^+$, such that

$$\left|\frac{\beta_{i,M}}{\sum_j \beta_{j,M}} - \alpha_i\right| < \iota$$

for all $i = 1, \ldots, n$. Thus, $\frac{\beta_{i,M}}{\sum_j \beta_{j,M}} > \alpha_i - \iota \geq -\log_\delta \mathbb{E}\left\{\lambda_i^{2T_1^*}\right\}$, which implies

$$\mathbb{E}\left\{\lambda_i^{2T_1^*}\right\}\delta^{\frac{\beta_{i,M}}{\sum_j \beta_{j,M}}} < 1, \quad i = 1, \ldots, n.$$

Since $\beta_{1,M}, \ldots, \beta_{n,M}$ are rational, there exist integers $\tau_1, \ldots, \tau_n, \overline{\tau}$ such that $\beta_{i,M} = \frac{\tau_i}{\overline{\tau}}$ and $\mathbb{E}\left\{\lambda_i^{2T_1^*}\right\}\delta^{\frac{\tau_i}{\sum_j \tau_j}} < 1$ for $i = 1, \ldots, n$, which implies

$$\mathbb{E}\left\{\lambda_i^{2\overline{T}_1}\delta^{\tau_i}\right\} = \mathbb{E}\left\{\lambda_i^{2T_1^*}\right\}^{\tau_1+\cdots+\tau_n}\delta^{\tau_i} < 1.$$

Similar to the proof of Theorem 3.4, we can show that the sampled sequence $\mathbb{E}\left\{e_{i,\check{T}_k}^2\right\}$ is bounded, and further $\mathbb{E}\left\{e_{i,t}^2\right\}$ is bounded. In view of Lemma 3.4, the sufficiency is proved. □

Remark 3.5 In view of Lemma 3.3, Theorem 3.5 can be equivalently stated as follows: if there exist α_is with $0 < \alpha_i \leq 1$ and $\sum_{i=1}^{d} \alpha_i = 1$, such that

$$\mathbb{E}\left\{ \lambda_i^{2T_1^*} \right\}^{\frac{v_i}{\alpha_i}} \delta < 1, \tag{3.46}$$

for $i = 1, \ldots, d$, the system is mean-square stabilizable. Then the existence of α_is in Theorem 3.5 can be determined as follows: let $\alpha_i^* = -v_i \log_\delta \mathbb{E}\left\{ \lambda_i^{2T_1^*} \right\}$, which is the lower bound for any feasible α_i from (3.46). If $\sum_i \alpha_i^* > 1$, there are no feasible α_is. Otherwise, one admissible α_i is given by $\alpha_i = \frac{\alpha_i^*}{\sum_j \alpha_j^*}$.

Remark 3.6 Theorem 3.3 can be equivalently expressed as follows: if there exist α_is with $0 < \alpha_i \leq 1$ and $\sum_{i=1}^{d} \alpha_i = 1$, such that

$$\lambda_i^{\frac{2v_i}{\alpha_i}} \rho(Q'D_1) < 1, \tag{3.47}$$

for $i = 1, \ldots, d$, the system is mean-square stabilizable. For power-constrained Markov lossy channels, in view of Lemma 3.3, (3.47) is equivalent to

$$\mathbb{E}\left\{ \lambda_i^{\frac{v_i}{\alpha_i} 2T_1^*} \right\} \delta < 1. \tag{3.48}$$

Since $\mathbb{E}\left\{ \lambda_i^{2T_1^*} \right\}^{\frac{v_i}{\alpha_i}} \leq \mathbb{E}\left\{ \lambda_i^{\frac{v_i}{\alpha_i} 2T_1^*} \right\}$ from Jensen's inequality, any λ_i that satisfies (3.48), must also satisfy (3.46). Thus, the adaptive TDMA scheduler outperforms the TDMA scheduler in the sense that it can tolerate more unstable systems.

When all the strictly unstable eigenvalues have the same magnitude, the sufficient condition in Theorem 3.5 coincides with the necessary condition in Theorem 3.2, as shown in the following corollary.

Corollary 3.2 *Suppose* $|\lambda_1| = \cdots = |\lambda_{d_u}| = \tilde{\lambda} > 1$ *and* $|\lambda_{d_u+1}| = \cdots = |\lambda_d| = 1$ *with* $1 \leq d_u \leq d$. *There exists an encoder/decoder pair* $\{\mathcal{E}_t(\cdot)\}_{t\geq 0}, \{\mathcal{D}_t(\cdot)\}_{t\geq 0}$, *such that the system* (3.1) *can be mean-square stabilized over the power- constrained Markov lossy channel if and only if*

$$(1 - \mathfrak{q})\tilde{\lambda}^{2} < 1,$$

$$\delta^{\frac{1}{v_1 + \cdots + v_{d_u}}} \tilde{\lambda}^{2} \left[1 + \frac{\mathfrak{p}(\tilde{\lambda}^{2} - 1)}{1 - (1 - \mathfrak{q})\tilde{\lambda}^{2}} \right] < 1.$$

Remark 3.7 As an application of the derived theorems, we have the following extensions.

- When $\mathfrak{p} = 0, \mathfrak{q} = 1$, the power-constrained Markov lossy channel degenerates to the AWGN channel, a necessary and sufficient condition to ensure mean-square stabilizability from Theorem 3.2 and Theorem 3.5 is $\sum_i \nu_i \ln |\lambda_i| < \frac{1}{2} \ln \left(1 + \frac{P}{\sigma_\omega^2}\right)$, which coincides with the stabilizability condition over AWGN channels in Braslavsky et al. [2007] and Freudenberg et al. [2010].
- If $\mathfrak{p} = 1 - \mathfrak{q}$, we can obtain the stabilizability condition over power-constrained lossy channels [Xu et al., 2016]. We can show that Theorems 3.2, 3.4, and 3.5 recover Lemma 1, Theorem 2 and Theorem 1 in Xu et al. [2016], respectively.
- For the power-constrained Markov lossy channel, taking the limit $P \to \infty, \sigma_\omega^2 \to 0$, we obtain the stabilizability condition for control over Markovian packet loss channel from Theorem 3.2 and Theorem 3.5 as $(1 - \mathfrak{q})|\lambda_1|^2 < 1$, which recovers the results in Xie and Xie [2009], Huang and Dey [2007], and Gupta et al. [2007]. Moreover, if $\mathfrak{p} = 1 - \mathfrak{q}$, we can further recover the stabilizability condition for control over i.i.d. erasure channels as in Elia [2005] and Sinopoli et al. [2004].

3.5.3 Numerical Illustrations

For two-dimensional systems controlled over power-constrained Markov lossy channels, suppose $P = 3, \sigma_\omega^2 = 1$, the regions for $(\ln |\lambda_1|, \ln |\lambda_2|)$ indicated by the derived necessary conditions and sufficient conditions are plotted in Figure 3.7 under different failure and recovery rates. We plot the necessary stabilizability region and sufficient stabilizability regions achieved with the optimal scheduler, the TDMA scheduler, and the adaptive TDMA scheduler for the case $\mathfrak{p} = 0.3, \mathfrak{q} = 0.6$. For the cases of $\mathfrak{p} = 0.6, \mathfrak{q} = 0.6$ and $\mathfrak{p} = 0.3, \mathfrak{q} = 0.9$, only the stabilizability region indicated by the necessity and sufficiency with the optimal scheduler is plotted. The other sufficient stabilizability regions are omitted for clarity but can be plotted in a similar way as in the case of $\mathfrak{p} = 0.3$, $\mathfrak{q} = 0.6$.

For the given failure and recovery rate, it is clear that the adaptive TDMA scheduler achieves a larger stabilizability region than the TDMA scheduler. When the two eigenvalues are with equal magnitude, the adaptive TDMA scheduler is optimal, which is implied in Corollary 3.2. Besides, the optimal scheduling Algorithm 3.2 is tight as proved in Theorem 3.4. Moreover, when we increase the failure rate $\mathfrak{p}$ or the recovery rate $\mathfrak{q}$, the stabilizability region is reduced or enlarged as expected due to the change in the communication channel's reliability.

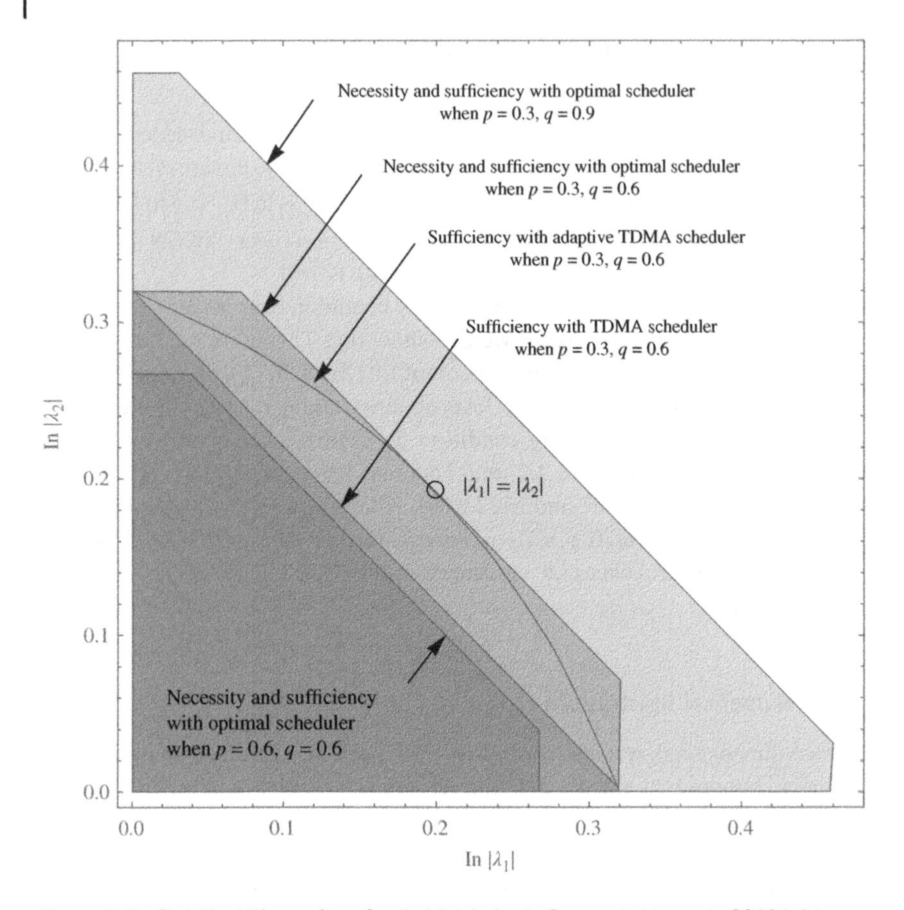

Figure 3.7 Stabilizability regions for $(\ln |\lambda_1|, \ln |\lambda_2|)$. Source: L. Xu et al., 2018/with permission of IEEE.

3.6 Conclusions

The chapter studies the mean-square stabilization problem of discrete-time LTI systems over Gaussian Markov channels, which suffer from both signal-to-noise ratio constraint and correlated channel fading modeled by a Markov process. The existence of a fundamental limitation for mean-square stabilization is first established. Sufficient stabilization conditions under a TDMA communication scheme are derived in terms of the stability of a MJLS. Moreover, a necessary and sufficient condition is presented for mean-square stabilization of two-dimensional systems controlled over power- constrained Markov lossy channels. Furthermore, improved sufficient stabilizability conditions are derived based on an adaptive TDMA communication scheme for general high-dimensional systems, which achieve a larger stabilizability region than the TDMA communication scheme.

Bibliography

R. B. Ash and C. A. Doléans-Dade. *Probability and measure theory.* Harcourt/Academic Press, San Diego, CA, 2000.

J. H. Braslavsky, R. H. Middleton, and J. S. Freudenberg. Feedback stabilization over signal-to-noise ratio constrained channels. *IEEE Transactions on Automatic Control*, 52(8):1391–1403, 2007.

L. Coviello, P. Minero, and M. Franceschetti. Stabilization over Markov feedback channels. In *Proceedings of the 50th IEEE Conference on Decision and Control*, pages 3776–3782, 2011.

N. Elia. Remote stabilization over fading channels. *Systems & Control Letters*, 54(3):237–249, 2005.

J. S. Freudenberg, R. H. Middleton, and V. Solo. Stabilization and disturbance attenuation over a Gaussian communication channel. *IEEE Transactions on Automatic Control*, 55(3):795–799, 2010.

A. Goldsmith. *Wireless communications.* Cambridge University Press, Cambridge, 2005.

A. J. Goldsmith and P. P. Varaiya. Capacity, mutual information, and coding for finite-state Markov channels. *IEEE Transactions on Information Theory*, 42(3):868–886, 1996.

V. Gupta, B. Hassibi, and R. M. Murray. Optimal LQG control across packet-dropping links. *Systems & Control Letters*, 56(6):439–446, 2007.

R. A. Horn and C. R. Johnson. *Matrix analysis.* Cambridge University Press, Cambridge, 1985.

M. Huang and S. Dey. Stability of Kalman filtering with Markovian packet losses. *Automatica*, 43(4):598–607, 2007.

J. Liu, N. Elia, and S. Tatikonda. Capacity-achieving feedback schemes for Gaussian finite-state Markov channels with channel state information. *IEEE Transactions on Information Theory*, 61(7):3632–3650, 2015.

P. Minero, M. Franceschetti, S. Dey, and G. N. Nair. Data rate theorem for stabilization over time-varying feedback channels. *IEEE Transactions on Automatic Control*, 54(2):243–255, 2009.

P. Minero, L. Coviello, and M. Franceschetti. Stabilization over Markov feedback channels: The general case. *IEEE Transactions on Automatic Control*, 58(2):349–362, 2013.

G. N. Nair and R. J. Evans. Stabilizability of stochastic linear systems with finite feedback data rates. *SIAM Journal on Control and Optimization*, 43(2):413–436, 2004.

B. Sinopoli, L. Schenato, M. Franceschetti, K. Poolla, M. I. Jordan, and S. S. Sastry. Kalman filtering with intermittent observations. *IEEE Transactions on Automatic Control*, 49(9):1453–1464, 2004.

S. Tatikonda, A. Sahai, and S. Mitter. Stochastic linear control over a communication channel. *IEEE Transactions on Automatic Control*, 49(9):1549–1561, 2004.

H. Viswanathan. Capacity of Markov channels with receiver CSI and delayed feedback. *IEEE Transactions on Information Theory*, 45(2):761–771, 1999.

L. Xie and L. Xie. Stability analysis of networked sampled-data linear systems with Markovian packet losses. *IEEE Transactions on Automatic Control*, 54(6):1375–1381, 2009.

L. Xu, Y. Mo, and L. Xie. Mean square stabilization of vector LTI systems over power constrained lossy channels. In *Proceedings of the 2016 American Control Conference*, pages 7129–7134, 2016.

K. You and L. Xie. Minimum data rate for mean square stabilizability of linear systems with Markovian packet losses. *IEEE Transactions on Automatic Control*, 56(4):772–785, 2011.

A. A. Zaidi, T. J. Oechtering, S. Yuksel, and M. Skoglund. Stabilization of linear systems over Gaussian networks. *IEEE Transactions on Automatic Control*, 59(9):2369–2384, 2014a.

A. A. Zaidi, T. J. Oechtering, S. Yüksel, and M. Skoglund. Stabilization and control over Gaussian networks. In *Information and Control in Networks*, Lecture Notes in Control and Information Sciences, G. Como, B. Bernhardsson, and A. Rantzer (eds), volume 450, pages 39–85. Springer, Cham, 2014b.

4

Linear-Quadratic Optimal Control of NCSs with Random Input Gains

4.1 Introduction

In this chapter, we will focus on the linear quadratic (LQ) optimal control problem of a discrete-time multi-input LTI system with random input gains, as depicted in Figure 4.1. Recall that in the solutions to the deterministic LQ optimal control problem, as well as to the deterministic optimal linear filtering problem, the well-known standard algebraic Riccati equations (AREs) arise and play an essential role. The solutions, properties, and applications of AREs are studied in a large number of research works [Lancaster and Rodman, 1995, Qiu, 1999, Zhou et al., 1996]. Likewise, in the study of the LQ optimal control of NCSs with random input gains, a modified algebraic Riccati equation (MARE), which plays a similar role to AREs, appears. In general, there are many solutions to the MARE. Among them, the one called mean-square stabilizing solution attracts our attention. With the static state feedback controller associated with the mean-square stabilizing solution, the stochastic LQ optimal control problem is solvable, i.e. the cost function is minimized and the closed-loop system is mean-square stable. Thus, the emphasis of this chapter is on how to address the existence problem of a mean-square stabilizing solution to the MARE. By employing the theory of cone-invariant operators, we propose an explicit necessary and sufficient condition ensuring the existence of the mean-square stabilizing solution to the MARE. Such a condition is compatible with the one ensuring the stabilizing solution to the standard ARE and it indicates that the common condition of observability or detectability of certain stochastic systems is unnecessary.

This chapter is organized as follows: the problem formulation is given in Section 4.2. The finite-horizon case is studied in Section 4.3. Some properties of cone-invariant operators and the solvability of the associated MARE are discussed

Control over Communication Networks: Modeling, Analysis, and Design of Networked Control Systems and Multi-Agent Systems over Imperfect Communication Channels, First Edition.
Jianying Zheng, Liang Xu, Qinglei Hu, and Lihua Xie.

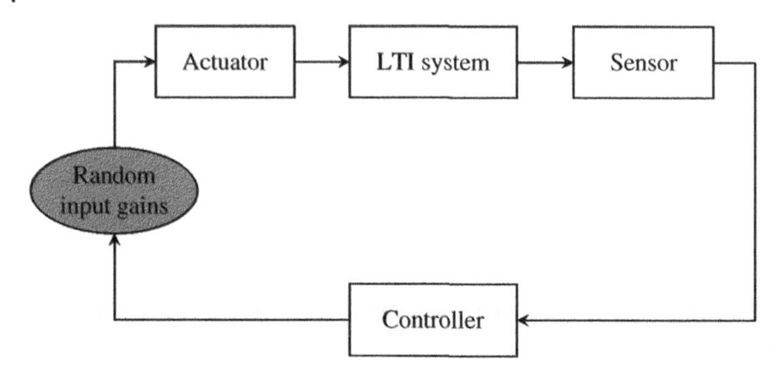

Figure 4.1 Linear time-invariant (LTI) systems with random input gains.

in Section 4.4. Then the infinite-horizon stochastic LQ optimal control problem is solved in Section 4.5. This chapter ends with some concluding remarks in Section 4.6.

4.2 Problem Formulation

A discrete-time LTI system with random input gains, as shown in Figure 4.2, is described by

$$\begin{aligned} x(k+1) &= Ax(k) + B\kappa(k)v(k), \\ z(k) &= Cx(k) + D\kappa(k)v(k), \end{aligned} \tag{4.1}$$

where $x(k) \in \mathbb{R}^n$ is the system state, $z(k) \in \mathbb{R}^p$ is the system output, and $v(k) \in \mathbb{R}^m$ is the control input, generated by a finite-dimensional LTI state feedback controller $\boldsymbol{K}$ with the form of

$$\begin{aligned} x_K(k+1) &= A_K x_K(k) + B_K x(k), \\ v(k) &= C_K x_K(k) + D_K x(k). \end{aligned} \tag{4.2}$$

There are random gains $\kappa(k)$ in the input channels. The parallel transmission strategy for the multi-input system is adopted, i.e. each element of $v(k)$ is sent

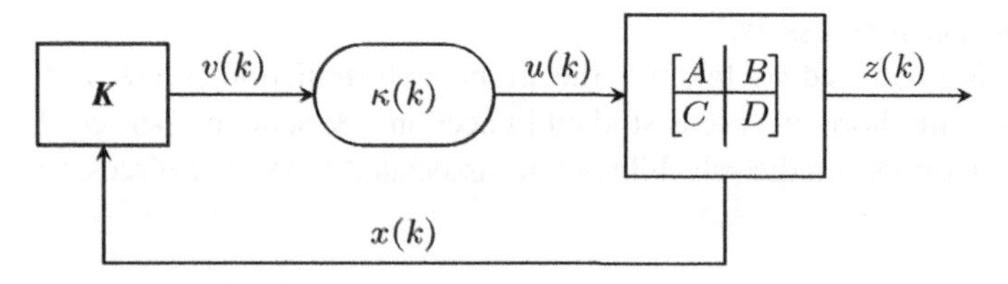

Figure 4.2 Discrete-time LTI systems with random input gains.

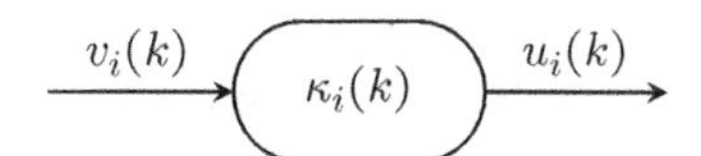

Figure 4.3 Parallel transmission strategy.

across an individual input channel (see Figure 4.3). Then $\kappa(k)$ can be given by a diagonal random matrix $\text{diag}\{\kappa_1(k), \dots, \kappa_m(k)\}$, whose diagonal elements $\kappa_i(k), i = 1, 2, \dots, m$, are mutually uncorrelated independent and identically distributed (i.i.d) random processes with mean $\mu_i = \mathbb{E}\left[\kappa_i(k)\right] \neq 0$ and variance $\sigma_i^2 = \mathbb{E}\left[(\kappa_i(k) - \mu_i)^2\right]$, respectively. The signal-to-noise ratio of the ith input channel is defined as $\text{SNR}_i \triangleq \frac{\mu_i^2}{\sigma_i^2}$. Denote

$$\begin{aligned} M &\triangleq \text{diag}\{\mu_1, \dots, \mu_m\}, \\ \Sigma^2 &\triangleq \text{diag}\{\sigma_1^2, \dots, \sigma_m^2\}, \\ \text{SNR} &\triangleq \text{diag}\{\text{SNR}_1, \dots, \text{SNR}_m\}, \end{aligned}$$

and

$$W \triangleq \begin{bmatrix} 1 + \text{SNR}_1^{-1} & 1 & \cdots & 1 \\ 1 & 1 + \text{SNR}_2^{-1} & \ddots & \vdots \\ \vdots & \ddots & \ddots & 1 \\ 1 & \cdots & 1 & 1 + \text{SNR}_m^{-1} \end{bmatrix} = \mathbf{1}_{m \times m} + \text{SNR}^{-1}.$$

The mean-square capacity of the ith input channel is defined as in Elia [2005]:

$$\mathfrak{C}_i \triangleq \frac{1}{2} \log(1 + \text{SNR}_i).$$

The overall mean-square capacity is given by

$$\mathfrak{C} = \sum_{i=1}^{m} \mathfrak{C}_i.$$

Clearly, the channel's capacity can be used to measure the information constraint. A larger $\mathfrak{C}_i$ indicates that there is less randomness in the ith channel and more reliable information can be transmitted. A larger $\mathfrak{C}$ indicates that we have more available information or disposable resources in total.

Denote the overall system state by $\hat{x}(k) = \left[x'(k) \;\; x'_K(k)\right]'$. Then the closed-loop system can be written as

$$\hat{x}(k+1) = \left\{ \begin{bmatrix} A & 0 \\ B_K & A_K \end{bmatrix} + \begin{bmatrix} B \\ 0 \end{bmatrix} \kappa(k) \begin{bmatrix} D_K & C_K \end{bmatrix} \right\} \hat{x}(k). \tag{4.3}$$

The state feedback controller $\boldsymbol{K}$ is said to be mean-square stabilizing if the corresponding closed-loop system (4.3) is stable in the mean-square sense, i.e. $\lim_{k \to \infty} \mathbb{E}[\hat{x}(k)\hat{x}'(k)] = 0$ for any initial state $\hat{x}(0)$. Then the objective of this stochastic LQ optimal control problem is to find a mean-square stabilizing controller $\boldsymbol{K}$ to minimize the cost function $J(x(0), v(\cdot))$, defined as

$$\begin{aligned} J(x(0), v(\cdot)) &= \mathbb{E}\left[\sum_{k=0}^{\infty} z'(k)z(k)\right] \\ &= \mathbb{E}\left\{\sum_{k=0}^{\infty}\begin{bmatrix} x(k) \\ \kappa(k)v(k)\end{bmatrix}'\begin{bmatrix} C' \\ D'\end{bmatrix}\begin{bmatrix} C & D\end{bmatrix}\begin{bmatrix} x(k) \\ \kappa(k)v(k)\end{bmatrix}\right\} \\ &= \mathbb{E}\left\{\sum_{k=0}^{\infty}\begin{bmatrix} x(k) \\ Mv(k)\end{bmatrix}'\begin{bmatrix} C'C & C'D \\ D'C & W\odot(D'D)\end{bmatrix}\begin{bmatrix} x(k) \\ Mv(k)\end{bmatrix}\right\}. \end{aligned}$$

It holds that $W \odot (D'D) = D'D + \mathrm{SNR}^{-1} \odot (D'D)$. We make the assumption on the system parameters that

$$W \odot (D'D) > 0,$$

instead of the stronger one that D has full column rank, which is a common assumption in many papers. For simplicity, in the proof of this chapter, we often let

$$\begin{bmatrix} Q & S \\ S' & R\end{bmatrix} \triangleq \begin{bmatrix} C'C & C'D \\ D'C & D'D\end{bmatrix}, \quad \tilde{R} \triangleq W \odot R.$$

One traditional way to handle this problem is to fix the individual channel capacities a priori and then find an optimal mean-square stabilizing controller. However, under this formulation, the problem is not always well posed for any given channel capacities, i.e. the cost function might always be infinity no matter what control signal is used or the control signal that minimizes the cost function does not stabilize the system. To tackle this difficulty, we put the problem under a novel framework, namely, the channel/controller codesign framework. In this case, the individual channel capacities $\mathfrak{C}_i$ are not given a priori. Instead, they are assumed to be designed or allocated under an overall capacity constraint $\mathfrak{C}$. The allocation of the overall capacity to the individual channels, called channel resource allocation, is formally given by the probability vector

$$\pi = \begin{bmatrix} \pi_1 & \pi_2 & \dots & \pi_m \end{bmatrix}',$$

where $0 \le \pi_i \le 1$, $\sum_{i=1}^{m} \pi_i = 1$, such that $\mathfrak{C}_i = \pi_i\mathfrak{C}$. With the help of the channel/controller codesign, our problem becomes easy to simultaneously design a feasible allocation π and a mean-square stabilizing controller $v(k)$ to minimize the cost function $J(x(0), v(\cdot))$ under a certain overall capacity constraint $\mathfrak{C}$.

4.3 Finite-Horizon LQ Optimal Control

When $\kappa(k)$ is deterministic, our problem is reduced to the classical LQ optimal control problem. In the deterministic LQ optimal control problem, the infinite-horizon case, if it is well posed, can be solved by taking the result to the finite-horizon case as the horizon length goes to infinity. Inspired by this, we first study the finite-horizon LQ optimal control of discrete-time LTI systems with random input gains in this section. The infinite-horizon case will be investigated in the subsequent sections.

In the finite-horizon case, the cost function with the horizon length N, the initial state $x(0)$, and the control signal $\upsilon_N \triangleq \{\upsilon(0), \upsilon(1), \dots, \upsilon(N)\}$ is given in the following quadratic form:

$$\begin{aligned} J(x(0), \upsilon_N) &= \mathbb{E}\left[\sum_{k=0}^{N} z'(k)z(k)\right] \\ &= \mathbb{E}\left\{\sum_{k=0}^{N} \begin{bmatrix} x(k) \\ M\upsilon(k) \end{bmatrix}' \begin{bmatrix} C'C & C'D \\ D'C & W \odot (D'D) \end{bmatrix} \begin{bmatrix} x(k) \\ M\upsilon(k) \end{bmatrix}\right\}. \end{aligned}$$

The objective is to find the optimal control signal

$$\upsilon_N^{\text{opt}} \triangleq \{\upsilon^{\text{opt}}(0), \upsilon^{\text{opt}}(1), \dots, \upsilon^{\text{opt}}(N)\}$$

to minimize $J_N(x(0), \upsilon_N)$. Note that for the finite-horizon LQ optimal control, the cost function is always bounded, and the system's stability is not concerned. Hence, the individual channel capacities can be arbitrarily given. The following theorem shows that the optimal control signal is a linear function of the state and depends on the mean-square capacity of each input channel.

Theorem 4.1 *For every initial state $x(0)$, the optimal control signal with the horizon length N is given by*

$$\upsilon^{\text{opt}}(k) = F(k)x(k), \tag{4.4}$$

where

$$\begin{aligned} F(k) &= -M^{-1}[W \odot (R + B'X_N(k+1)B)]^{-1}[B'X_N(k+1)A + S'], \\ X_N(k) &= A'X_N(k+1)A + Q - [A'X_N(k+1)B + S] \\ &\quad \times [W \odot (B'X_N(k+1)B + R)]^{-1}[B'X_N(k+1)A + S'], \\ X_N(N+1) &= 0, \end{aligned} \tag{4.5}$$

for $k = N, N-1, \dots, 0$, and the optimal cost is given by $x'(0)X_N(0)x(0)$.

Proof: The proof is based on dynamic programming. Define the initial condition and cost-to-go function as

$$X_N(N+1) = 0, \quad L(N+1) = 0,$$

$$L(k) = \min_{\upsilon(k)} \mathbb{E}\left\{\begin{bmatrix} x(k) \\ M\upsilon(k) \end{bmatrix}' \begin{bmatrix} Q & S \\ S' & W \odot R \end{bmatrix} \begin{bmatrix} x(k) \\ M\upsilon(k) \end{bmatrix} + L(k+1)\right\},$$

for $k = N, \dots, 1, 0$.

First, it is easy to see that

$$L(N+1) = x'(N+1)X_N(N+1)x(N+1) = 0.$$

Next, we will show that if $L(k+1)$ is a quadratic function of the state $x(k+1)$, i.e. there exists $X_N(k+1) \geq 0$ such that $L(k+1) = x'(k+1)X_N(k+1)x(k+1)$, then $L(k)$ is also a quadratic function of the state $x(k)$. From the definition of $L(k)$, we have

$$L(k) = \min_{\upsilon(k)} \mathbb{E}\left\{ \begin{bmatrix} x(k) \\ M\upsilon(k) \end{bmatrix}' \begin{bmatrix} Q & S \\ S' & W \odot R \end{bmatrix} \begin{bmatrix} x(k) \\ M\upsilon(k) \end{bmatrix} \right.$$
$$\left. + \begin{bmatrix} x(k) \\ \kappa(k)\upsilon(k) \end{bmatrix}' \begin{bmatrix} A' \\ B' \end{bmatrix} X_N(k+1) \begin{bmatrix} A & B \end{bmatrix} \begin{bmatrix} x(k) \\ \kappa(k)\upsilon(k) \end{bmatrix} \right\}.$$

With some simple calculations, we have

$$\begin{aligned} &\mathbb{E}[\upsilon'(k)\kappa(k)B'X_N(k+1)B\kappa(k)\upsilon(k)] \\ &\quad = \upsilon'(k)[MB'X_N(k+1)BM + \Sigma^2 \odot B'X_N(k+1)B]\upsilon(k). \end{aligned}$$

Then it holds that

$$\begin{aligned} L(k) &= \min_{\upsilon(k)} \left\{ \begin{bmatrix} x(k) \\ M\upsilon(k) \end{bmatrix}' \begin{bmatrix} Q & S \\ S' & W \odot R \end{bmatrix} \begin{bmatrix} x(k) \\ M\upsilon(k) \end{bmatrix} \right. \\ &\quad + \begin{bmatrix} x(k) \\ M\upsilon(k) \end{bmatrix}' \begin{bmatrix} A' \\ B' \end{bmatrix} X_N(k+1) \begin{bmatrix} A & B \end{bmatrix} \begin{bmatrix} x(k) \\ M\upsilon(k) \end{bmatrix} \\ &\quad \left. + \upsilon'(k)[\Sigma^2 \odot B'X_N(k+1)B]\upsilon(k) \right\} \\ &= \min_{\upsilon(k)} \left\{ \begin{bmatrix} x(k) \\ M\upsilon(k) \end{bmatrix}' \begin{bmatrix} T_1 & T_2' \\ T_2 & T_3 \end{bmatrix} \begin{bmatrix} x(k) \\ M\upsilon(k) \end{bmatrix} \right\} \\ &= \min_{\upsilon(k)} \left\{ [\upsilon(k) - F(k)x(k)]'M'T_3M[\upsilon(k) - F(k)x(k)] \right\} \\ &\quad + x'(k)(T_1 - T_2'T_3^{-1}T_2)x(k), \end{aligned}$$

where

$$\begin{bmatrix} T_1 & T_2' \\ T_2 & T_3 \end{bmatrix} = \begin{bmatrix} A'X_N(k+1)A + Q & A'X_N(k+1)B + S \\ B'X_N(k+1)A + S' & W \odot (B'X_N(k+1)B + R) \end{bmatrix}$$

and

$$F(k) = -M^{-1}T_3^{-1}T_2.$$

From this, we can obtain the optimal control law, which is given as

$$\begin{aligned} \upsilon^{\mathrm{opt}}(k) &= F(k)x(k) \\ &= -M^{-1}[W \odot (B'X_N(k+1)B + R)]^{-1}(B'X_N(k+1)A + S')x(k). \end{aligned}$$

Moreover, the cost-to-go function $L(k)$ is given by

$$L(k) = x'(k)(T_1 - T_2'T_3^{-1}T_2)x(k) = x'(k)X_N(k)x(k),$$

where $X_N(k)$ is given by (4.5), as a result of which, the optimal control cost is given by

$$L(0) = x'(0)X_N(0)x(0).$$

The proof is completed. □

4.4 Solvability of Modified Algebraic Riccati Equation

With the finite-horizon result as shown in Theorem 4.1, the infinite-horizon case can be solved by taking the horizon length $N \to \infty$. However, this requires that as $N \to \infty$, the matrix $X_N(0)$ solved by the backward iteration of $X_N(k)$ in (4.5) converges to a positive semi-definite matrix X which satisfies the following discrete-time MARE:

$$\begin{aligned} &A'XA - X + C'C \\ &\quad - (A'XB + C'D)\left[W \odot (B'XB + D'D)\right]^{-1}(B'XA + D'C) = 0. \end{aligned} \tag{4.6}$$

Moreover, the matrix X is required to be mean-square stabilizing in the sense that the associated state feedback gain

$$F(X) \triangleq -M^{-1}[W \odot (B'XB + D'D)]^{-1}(B'XA + D'C) \tag{4.7}$$

is mean-square stabilizing with $\kappa(k)$. Thus, this section is dedicated to studying the existence of a mean-square stabilizing solution to the MARE (4.6).

4.4.1 Cone-Invariant Operators

Before stating the results, we first introduce some preliminary knowledge from cone-invariant operators theory.

Let $\mathcal{V}$ be a finite-dimensional real vector space. A nonempty subset $\mathcal{V}_+ \subset \mathcal{V}$ is said to be a convex cone if $\mathcal{V}_+ + \mathcal{V}_+ = \mathcal{V}_+$ and $a\,\mathcal{V}_+ \subset \mathcal{V}_+$ for all real nonnegative numbers a. A convex cone is pointed if $\mathcal{V}_+ \cap -\mathcal{V}_+ = \{0\}$ and solid if int $\mathcal{V}_+$ is not empty. A convex cone is said to be proper if it is closed, pointed, and solid. A proper cone induces a partial order on the vector space as follows: for $X, Y \in \mathcal{V}$, we write $X \geq_{\mathcal{V}_+} Y$ if $X - Y \in \mathcal{V}_+$, and $X >_{\mathcal{V}_+} Y$ if $X - Y \in \text{int } \mathcal{V}_+$.

Example 4.1

- *The set of nonnegative vectors* $\mathbb{R}_+^n$ *is a proper cone and induces the entrywise partial order in* $\mathbb{R}^n$.

- *The set of nonnegative polynomials on* $[0, 1]$

$$\{x \in \mathbb{R}^n : x_1 t^{n-1} + x_2 t^{n-2} + \cdots + x_n \geq 0 \, for \; t \in [0, 1]\}$$

is also a proper cone in $\mathbb{R}^n$.

- *The space of symmetric matrices* S_n *is a real Hilbert space endowed with the Frobenius inner product* $\langle X, Y \rangle = trXY$ *and the Frobenius norm*

$$\|X\| = \sqrt{\sum_{i=1}^{n}\sum_{i=1}^{n} x_{ij}^2}.$$

Then $\mathcal{P}_n$ *is a proper cone in* S_n *and induces the partial order as follows: given* $X, Y \in S_n$, $X \geq_{\mathcal{P}_n} Y$ *if and only if* $v'(X - Y)v \geq 0, \; \forall v \in \mathbb{R}^n$.

Definition 4.1 Let $\mathcal{U}, \mathcal{V}$ be finite-dimensional real vector spaces together with the proper cones $\mathcal{U}_+$ and $\mathcal{V}_+$, respectively. A linear operator $\boldsymbol{T} : \mathcal{U} \to \mathcal{V}$ is said to be $(\mathcal{U}_+, \mathcal{V}_+)$-invariant if $\boldsymbol{T}\mathcal{U}_+ \subset \mathcal{V}_+$. When $\mathcal{U}_+ = \mathcal{V}_+$, we simply write $(\mathcal{U}_+, \mathcal{V}_+)$-invariant as $\mathcal{U}_+$-invariant.

A typical example of cone-invariant operators is nonnegative matrices. To be specific, an $m \times n$ nonnegative matrix $\mathbb{R}_+^{m \times n}$ is a $(\mathbb{R}_+^n, \mathbb{R}_+^m)$-invariant operator. Another class of cone-invariant operators is positive semidefinite cone-invariant operators, which leave the positive semidefinite cones invariant. For example, the operator

$$\boldsymbol{T} : X \in S_n \mapsto AXA' \in S_m$$

with $A \in \mathbb{R}^{m \times n}$ is $(\mathcal{P}_n, \mathcal{P}_m)$-invariant.

For the nonnegative square matrices, we have the famous Perron–Frobenius theorem to characterize their spectral properties. It was first extended to a compact cone-invariant operator in a Banach space by Krein and Rutman [1950]. A finite-dimensional Krein–Rutman theorem is stated below.

Theorem 4.2 *Let* $\mathcal{V}_+$ *be a proper cone in the finite-dimensional real vector space* $\mathcal{V}$ *and* $\boldsymbol{T} : \mathcal{V} \to \mathcal{V}$ *be a linear* $\mathcal{V}_+$*-invariant operator. Then* $\rho(\boldsymbol{T})$ *is an eigenvalue of* $\boldsymbol{T}$ *with a corresponding eigenvector* X *in* $\mathcal{V}_+ \backslash \{0\}$, *i.e.* $\boldsymbol{T}X = \rho(\boldsymbol{T})X$.

Besides the spectral radius, a cone-invariant operator may have other eigenvalues with some corresponding eigenvectors in the cone. We call these eigenvalues distinguished eigenvalues and the corresponding eigenvectors in the cone distinguished eigenvectors. It is easy to see that distinguished eigenvalues are always real and nonnegative, and the spectral radius is the largest one. To tell whether an eigenvalue is distinguished or not, it is equivalent to solve a conic feasibility problem. Particularly, when the underlying cone is $\mathbb{R}^n_+$, the conic feasibility problem becomes a linear programming (LP) feasibility problem and when the underlying cone is $\mathcal{P}_n$, it corresponds to a semidefinite programming (SDP) feasibility problem.

Analogous to the definition of generalized eigenvalues of a matrix pair, we can define generalized distinguished eigenvalues of a cone-invariant operator pair. Given two $\mathcal{V}_+$-invariant operators $\boldsymbol{T}_1$ and $\boldsymbol{T}_2$, λ is a generalized distinguished eigenvalue of the operator pair $(\boldsymbol{T}_1, \boldsymbol{T}_2)$ if there exists $X \in \mathcal{V}_+$ such that $\boldsymbol{T}_1(X) = \lambda \boldsymbol{T}_2(X)$, and X is called a generalized distinguished eigenvector corresponding to λ.

Let $\mathcal{X}$, $\mathcal{U}$, and $\mathcal{Y}$ be finite-dimensional real vector spaces together with the proper cones $\mathcal{X}_+$, $\mathcal{U}_+$, and $\mathcal{Y}_+$, respectively. Then a discrete-time LTI system with the following form:

$$\begin{aligned} X(k+1) &= \boldsymbol{A}X(k) + \boldsymbol{B}U(k), \\ Y(k) &= \boldsymbol{C}X(k) + \boldsymbol{D}U(k), \end{aligned} \tag{4.8}$$

where $X(k) \in \mathcal{X}$, $U(k) \in \mathcal{U}$, and $Y(k) \in \mathcal{Y}$, denoted by $\left[\begin{array}{c|c} \boldsymbol{A} & \boldsymbol{B} \\ \hline \boldsymbol{C} & \boldsymbol{D} \end{array}\right]$ for brevity, is said to be a cone-invariant system if the partitioned linear operator

$$\begin{bmatrix} \boldsymbol{A} & \boldsymbol{B} \\ \boldsymbol{C} & \boldsymbol{D} \end{bmatrix} : \mathcal{X} \times \mathcal{U} \to \mathcal{X} \times \mathcal{Y}$$

is $(\mathcal{X}_+ \times \mathcal{U}_+, \mathcal{X}_+ \times \mathcal{Y}_+)$-invariant. When $\mathcal{X} = \mathbb{R}^n$ with $\mathcal{X}_+ = \mathbb{R}^n_+$, $\mathcal{U} = \mathbb{R}^m$ with $\mathcal{U}_+ = \mathbb{R}^m_+$, and $\mathcal{Y} = \mathbb{R}^p$ with $\mathcal{Y}_+ = \mathbb{R}^p_+$, this cone-invariant system reduces to a positive system [Farina and Rinaldi, 2011], and the partitioned linear operator is exactly a nonnegative matrix with dimension $(n+p) \times (n+m)$.

For a cone-invariant system, weaker notions of cone-stability, cone-observability, and cone-detectability can be defined by restricting the states, inputs, and outputs to their respective cones, rather than adopting the standard stability, observability, and detectability of linear systems.

Definition 4.2 The autonomous cone-invariant system $X(k+1) = \boldsymbol{A}X(k)$ is said to be $\mathcal{X}_+$-stable if for any initial state $X(0) \in \mathcal{X}_+$, it holds that $\lim_{k\to\infty} X(k) = 0$.

Theorem 4.3 *The following statements are equivalent.*

(i) The cone-invariant system $X(k+1) = \boldsymbol{A}X(k)$ is $\mathcal{X}_+$-stable.
(ii) $\rho(\boldsymbol{A}) < 1$.
(iii) There exists $X \in int\ \mathcal{X}_+$ such that $\boldsymbol{A}X <_{\mathcal{X}_+} X$.

Proof: The equivalence of (i) and (ii) can be easily verified by the definition of $\mathcal{X}_+$-stability. Here we show the equivalence of (ii) and (iii).

(ii) ⇒ (iii): By Theorem 4.2, there exists $Y \in \mathcal{X}_+ \setminus \{0\}$ such that $\boldsymbol{A}Y = \rho(\boldsymbol{A})Y$. Then we have $\boldsymbol{A}Y < Y$ due to $\rho(\boldsymbol{A}) < 1$. This implies that $U = Y - \boldsymbol{A}Y \in \text{int } \mathcal{X}_+$. Hence, the linear iteration

$$X(k+1) = \boldsymbol{A}X(k) + U,$$

with $X(0) = 0$, converges, and the solution $X = \lim_{k\to\infty} X(k)$ satisfies $X \in \text{int } \mathcal{X}_+$ and $\boldsymbol{A}X <_{\mathcal{X}_+} X$.

(ii) ⇐ (iii): Let $Y \in \mathcal{X}_+^* \setminus \{0\}$ be a distinguished eigenvector of the adjoint operator $\boldsymbol{A}^*$ corresponding to $\rho(\boldsymbol{A}^*)$ and $X \in \text{int } \mathcal{X}_+$ satisfy that $\boldsymbol{A}X <_{\mathcal{X}_+} X$. Then by the definition of adjoint operator, we have that

$$\begin{aligned}\boldsymbol{A}X <_{\mathcal{X}_+} X &\Rightarrow \langle \boldsymbol{A}X, Y\rangle < \langle X, Y\rangle \\ &\Rightarrow \langle X, \boldsymbol{A}^*Y\rangle < \langle X, Y\rangle \\ &\Rightarrow \langle X, \rho(\boldsymbol{A})Y\rangle < \langle X, Y\rangle \\ &\Rightarrow \rho(\boldsymbol{A})\langle X, Y\rangle < \langle X, Y\rangle \Rightarrow \rho(\boldsymbol{A}) < 1,\end{aligned}$$

which completes the proof. □

Implied by the solidness of $\mathcal{X}_+$, which is equivalent to that $\mathcal{X}_+ - \mathcal{X}_+ = \mathcal{X}$, we readily see that the autonomous cone-invariant system is stable in the whole vector space if and only if it is $\mathcal{X}_+$-stable. Therefore, the stability property is cone-independent. However, such conclusion does not hold for cone-observability and cone-detectability.

Consider a cone-invariant system

$$\begin{aligned}x(k+1) &= \boldsymbol{A}x(k), \\ y(k) &= \boldsymbol{C}x(k),\end{aligned}$$

which is denoted by $\left[\frac{\boldsymbol{A}}{\boldsymbol{C}}\right]$. The cone-observability and cone-detectability are defined as follows:

Definition 4.3

(i) $\left[\frac{\boldsymbol{A}}{\boldsymbol{C}}\right]$ is said to be $\mathcal{X}_+$-observable if for any $X(0) \in \mathcal{X}_+$ such that $Y(k) \equiv 0$ for all $k \geq 0$, it holds that $X(0) = 0$.

(ii) $\left[\frac{\boldsymbol{A}}{\boldsymbol{C}}\right]$ is said to be $\mathcal{X}_+$-detectable if for any $X(0) \in \mathcal{X}_+$ such that $Y(k) \equiv 0$ for all $k \geq 0$, it holds that $\lim_{k\to\infty} X(k) = 0$.

The distinguished eigenvalues of the $\mathcal{X}_+$-invariant operator $\boldsymbol{A}$ play an important role in characterizing the system's $\mathcal{X}_+$-observability and $\mathcal{X}_+$-detectability.

Definition 4.4 A distinguished eigenvalue λ of $\boldsymbol{A}$ is said to be observable with respect to $\boldsymbol{C}$ if

$$\boldsymbol{C}X \neq 0,$$

for any distinguished eigenvector X of $\boldsymbol{A}$ corresponding to λ. We simply say that λ is an observable distinguished eigenvalue of $\left[\frac{\boldsymbol{A}}{\boldsymbol{C}}\right]$. Otherwise, it is said to be an unobservable distinguished eigenvalue.

Theorem 4.4

(i) $\left[\frac{\boldsymbol{A}}{\boldsymbol{C}}\right]$ *is $\mathcal{X}_+$-observable if and only if all the distinguished eigenvalues of* $\left[\frac{\boldsymbol{A}}{\boldsymbol{C}}\right]$ *are observable.*

(ii) $\left[\frac{\boldsymbol{A}}{\boldsymbol{C}}\right]$ *is $\mathcal{X}_+$-detectable if and only if all the unstable distinguished eigenvalues of* $\left[\frac{\boldsymbol{A}}{\boldsymbol{C}}\right]$ *are observable.*

Proof: We first prove the criterion for cone-observability. To show the necessity, suppose that there exists a distinguished eigenvalue $\lambda \geq 0$ with $X \in \mathcal{X}_+\backslash\{0\}$ such that $\boldsymbol{A}X = \lambda X,\ \boldsymbol{C}X = 0$. Let $X(0) = X \neq 0$. Then

$$Y(k) = \boldsymbol{C}\boldsymbol{A}^k X = \lambda^k \boldsymbol{C}X = 0,$$

for any $k \geq 0$. Hence, $\left[\frac{\boldsymbol{A}}{\boldsymbol{C}}\right]$ is not $\mathcal{X}_+$-observable, which causes confliction. This shows the necessity.

For the sufficiency, assume that there exists $X(0) = X_0 \in \mathcal{X}_+$ such that $Y(k) = \boldsymbol{C}\boldsymbol{A}^k X_0 \equiv 0$ for any $k \geq 0$. It suffices to show that $X_0 = 0$. Denote by S_{X_0} the linear subspace $\text{span}\{X_0, \boldsymbol{A}X_0, \boldsymbol{A}^2X_0, \dots\}$. Then $\boldsymbol{C}X = 0$ for any $X \in S_{X_0}$. By Theorem 2.2 in Tam and Wu [1989], we have that $\rho(\boldsymbol{A}|_{S_{X_0}})$, where $\boldsymbol{A}|_{S_{X_0}}$ denotes the restriction of $\boldsymbol{A}$ to the invariant subspace S_{X_0}, is a distinguished eigenvalue of $\boldsymbol{A}$ with a distinguished eigenvector X in $S_{X_0} \cap \mathcal{X}_+$, i.e. $\boldsymbol{A}X = \rho(\boldsymbol{A}|_{S_{X_0}})X$. At the same time, $\boldsymbol{C}X = 0$. In order not to conflict with the cone-observability criterion, S_{X_0} has to be $\{0\}$, which implies $X_0 = 0$. Hence, $\left[\frac{\boldsymbol{A}}{\boldsymbol{C}}\right]$ is $\mathcal{X}_+$-observable.

The proof of the cone-detectability criterion is similar. We only take a look at the sufficiency part. We need to show that $\lim_{k\to\infty} X(k) = 0$ for $X(0) = X_0 \in \mathcal{X}_+$ satisfying $Y(k) \equiv 0$. In this case, we also have that $\boldsymbol{A}X = \rho(\boldsymbol{A}|_{S_{X_0}})X$, $\boldsymbol{C}X = 0$ for some $X \in S_{X_0} \cap \mathcal{X}_+$. By the cone-detectability criterion, $\rho(\boldsymbol{A}|_{S_{X_0}})$ should be less than 1. Hence, $\left[\frac{\boldsymbol{A}}{\boldsymbol{C}}\right]$ is stable with X_0. This implies the cone-detectability of $\left[\frac{\boldsymbol{A}}{\boldsymbol{C}}\right]$. □

Remark 4.1 Theorem 4.4 can be seen as PBH tests for cone-invariant systems. We can also define the observability and detectability of a cone-invariant system in the whole vector space $\mathcal{X}$. Take the observability as an example. $\left[\frac{\boldsymbol{A}}{\boldsymbol{C}}\right]$ is said to be observable if, for any $X(0) \in \mathcal{X}$ such that $Y(k) \equiv 0$ for all $k > 0$, it holds that $X(0) = 0$. However, unlike stability, cone-observability cannot imply observability. We use a simple example to declare this.

Example 4.2 *Consider the following positive system:*

$$x(k+1) = \begin{bmatrix} 1 & 3 \\ 2 & 2 \end{bmatrix} x(k),$$
$$y(k) = \begin{bmatrix} 2 & 3 \end{bmatrix} x(k),$$

where $x(k) \in \mathbb{R}^2$ *and* $y(k) \in \mathbb{R}$. *The spectrum of* $\begin{bmatrix} 1 & 3 \\ 2 & 2 \end{bmatrix}$ *is* $\{4, -1\}$, *in which 4 is a distinguished eigenvalue. A basis of the eigenspace corresponding to 4 is given by* $\begin{bmatrix} 1 \\ 1 \end{bmatrix}$. *Then* $\begin{bmatrix} 2 & 3 \end{bmatrix}\begin{bmatrix} 1 \\ 1 \end{bmatrix} = 5 \neq 0$. *Hence, by Theorem 4.4, this system is* $\mathbb{R}^2_+$*-observable. On the other hand,* $\begin{bmatrix} 3 \\ -2 \end{bmatrix}$ *is an eigenvector corresponding to* -1 *and* $\begin{bmatrix} 2 & 3 \end{bmatrix}\begin{bmatrix} 3 \\ -2 \end{bmatrix} = 0$. *Then by the observability criterion for an LTI system, this system is not observable. Hence, cone-observability is not equivalent to observability. Likewise, cone-detectability cannot imply detectability, which is defined in a similar way to observability.*

Remark 4.2 Similarly, given two $\mathcal{X}_+$-invariant operators $\boldsymbol{A}_1$ and $\boldsymbol{A}_2$, a generalized distinguished eigenvalue λ of $(\boldsymbol{A}_1, \boldsymbol{A}_2)$ is said to be an observable generalized distinguished eigenvalue of $\left[\frac{(\boldsymbol{A}_1, \boldsymbol{A}_2)}{\boldsymbol{C}}\right]$, if for any generalized distinguished eigenvector X corresponding to λ, it holds that $\boldsymbol{C}X \neq 0$.

4.4.2 Solvability

Since the optimal controller to the stochastic LQ optimal control, if it exists, is a static one, we limit our attention to the static state feedback in the rest of this chapter. Let $v(k)$ be generated by a static state feedback gain F. Then the closed-loop system (4.3) is given by

$$\begin{aligned} x(k+1) &= (A + B\kappa(k)F)x(k), \\ z(k) &= (C + D\kappa(k)F)x(k). \end{aligned} \tag{4.9}$$

Let $X(k) \triangleq \mathbb{E}[x(k)x'(k)] \in \mathcal{P}_n$ and $Z(k) \triangleq \mathbb{E}[z(k)z'(k)] \in \mathcal{P}_p$. Then we obtain the following cone-invariant system:

$$\begin{aligned} X(k+1) &= (A + BMF)X(k)(A + BMF)' + B[\Sigma^2 \odot (FX(k)F')]B', \\ Z(k) &= (C + DMF)X(k)(C + DMF)' + D[\Sigma^2 \odot (FX(k)F')]D'. \end{aligned}$$

Denote the associated $\mathcal{P}_n$-invariant and $(\mathcal{P}_n, \mathcal{P}_p)$-invariant operators as follows:

$$\begin{aligned} \boldsymbol{A}_F &: X \in \mathcal{S}_n \mapsto (A + BMF)X(A + BMF)' + B[\Sigma^2 \odot (FXF')]B' \in \mathcal{S}_n, \\ \boldsymbol{C}_F &: X \in \mathcal{S}_n \mapsto (C + DMF)X(C + DMF)' + D[\Sigma^2 \odot (FXF')]D' \in \mathcal{S}_p. \end{aligned}$$

Then the above cone-invariant system can be denoted by $\left[\begin{array}{c} \boldsymbol{A}_F \\ \hline \boldsymbol{C}_F \end{array}\right]$ for simplicity. The adjoint operator $\boldsymbol{A}_F^*$ satisfying

$$\langle \boldsymbol{A}_F X, Y \rangle = \langle X, \boldsymbol{A}_F^* Y \rangle$$

for all $X, Y \in \mathcal{S}_n$ is given by

$$\boldsymbol{A}_F^* : X \in \mathcal{S}_n \mapsto (A + BMF)'X(A + BMF) + F'[\Sigma^2 \odot (B'XB)]F \in \mathcal{S}_n.$$

System (4.1) is said to be mean-square stabilizable with the random input gains $\kappa(k)$ if there exists F such that the closed-loop system (4.9) is mean-square stable. We can simply say that $\{[A|B], \kappa(k)\}$ is mean-square stabilizable. By Lemma 1 in Xiao et al. [2012], we have the following result:

Lemma 4.1

(i) $\{[A|B], \kappa(k)\}$ *is mean-square stabilizable if and only if, for any* $Y > 0$, *there exist* F *and* $X > 0$ *such that*

$$X = \boldsymbol{A}_F^* X + Y. \tag{4.10}$$

(ii) When $\{[A|B], \kappa(k)\}$ *is mean-square stabilizable, then for any* $Y \geq 0$, *there exist* F *and* $X \geq 0$ *such that Eq. (4.10) holds.*

In Dragan et al. [2010], one numerical necessary and sufficient condition ensuring the existence of a mean-square stabilizing solution to the MARE (4.6) is given in terms of the feasibility of some linear matrix inequalities (LMIs). However, such a condition has no explicit interpretation with respect to the dynamical properties of the underlying stochastic system. We hope to obtain a necessary and sufficient condition directly in terms of the system parameters with an explicit interpretation of the dynamical properties of the underlying system.

We first consider a very special case when $\begin{bmatrix} C'C & C'D \\ D'C & D'D \end{bmatrix}$ is positive definite. Then a necessary and sufficient condition ensuring the existence of a mean-square stabilizing solution to the MARE (4.6) is provided.

Theorem 4.5 *When* $\begin{bmatrix} C'C & C'D \\ D'C & D'D \end{bmatrix} > 0$, *there exists a mean-square stabilizing solution to the MARE* (4.6) *if and only if* $\{[A|B], \kappa(k)\}$ *is mean-square stabilizable.*

Proof: The necessity is quite straightforward by the definition of the mean-square stabilizing solution. It suffices to show the sufficiency.

First, we will show the existence of a solution $X > 0$ to the MARE (4.6). Denote

$$V_N(x(0)) \triangleq \min_{v_N} (J_N(x(0), v_N).$$

Then by Theorem 4.1, $V_N(x(0)) = x'(0)X_N(0)x(0)$. For $N_2 > N_1 \geq 0$,

$$V_{N_2}(x(0)) \geq V_{N_1}(x(0)),$$

which implies

$$X_{N_2}(0) \geq X_{N_1}(0),$$

i.e. $X_N(0)$ is monotonically increasing with respect to N. Furthermore, when $\begin{bmatrix} Q & S \\ S' & R \end{bmatrix} > 0$, by Schur complements, we have $Q - SR^{-1}S' > 0$, which implies

$$X_1(0) = Q - S(W \odot R)^{-1}S' > 0.$$

Then we have

$$X_{N_2}(0) \geq X_{N_1}(0) > 0$$

for $N_2 > N_1 \geq 1$.

Since $\{[A|B], \kappa(k)\}$ is mean-square stabilizable, by Lemma 4.1, there exist a feedback gain F and $Y > 0$ such that

$$Y = (A + BMF)'Y(A + BMF) + F'[\Sigma^2 \odot (B'YB)]F + I. \tag{4.11}$$

It is easy to obtain

$$\begin{aligned} \operatorname{tr}[X(k+1)Y] &= \operatorname{tr}\{\mathbb{E}[(A + BK(k)F)x(k)x'(k)(A + BK(k)F)']Y\} \\ &= \operatorname{tr}\{(A + BMF)X(k)(A + BMF)'Y + B[\Sigma^2 \odot (FG(k)F')]B'Y\} \end{aligned}$$

$$= \text{tr}\{X(k)(A + BMF)'Y(A + BMF) + X(k)F'[\Sigma^2 \odot (B'YB)]F\}$$
$$= \text{tr}[X(k)Y] - \text{tr}[X(k)],$$

where Eq. (4.11) is used to derive the last equality. By summing over $k = 0, 1, \dots, N$, we obtain

$$\sum_{k=0}^{N} \text{tr}[X(k)] = \text{tr}\,[X(0)Y] - \text{tr}\,[X(N+1)Y].$$

Since the closed-loop system (4.9) is mean-square stable with F, i.e. $\lim_{k\to\infty} X(k) = 0$, we obtain

$$\sum_{k=0}^{\infty} \mathbb{E}[\|x(k)\|^2] = \sum_{k=0}^{\infty} \text{tr}\left[X(k)\right] = \text{tr}[X(0)Y] \le r_1 \|x(0)\|^2,$$

where $r_1 = \lambda_{\max}(Y)$.

With $v(k) = Fx(k)$, the cost function becomes

$$J(x(0), v_\cdot) = \sum_{k=0}^{\infty} \text{tr}\{\mathbb{E}[z(k)z(k)']\} = \sum_{k=0}^{\infty} \text{tr}[X(k)U] \le r_2 \|x(0)\|^2$$

with

$$U \triangleq (C + DMF)'(C + DMF) + F'[\Sigma^2 \odot R]F$$

and $r_2 = \lambda_{\max}(U)\lambda_{\max}(Y)$. Then

$$x'(0)X_N(0)x(0) = V_N(x(0)) \le V_\infty(x(0)) \le J_\infty(x(0), v_\infty) \le r_2\|x(0)\|^2,$$

which implies

$$0 \le X_N(0) \le r_2 I.$$

By the Monotone Convergence Theorem, $X = \lim_{k\to\infty} X_N(0) > 0$ exists.

Next, we show that X is a mean-square stabilizing solution. Notice that with $F_X \triangleq F(X)$ defined in (4.7), the MARE (4.6) becomes

$$\begin{aligned} X &= (A + BMF_X)'X(A + BMF_X) + F_X'[\Sigma^2 \odot (B'XB)]F_X + U \\ &= \boldsymbol{A}_{F_X}^* + U_F, \end{aligned} \tag{4.12}$$

where

$$U_F = Q - SR^{-1}S' + (SR^{-1} + F_X'M')R(SR^{-1} + F_X'M')' + F_X'[\Sigma^2 \odot R]F_X.$$

Since $Q - SR^{-1}S' > 0$, $U_F > 0$. Therefore, from (4.12), $X > \boldsymbol{A}_{F_X}^*$. By Lemma 4.1, the closed-loop system (4.9) is mean-square stable with feedback gain F_X, which shows that X is a mean-square stabilizing solution to the MARE (4.6). □

Next, we will consider the general case with only $W \odot (D'D) > 0$. Denote

$$\mathcal{M}(X) \triangleq A'X\,A + C'C - X,$$

$$\mathcal{L}(X) \triangleq A'XB + C'D,$$
$$\mathcal{N}(X) \triangleq W \odot (B'XB + D'D).$$

The following set is defined:

$$\Gamma = \left\{ X \in \mathcal{S}^n \middle| \begin{bmatrix} \mathcal{M}(X) & \mathcal{L}(X) \\ \mathcal{L}'(X) & \mathcal{N}(X) \end{bmatrix} \geq 0,\ \mathcal{N}(X) > 0 \right\}.$$

Note that the left-hand side of the MARE (4.6) is the Schur complement of the block $\mathcal{N}(X)$ of $\begin{bmatrix} \mathcal{M}(X) & \mathcal{L}(X) \\ \mathcal{L}'(X) & \mathcal{N}(X) \end{bmatrix}$. Hence, a solution to the MARE (4.6) is a member of Γ. Some other important solutions to the MARE (4.6) are defined as follows:

Definition 4.5 A solution $X \in \mathcal{S}_n$ to the MARE (4.6) is said to be a maximal solution if $X \geq \tilde{X}$ for all $\tilde{X} \in \Gamma$. It is said to be a strong solution if $\rho(\boldsymbol{A}_{F(X)}) \leq 1$.

The maximal solution can be numerically computed by solving the following convex optimization problem:

$$\begin{aligned} \max \quad & \operatorname{tr}(X), \\ \text{subject to} \quad & X \in \Gamma. \end{aligned} \tag{4.13}$$

Lemma 4.2

(i) *When $\{[A|B], \kappa(k)\}$ is mean-square stabilizable, there exists the unique maximal solution $X_+ \in \mathcal{P}_n$ to the MARE (4.6) and the minimal cost is given by $x'(0)X_+x(0)$. Moreover, X_+ is a strong solution.*

(ii) *When $\{[A|B], \kappa(k)\}$ is mean-square stabilizable, the mean-square stabilizing solution, if it exists, coincides with the maximal solution.*

Two lemmas are needed to prove Lemma 4.2.

Lemma 4.3 *Let $v(k)$ be generated by a mean-square stabilizing controller. Then for any $X \in \mathcal{S}_n$,*

$$J(x(0), v(\cdot)) = \mathbb{E}\left\{ \sum_{k=0}^{\infty} \begin{bmatrix} x(k) \\ Mv(k) \end{bmatrix}' \begin{bmatrix} \mathcal{M}(X) & \mathcal{L}(X) \\ \mathcal{L}'(X) & \mathcal{N}(X) \end{bmatrix} \begin{bmatrix} x(k) \\ Mv(k) \end{bmatrix} \right\} + x'(0)Xx(0).$$

Moreover, when $X \in \Gamma$, $J(x(0), v(\cdot)) \geq x'(0)Xx(0)$.

Proof: It is easy to see that

$$\begin{aligned} &\mathbb{E}\left[x'(N)Xx(N)\right] - x'(0)Xx(0) \\ &= \mathbb{E}\left\{ \sum_{k=0}^{N-1} \begin{bmatrix} x(k) \\ Mv(k) \end{bmatrix}' \begin{bmatrix} A'XA - X & A'XB \\ B'XA & W \odot (B'XB) \end{bmatrix} \begin{bmatrix} x(k) \\ Mv(k) \end{bmatrix} \right\}. \end{aligned}$$

When $v(k)$ is generated by a mean-square stabilizing controller, $\mathbb{E}[x'(N)Xx(N)]$ vanishes as $N \to \infty$, i.e.

$$\mathbb{E}\left\{\sum_{k=0}^{\infty}\begin{bmatrix} x(k) \\ Mv(k)\end{bmatrix}'\begin{bmatrix} A'XA-X & A'XB \\ B'XA & W\odot(B'XB)\end{bmatrix}\begin{bmatrix} x(k) \\ Mv(k)\end{bmatrix}\right\} + x'(0)Xx(0) = 0.$$

Add the left-hand side of the above equality to the right-hand side of the cost function $J(x(0), v(\cdot))$ to obtain the desired result. □

Lemma 4.4 *Given*

$$\begin{bmatrix} Q_1 & S_1 \\ S_1' & R_1\end{bmatrix} \geq \begin{bmatrix} Q_2 & S_2 \\ S_2' & R_2\end{bmatrix} \geq 0$$

with $W \odot R_2 > 0$, suppose that the mean-square stabilizing solutions to the associated MAREs

$$X = A'XA + Q_1 - (A'XB + S_1)[W\odot(R_1 + B'XB)]^{-1}(B'XA + S_1'),$$
$$X = A'XA + Q_2 - (A'XB + S_2)[W\odot(R_2 + B'XB)]^{-1}(B'XA + S_2')$$

exist, denoted by X_1 and X_2, respectively. Then $X_1 \geq X_2$.

Proof: Denote the corresponding mean-square stabilizing controllers of these two MAREs by F_1 and F_2, respectively. Then the associated MAREs can be rewritten as

$$\begin{aligned} X_1 &= (A+BMF_1)'X_1(A+BMF_1) + F_1'[\Sigma^2\odot(B'X_1B)]F_1 \\ &\quad + Q_1 + S_1MF_1 + F_1'MS_1' + F_1'M(W\odot R_1)MF_1 \\ &= \begin{bmatrix} I \\ MF_1\end{bmatrix}'\begin{bmatrix} A'X_1A+Q_1 & A'X_1B+S_1 \\ B'X_1A+S_1' & W\odot(B'X_1B+R_1)\end{bmatrix}\begin{bmatrix} I \\ MF_1\end{bmatrix}, \\ X_2 &= (A+BMF_2)'X_2(A+BMF_2) + F_2'[\Sigma^2\odot(B'X_2B)]F \\ &\quad + Q_2 + S_2MF_2 + F_2'MS_2' + F_2'M(W\odot R_2)MF_2 \\ &= \begin{bmatrix} I \\ MF_2\end{bmatrix}'\begin{bmatrix} A'X_2A+Q_2 & A'X_2B+S_2 \\ B'X_2A+S_2' & W\odot(B'X_2B+R_2)\end{bmatrix}\begin{bmatrix} I \\ MF_2\end{bmatrix} \\ &= \begin{bmatrix} I \\ MF_1\end{bmatrix}'\begin{bmatrix} A'X_2A+Q_2 & A'X_2B+S_2 \\ B'X_2A+S_2' & W\odot(B'X_2B+R_2)\end{bmatrix}\begin{bmatrix} I \\ MF_1\end{bmatrix} \\ &\quad - (F_2-F_1)'M[W\odot(B'X_2B+R_2)]M(F_2-F_1). \end{aligned}$$

Denote $\Delta X = X_1 - X_2$. Then

$$\Delta X = \begin{bmatrix} I \\ MF_1\end{bmatrix}'\begin{bmatrix} A'\Delta XA & A'\Delta XB \\ B'\Delta XA & W\odot(B'\Delta XB)\end{bmatrix}\begin{bmatrix} I \\ MF_1\end{bmatrix} + \hat{R},$$

where

$$\hat{R} = (F_2 - F_1)'M[W \odot (B'X_2B + R_2)]M(F_2 - F_1) \\ + \begin{bmatrix} I \\ MF_1 \end{bmatrix}' \begin{bmatrix} Q_1 - Q_2 & S_1 - S_2 \\ S_1' - S_2' & W \odot (R_1 - R_2) \end{bmatrix} \begin{bmatrix} I \\ MF_1 \end{bmatrix} \geq 0.$$

Because F_1 is a mean-square stabilizing controller, we have $\Delta X = X_1 - X_2 \geq 0$, which completes the proof. □

Proof of Lemma 4.2: (i) Because $\begin{bmatrix} Q & S \\ S' & R \end{bmatrix} \geq 0$, $\tilde{R} > 0$, Γ is always nonempty with at least one element $\overline{X} = 0$. Given an arbitrary $X \in \Gamma$, for any $\delta > 0$, it holds that

$$\begin{bmatrix} Q_0 & S_0 \\ S_0' & R_0 \end{bmatrix} \triangleq \begin{bmatrix} \mathcal{M}(X) + \delta I & \mathcal{L}(X) \\ \mathcal{L}'(X) & \mathcal{N}(X) \end{bmatrix} > 0.$$

By Theorem 4.5, a mean-square stabilizing solution $X_0 \in \mathcal{P}_n$ to the MARE

$$X_0 = A'X_0A + Q_0 - (A'X_0B + S_0)[W \odot (R_0 + B'X_0B)]^{-1}(B'X_0A + S_0')$$

exists with the mean-square stabilizing controller given by

$$F_0 = -M^{-1}[W \odot (R_0 + B'X_0B)]^{-1}(B'X_0A + S_0').$$

The above MARE can be rewritten as

$$X + X_0 = A'(X + X_0)A + Q + \delta I - [A'(X + X_0)B + S] \\ \times \{W \odot [R + B'(X + X_0)B]\}^{-1}[B'(X + X_0)A + S'].$$

Given that

$$F_0 = -M^{-1}\{W \odot [R + B'(X + X_0)B]\}^{-1}[B'(X + X_0)A + S'],$$

it follows that $X_\delta = X + X_0$ is the mean-square stabilizing solution to the following MARE:

$$\mathcal{M}(X_\delta) + \delta I - \mathcal{L}(X_\delta)\mathcal{N}(X_\delta)^{-1}\mathcal{L}'(X_\delta) = 0,$$

and $X_\delta \geq 0$. Define a positive decreasing sequence δ_i with $\delta_i \to 0$ as $i \to \infty$. Then denote by X_{δ_i} the mean-square stabilizing solution to the associated MARE

$$\mathcal{M}(X_{\delta_i}) + \delta_i I - \mathcal{L}(X_{\delta_i})\mathcal{N}(X_{\delta_i})^{-1}\mathcal{L}'(X_{\delta_i}) = 0.$$

Hence, the associated controller

$$F_{\delta_i} = -M^{-1}\mathcal{N}(X_{\delta_i})^{-1}\mathcal{L}'(X_{\delta_i})$$

minimizes the following cost function:

$$J_{\delta_i}(x(0), v(\cdot)) = \mathbb{E}\left\{\sum_{k=0}^{\infty} \begin{bmatrix} x(k) \\ \kappa(k)v(k) \end{bmatrix}' \begin{bmatrix} Q + \delta_i I & S \\ S' & R \end{bmatrix} \begin{bmatrix} x(k) \\ \kappa(k)v(k) \end{bmatrix}\right\}.$$

The minimal cost is given by $V_{\delta_i}(x(0)) \triangleq \inf_{K} J_{\delta_i}(x(0), v(\cdot)) = x'(0)X_{\delta_i}x(0)$.

According to Lemma 4.4,

$$X_{\delta_0} \geq \cdots \geq X_{\delta_i} \geq \cdots \geq X,$$

where the last inequality holds as $V_{\delta_i}(x(0)) \geq x'(0)Xx(0)$ by Lemma 4.3. By the Monotone Convergence Theorem, X_{δ_i} converges to a unique matrix X_+ as $i \to \infty$. Then X_+ satisfies

$$\mathcal{M}(X_+) - \mathcal{L}(X_+)\mathcal{N}(X_+)^{-1}\mathcal{L}'(X_+) = 0$$

and $X_+ \geq X$. Because X is any element in the set Γ, X_+ is exactly the unique maximal solution. Moreover, $X_+ \geq \overline{X} = 0$. By the continuity of the eigenvalues, X_+ is a strong solution.

Because

$$V(x(0)) \triangleq \inf_{K} J(x(0), v(\cdot)) \leq V_{\delta_i}(x(0)) = x'(0)X_{\delta_i}x(0) \leq \cdots \leq V_{\delta_0}(x(0)),$$

it follows that $V(x(0)) \leq x'(0)X_+x(0)$ due to $\lim_{i\to\infty} X_{\delta_i} = X_+$. Meanwhile, by Lemma 4.3,

$$V(x(0)) \geq x'(0)X_+x(0).$$

Therefore, $V(x(0)) = x'(0)X_+x(0)$.

(ii) Suppose that the mean-square stabilizing solution exists, denoted by X. Denote the maximal solution by X_+. Then

$$V(x(0)) = x'(0)Xx(0) = x'(0)X_+x(0),$$

which implies that $X = X_+$. □

Recall the standard definite ARE theory for some insight. An explicit, necessary, and sufficient condition ensuring the existence of a stabilizing solution to the ARE

$$A'XA - X + C'C - (A'XB + C'D)(B'XB + D'D)^{-1}(B'XA + D'C) = 0 \tag{4.14}$$

is that

(i) $[A|B]$ is stabilizable;
(ii) $\begin{bmatrix} A - e^{j\omega}I & B \\ C & D \end{bmatrix}$ has full column rank for all $\omega \in [0, 2\pi)$.

The above condition can be found in Qiu [1999]. Moreover, as long as the stabilizing solution exists, it is unique. When D has full column rank, condition (ii) is equivalent in that

$$\left[\frac{A - B(D'D)^{-1}D'C}{C - D(D'D)^{-1}D'C}\right]$$

has no unobservable eigenvalues on the unit circle [Zhou et al., 1996].

Inspired by the above results, we define the following stochastic system:

$$\begin{aligned} x(k+1) &= \{A - B\kappa(k)M^{-1}[W \odot (D'D)]^{-1}D'C\}x(k), \\ z(k) &= \{C - D\kappa(k)M^{-1}[W \odot (D'D)]^{-1}D'C\}x(k). \end{aligned} \tag{4.15}$$

The corresponding cone-invariant system is exactly $\begin{bmatrix} \boldsymbol{A}_{-M^{-1}[W\odot (D'D)]^{-1}D'C} \\ \boldsymbol{C}_{-M^{-1}[W\odot (D'D)]^{-1}D'C} \end{bmatrix}$.

The following key lemma, which itself not only is an interesting result but also plays an essential role in deriving the necessary and sufficient condition, is obtained.

Lemma 4.5 *Given a solution $X \in \mathcal{P}_n$ to the MARE (4.6), any unstable distinguished eigenvalue of $\boldsymbol{A}_{F(X)}$ is an unobservable distinguished eigenvalue of* $\begin{bmatrix} \boldsymbol{A}_{-M^{-1}[W\odot (D'D)]^{-1}D'C} \\ \boldsymbol{C}_{-M^{-1}[W\odot (D'D)]^{-1}D'C} \end{bmatrix}$.

Proof: The MARE (4.6) can be rewritten as

$$X = \boldsymbol{A}_F^* X + H,$$

where $F = F(X)$ and $H \triangleq (C + DMF)'(C + DMF) + F'[\Sigma^2 \odot (D'D)]F$.

Suppose that $\lambda \geq 1$ is an unstable distinguished eigenvalue of $\boldsymbol{A}_F$, and $Y \geq 0$ is a distinguished eigenvector, i.e. $\boldsymbol{A}_F Y = \lambda Y$. Then

$$0 \leq \mathrm{tr}(HY) = \mathrm{tr}[(X - \boldsymbol{A}_F^* X)Y] = \mathrm{tr}[X(Y - \boldsymbol{A}_F Y)] = (1 - \lambda)\mathrm{tr}(XY) \leq 0,$$

i.e. $\mathrm{tr}(HY) = 0$. Meanwhile,

$$\begin{aligned} H &= Q + SMF + F'MS' + F'M\tilde{R}MF \\ &= Q - S\tilde{R}^{-1}S' + (\tilde{R}^{-1}S' + MF)'\tilde{R}(\tilde{R}^{-1}S' + MF). \end{aligned}$$

We have that

$$\mathrm{tr}\{[Q - S\tilde{R}^{-1}S' + (\tilde{R}^{-1}S' + MF)'\tilde{R}(\tilde{R}^{-1}S' + MF)]Y\} = 0.$$

Because $\begin{bmatrix} Q & S \\ S' & R \end{bmatrix} \geq 0$ and $\tilde{R} = R + \mathrm{SNR}^{-1} \odot R > 0$, it follows that $\begin{bmatrix} Q & S \\ S' & \tilde{R} \end{bmatrix} \geq$ 0and the Schur complement $Q - S\tilde{R}^{-1}S' \geq 0$. Therefore, we have

$$\begin{aligned} &\mathrm{tr}\left[(Q - S\tilde{R}^{-1}S')Y\right] = 0, \\ &\mathrm{tr}\left[(\tilde{R}^{-1}S' + MF)'\tilde{R}(\tilde{R}^{-1}S' + MF)Y\right] = 0. \end{aligned}$$

Moreover, with the assumption $\tilde{R} > 0$, it holds that $(\tilde{R}^{-1}S' + MF)Y = 0$, i.e.

$$MFY = -\tilde{R}^{-1}S'Y.$$

It follows that $\lambda Y = \boldsymbol{A}_{-M^{-1}\tilde{R}^{-1}S'}(Y)$, i.e. λ is an unstable distinguished eigenvalue of $\boldsymbol{A}_{-M^{-1}\tilde{R}^{-1}S'}$. At the same time,

$$\begin{aligned}
\mathrm{tr}(\boldsymbol{C}_{-M^{-1}\tilde{R}^{-1}S'}) &= \mathrm{tr}\left\{\mathbb{E}\left[(C - D\kappa(k)M^{-1}\tilde{R}^{-1}S')Y(C - D\kappa(k)M^{-1}\tilde{R}^{-1}S')'\right]\right\} \\
&= \mathrm{tr}\left\{\mathbb{E}\left[(C - D\kappa(k)M^{-1}\tilde{R}^{-1}S')'(C - D\kappa(k)M^{-1}\tilde{R}^{-1}S')Y\right]\right\} \\
&= \mathrm{tr}(Q - S\tilde{R}^{-1}S')Y \\
&= 0,
\end{aligned}$$

which implies $\boldsymbol{C}_{-M^{-1}\tilde{R}^{-1}S'} = 0$. Therefore, λ is an unobservable distinguished eigenvalue of $\left[\begin{array}{c}\boldsymbol{A}_{-M^{-1}\tilde{R}^{-1}S'} \\ \hline \boldsymbol{C}_{-M^{-1}\tilde{R}^{-1}S'}\end{array}\right]$. □

By virtue of Lemma 4.5, we can obtain an explicit, necessary, and sufficient condition ensuring the existence of a mean-square stabilizing solution to the MARE (4.6).

Theorem 4.6 *When $W \odot (D'D) > 0$, a mean-square stabilizing solution to the MARE (4.6) exists if and only if*

(i) $\{[A|B], \kappa(k)\}$ *is mean-square stabilizable;*

(ii) $\left[\begin{array}{c}\boldsymbol{A}_{-M^{-1}[W\odot(D'D)]^{-1}D'C} \\ \hline \boldsymbol{C}_{-M^{-1}[W\odot(D'D)]^{-1}D'C}\end{array}\right]$ *has no unobservable distinguished eigenvalue at point 1.*

Moreover, if the mean-square stabilizing solution exists, it is unique.

Proof: We first start with the sufficiency part. By Lemma 4.2, the maximal solution X_+ exists and $\rho(\mathcal{L}_{F(X_+)}) \leq 1$. Because $\lambda = 1$ is not an unobservable distinguished eigenvalue of $\left[\begin{array}{c}\boldsymbol{A}_{-M^{-1}\tilde{R}^{-1}S'} \\ \hline \boldsymbol{C}_{-M^{-1}\tilde{R}^{-1}S'}\end{array}\right]$, by Lemma 4.5, $\rho(\mathcal{L}_{F(X_+)}) < 1$. Therefore, the maximal solution is exactly the mean-square stabilizing solution.

Now, take a look at the necessity part. Evidently, the mean-square stabilizabity of $\{[A|B], \kappa(k)\}$ is necessary. Suppose that $\left[\begin{array}{c}\boldsymbol{A}_{-M^{-1}\tilde{R}^{-1}S'} \\ \hline \boldsymbol{C}_{-M^{-1}\tilde{R}^{-1}S'}\end{array}\right]$ has an unobservable distinguished eigenvalue at 1, i.e. there exists $Y \geq 0$ such that

$$\begin{aligned}
&\boldsymbol{A}_{-M^{-1}\tilde{R}^{-1}S'}Y = Y, \\
&\boldsymbol{C}_{-M^{-1}\tilde{R}^{-1}S'}Y = 0.
\end{aligned}$$

We show that in this case, for any $X \in \mathcal{P}_n$ satisfying the MARE (4.6), it is not a mean-square stabilizing solution. Denote

$$T = \begin{bmatrix} T_{11} & T_{12} \\ T'_{12} & T_{22} \end{bmatrix} \triangleq \begin{bmatrix} I & -S\tilde{R}^{-1} \\ 0 & I \end{bmatrix} \begin{bmatrix} \mathcal{M}(X) & \mathcal{L}(X) \\ \mathcal{L}'(X) & \mathcal{N}(X) \end{bmatrix} \begin{bmatrix} I & 0 \\ -\tilde{R}^{-1}S' & I \end{bmatrix}.$$

Hence,

$$\begin{aligned} T_{11} &= \boldsymbol{A}^*_{-M^{-1}\tilde{R}^{-1}S'} X - X + Q - S\tilde{R}^{-1}S', \\ T_{12} &= (A - B\tilde{R}^{-1}S')'XB - S\tilde{R}^{-1}[\mathrm{SNR}^{-2} \odot (B'XB)], \\ T_{22} &= \mathcal{N}(X). \end{aligned}$$

With some computations, we have

$$\begin{aligned} &\mathcal{M}(X) - \mathcal{L}(X)\mathcal{N}(X)^{-1}\mathcal{L}'(X) \\ &= \begin{bmatrix} I \\ -\mathcal{N}(X)^{-1}\mathcal{L}'(X) \end{bmatrix}' \begin{bmatrix} \mathcal{M}(X) & \mathcal{L}(X) \\ \mathcal{L}'(X) & \mathcal{N}(X) \end{bmatrix} \begin{bmatrix} I \\ -\mathcal{N}(X)^{-1}\mathcal{L}'(X) \end{bmatrix} \\ &= \begin{bmatrix} I \\ -T_{22}^{-1}T'_{12} \end{bmatrix}' T \begin{bmatrix} I \\ -T_{22}^{-1}T'_{12} \end{bmatrix} \\ &= T_{11} - T_{12}T_{22}^{-1}T'_{12}, \end{aligned}$$

and

$$\begin{aligned} &\mathrm{tr}\left[(Q - S\tilde{R}^{-1}S')Y\right] \\ &= \mathrm{tr}\left\{\mathbb{E}\left[(C - D\kappa(k)M^{-1}\tilde{R}^{-1}S')'(C - D\kappa(k)M^{-1}\tilde{R}^{-1}S')Y\right]\right\} \\ &= \mathrm{tr}\left\{\mathbb{E}\left[(C - D\kappa(k)M^{-1}\tilde{R}^{-1}S')Y(C - D\kappa(k)M^{-1}\tilde{R}^{-1}S')'\right]\right\} \\ &= \mathrm{tr}(\boldsymbol{C}_{-M^{-1}\tilde{R}^{-1}S'}Y) = 0. \end{aligned}$$

It follows that

$$\begin{aligned} \mathrm{tr}(T_{11}Y) &= \mathrm{tr}\left[(\boldsymbol{A}^*_{-M^{-1}\tilde{R}^{-1}S'}X - X)Y\right] + \mathrm{tr}\left[(Q - S\tilde{R}^{-1}S')Y\right] \\ &= \mathrm{tr}\left[X(\boldsymbol{A}_{-M^{-1}\tilde{R}^{-1}S'}Y - Y)\right] \\ &= 0. \end{aligned}$$

Then for any solution $X \in \mathcal{P}_n$ to the MARE (4.6), we have

$$\begin{aligned} 0 &= \mathrm{tr}\left[(\mathcal{M}(X) - \mathcal{L}(X)\mathcal{N}(X)^{-1}\mathcal{L}'(X))Y\right] \\ &= \mathrm{tr}\left[(T_{11} - T_{12}T_{22}^{-1}T'_{12})Y\right] \\ &= -\mathrm{tr}(T_{12}T_{22}^{-1}T'_{12}Y), \end{aligned}$$

which implies $T_{12}T_{22}^{-1}T'_{12}Y = 0$, as $T_{22} > 0$ and $Y \geq 0$. It follows that $T'_{12}Y = 0$. Therefore,

$$\begin{aligned} &[\tilde{R}^{-1}S' + MF(X)]Y \\ &= [\tilde{R}^{-1}S' - \mathcal{N}(X)^{-1}(B'XA + S')]Y \end{aligned}$$

$$
\begin{aligned}
&= \mathcal{N}(X)^{-1}\{[W \odot (B'XB)]\tilde{R}^{-1}S' - B'XA\}Y \\
&= \mathcal{N}(X)^{-1}\{[\mathrm{SNR}^{-2} \odot (B'XB)]\tilde{R}^{-1}S' - B'X(A - B\tilde{R}^{-1}S')\}Y \\
&= -\mathcal{N}(X)^{-1}T'_{12}Y = 0,
\end{aligned}
$$

which implies that $MF(X)Y = -\tilde{R}^{-1}S'Y$. Hence,

$$\mathcal{L}_{F(X)}(Y) = \boldsymbol{A}_{-M^{-1}\tilde{R}^{-1}S'}Y = Y,$$

i.e. $1 \in \sigma_{\mathcal{P}}(\mathcal{L}_{F(X)})$, and then $\rho_{\mathcal{P}}(\mathcal{L}_{F(X)}) \geq 1$. Therefore, the closed-loop system (4.9) is not mean-square stabilizing with the controller $F(X)$. This shows the necessity. □

Remark 4.3 Theorem 4.6 solves the question of whether there is a necessary and sufficient condition analogous to the one ensuring the stabilizing solution to the standard ARE (4.14). It indicates that the common assumption or condition of the observability or detectability of certain stochastic or cone-invariant system is unnecessary. Moreover, it shows that the distinguished eigenvalue at 1 of a cone-invariant system plays a similar role to the eigenvalues of an LTI system on the unit circle.

Consider a special case in which $C'D = 0$, as is employed in many research papers. Then the MARE (4.6) becomes

$$A'XA - X + C'C - A'XB[W \odot (B'XB + D'D)]^{-1}B'XA = 0. \tag{4.16}$$

By the definition of distinguished eigenvalues, we can make the following conclusion.

Corollary 4.1 *There is a unique mean-square stabilizing solution to the MARE (4.16) if and only if*

(i) $\{[A|B], \kappa(k)\}$ *is mean-square stabilizable;*

(ii) $\left[\dfrac{A}{C}\right]$ *has no unobservable eigenvalues on the unit circle.*

The result for the MARE (4.6) can be easily generalized to a more general class of MAREs. Consider a MARE of the following form:

$$
\begin{aligned}
&A'XA - X + Q + \sum_{i=1}^{N} A_i'XA_i - \left(A'XB + S + \sum_{i=1}^{N} A_i'XB_i\right) \\
&\times \left(B'XB + R + \sum_{i=1}^{N} B_i'XB_i\right)^{-1} \left(A'XB + S + \sum_{i=1}^{N} A_i'XB_i\right)' = 0,
\end{aligned}
\tag{4.17}
$$

where $A, A_i \in \mathbb{R}^{n\times n}$, $B, B_i \in \mathbb{R}^{n\times m}$ for $i = 1, \ldots, N$, and $\begin{bmatrix} Q & S \\ S' & R \end{bmatrix} \in \mathcal{P}_{n+m}$ with $R > 0$. Such a MARE often arises in the stochastic LQ optimal control problem with multiple multiplicative white noises on both the system state and control input.

The associated cone-invariant operators are given by

$$\hat{\mathbf{A}}_F : X \in \mathbb{S}_n \mapsto (A + BF)X(A + BF)' + \sum_{i=1}^{N} (A_i + B_iF)X(A_i + B_iF)' \in \mathbb{S}_n,$$

$$\hat{\mathbf{C}} : X \in \mathbb{S}_n \mapsto (Q - SR^{-1}S')^{1/2}X(Q - SR^{-1}S')^{1/2} \in \mathbb{S}_n.$$

Then we have the following result.

Corollary 4.2 *There exists a mean-square stabilizing solution to MARE* (4.17) *if and only if*

(i) *there exists F such that $\hat{\mathbf{A}}_F$ is mean-square stable;*

(ii) $\begin{bmatrix} \hat{\mathbf{A}}_{-R^{-1}S'} \\ \hat{\mathbf{C}} \end{bmatrix}$ *has no unobservable distinguished eigenvalue at 1.*

Corollary 4.2 is also stated in Ungureanu [2015], which extends our result to a class of MAREs in infinite dimensions.

Remark 4.4 Sometimes we may encounter NCSs with $W \odot (D'D)$ being singular. In this situation, the cone-invariant system $\left[\frac{\mathbf{A}_{-M^{-1}[W \odot (D'D)]^{-1}D'C}}{\mathbf{C}_{-M^{-1}[W \odot (D'D)]^{-1}D'C}}\right]$ is not well defined. In Zheng and Qiu [2018], a sufficient condition and a necessary condition are provided by further employing the tool of cone-invariant operators and some classical results from the theory of standard AREs. These two conditions coincide when D has full column rank. Interesting readers can refer to Zheng and Qiu [2018] for details.

4.5 LQ Optimal Control

We have discussed the existence issue of the mean-square stabilizing solution to the MARE (4.6) and obtained an analytical necessary and sufficient condition ensuring the existence. Now, we are well prepared to complete the study of the LQ optimal control of LTI systems with random input gains under the channel/controller codesign framework, i.e. the individual channel capacities $\mathfrak{C}_i$ are assumed to be designed under a given overall capacity constraint $\mathfrak{C}$.

Under the channel/controller codesign framework, $[A|B]$ is said to be stabilizable with the overall capacity $\mathfrak{C}$ if there is an allocation π such that $\mathfrak{C}_i = \pi_i\mathfrak{C}$ and $\{[A|B], \kappa(k)\}$ is mean-square stabilizable.

Recall the following two concepts, which were introduced to dynamical systems theory a long time ago but only appeared in the control literature recently. One is the Mahler measure [Mahler, 1960] of an $n \times n$ matrix A, denoted by $M(A)$, which is the product of the absolute value of the unstable eigenvalues of A, i.e.

$$M(A) = \prod_{i=1}^{n} \max\{1, |\lambda_i(A)|\}.$$

The second is the topological entropy [Bowen, 1971] of A, denoted by $h(A)$, which is the logarithm of $M(A)$, i.e.

$$h(A) = \log M(A).$$

In Xiao et al. [2012], a result on stabilizability with a given overall capacity is given.

Lemma 4.6 *$[A|B]$ is stabilizable with capacity $\mathfrak{C}$ if and only if $\mathfrak{C} > h(A)$.*

Remark 4.5 Lemma 4.6 indicates that the minimum total channel capacity required for mean-square stabilization with channel resource allocation is equal to the topological entropy of the open-loop plant. It is worth emphasizing the basic idea in the proof of Lemma 4.6. Without loss of generality, $[A|B]$ can be assumed to have the following Wonham decomposition [Wonham, 1967]:

$$A = \begin{bmatrix} A_1 & * & \cdots & * \\ 0 & A_2 & \ddots & \vdots \\ \vdots & \ddots & \ddots & * \\ 0 & \cdots & 0 & A_m \end{bmatrix}, \quad B = \begin{bmatrix} b_1 & * & \cdots & * \\ 0 & b_2 & \ddots & \vdots \\ \vdots & \ddots & \ddots & * \\ 0 & \cdots & 0 & b_m \end{bmatrix},$$

where each pair $[A_i|b_i]$ is stabilizable. It is clear from the Wonham decomposition that A_i contains all the unstable eigenvalues of A, which are controllable by the ith input but not controllable by any previous inputs. For a given overall capacity $\mathfrak{C}$, a feasible allocation π can always be found such that $\mathfrak{C}_i = \pi_i \mathfrak{C} > h(A_i)$. With this allocation, we sequentially design f_i such that $[A_i|b_i]$ is stabilized with capacity $\mathfrak{C}_i$. The existence of such an f_i is guaranteed by the result in Elia [2005] for the state feedback mean-square stabilization of a single-input system over a fading channel. By such a sequential design, $[A|B]$ can be stabilized with capacity $\mathfrak{C}$.

As mentioned in the previous section, the solvability of our LQ optimal control problem depends on the existence of the mean-square stabilizing solution to the MARE (4.6). It is formally shown by the following lemma.

Lemma 4.7 *Under the channel/controller codesign framework, the LQ optimal control problem of system (4.1) is solvable with the given overall capacity $\mathfrak{C}$ if and only if the MARE (4.6) has the unique mean-square stabilizing solution X together with a feasible allocation π. Then, the associated static state feedback gain $F(X)$ is the optimal controller with the minimal cost $x'(0)Xx(0)$.*

Proof: Suppose that a feasible allocation π is given. Therefore, W is determined. Let $v(\cdot)$ be generated by a mean-square stabilizing controller and $X \in \mathcal{S}_n$. Then,

$$\begin{aligned} J(x(0), v(\cdot)) &= \mathbb{E}\left\{\sum_{k=0}^{\infty}\begin{bmatrix} x(k) \\ Mv(k)\end{bmatrix}'\begin{bmatrix}\mathcal{M}(X) & \mathcal{L}(X)\\ \mathcal{L}'(X) & \mathcal{N}(X)\end{bmatrix}\begin{bmatrix} x(k) \\ Mv(k)\end{bmatrix}\right\} + x'(0)Xx(0) \\ &= \mathbb{E}\left\{\sum_{k=0}^{\infty}\left[v(k) + M^{-1}\mathcal{N}(X)^{-1}\mathcal{L}'(X)x(k)\right]'M'\mathcal{N}(X)M\right. \\ &\quad \times \left[v(k) + M^{-1}\mathcal{N}(X)^{-1}\mathcal{L}'(X)x(k)\right] \\ &\quad \left. + x'(k)\left[\mathcal{M}(X) - \mathcal{L}(X)\mathcal{N}(X)^{-1}\mathcal{L}'(X)\right]x(k)\right\} + x'(0)Xx(0). \end{aligned}$$

Sufficiency: Let X be the mean-square stabilizing solution to the MARE (4.6). Then the cost function becomes

$$\begin{aligned} J(x(0), v(\cdot)) &= \mathbb{E}\left\{\sum_{k=0}^{\infty}\left[v(k) + M^{-1}\mathcal{N}(X)^{-1}\mathcal{L}'(X)x(k)\right]'M'\mathcal{N}(X)M\right. \\ &\quad \times \left[v(k) + M^{-1}\mathcal{N}(X)^{-1}\mathcal{L}'(X)x(k)\right] \\ &\quad \left. + x'(k)\left[\mathcal{M}(X) - \mathcal{L}(X)\mathcal{N}(X)^{-1}\mathcal{L}'(X)\right]x(k)\right\} + x'(0)Xx(0). \end{aligned}$$

Evidently, when $v(k) = -M^{-1}\mathcal{N}(X)^{-1}\mathcal{L}'(X)x(k) = F(X)x(k)$, the minimum cost is achieved, i.e.

$$\inf_{K} J(x(0), v(\cdot)) = J(x(0), F(X)x(\cdot)) = x'(0)Xx(0).$$

Necessity: Suppose $v(\cdot)$ is generated by the optimal mean-square stabilizing controller $\hat{K}$ and X_+ is the maximal solution to the MARE (4.6) under the allocation π. Then, the minimal cost is given as $x'(0)X_+x(0)$ by Lemma 4.2. Hence,

$$\begin{aligned} &\inf_{K} J(x(0), v(\cdot)) = J(x(0), \hat{K}x(\cdot)) \\ &= \mathbb{E}\left\{\sum_{k=0}^{\infty}\left[v(k) + M^{-1}\mathcal{N}(X_+)^{-1}\mathcal{L}'(X_+)x(k)\right]'M'\mathcal{N}(X_+)M\right. \\ &\quad \times \left[v(k) + M^{-1}\mathcal{N}(X_+)^{-1}\mathcal{L}'(X_+)x(k)\right] \\ &\quad \left. + x'(k)\left[\mathcal{M}(X_+) - \mathcal{L}(X_+)\mathcal{N}(X_+)^{-1}\mathcal{L}'(X_+)\right]x(k)\right\} + x'(0)X_+x(0) \\ &= \mathbb{E}\left\{\sum_{k=0}^{\infty}\left[v(k) + M^{-1}\mathcal{N}(X_+)^{-1}\mathcal{L}'(X_+)x(k)\right]'M'\mathcal{N}(X_+)M\right. \\ &\quad \left. \times \left[v(k) + M^{-1}\mathcal{N}(X_+)^{-1}\mathcal{L}'(X_+)x(k)\right]\right\} + x'(0)X_+x(0) \\ &= x'(0)X_+x(0), \end{aligned}$$

which implies that

$$v(k) = -M^{-1}\mathcal{N}(X_+)^{-1}\mathcal{L}'(X_+)x(k) = \hat{K}x(k),$$

i.e. the maximal solution is mean-square stabilizing. □

The following ultimate conclusion is obtained by combining Theorem 4.6, Lemma 4.6, and Lemma 4.7.

Theorem 4.7 *Under the channel/controller codesign framework, the LQ optimal control problem of system* (4.1) *is solvable if and only if*

(i) $\mathfrak{C} > h(A)$;

(ii) $\left[\dfrac{\boldsymbol{A}_{-M^{-1}[W\odot(D'D)]^{-1}D'C}}{\boldsymbol{C}_{-M^{-1}[W\odot(D'D)]^{-1}D'C}}\right]$ *has no unobservable distinguished eigenvalue at point 1.*

The mean-square stabilization problem is analytically solved under the channel/controller codesign framework. The observability of the distinguished eigenvalue at 1 can be checked by solving a classical LMI problem. Once the above necessary and sufficient condition holds, we can get the mean-square stabilizing solution by obtaining the maximal solution, which can be numerically computed by solving the convex optimization problem (4.13). Note that when $C'D = 0$, the condition becomes simpler and is very easy to verify.

Corollary 4.3 *When* $C'D = 0$, *this LQ problem is solvable under the framework of channel/controller codesign if and only if*

(i) $\mathfrak{C} > h(A)$.

(ii) $\left[\dfrac{A}{C}\right]$ *has no unobservable eigenvalues on the unit circle.*

A simple example is given to illustrate the main theorem.

Example 4.3 *Consider an LTI system* $\left[\begin{array}{c|c} A & B \\ \hline C & D \end{array}\right]$ *with*

$$A = \begin{bmatrix} 2 & 0 & 0 \\ 0 & 1 & 0 \\ 0 & 0 & 1 \end{bmatrix}, \quad B = \begin{bmatrix} 1 & 0 \\ 1 & 1 \\ 0 & 1 \end{bmatrix}.$$

The overall capacity of the input channels is $\mathfrak{C} = \frac{1}{2}\log 10$. *Note that* $[A|B]$ *is already in the Wonham decomposition form, with*

$$A_1 = \begin{bmatrix} 2 & 0 \\ 0 & 1 \end{bmatrix}, \ b_1 = \begin{bmatrix} 1 \\ 1 \end{bmatrix}, \ A_2 = 1, \text{ and } b_2 = 1.$$

The topological entropy of the plant is

$$h(A) = h(A_1) + h(A_2) = \log(2 * 1) + \log 1 = \log 2.$$

Clearly, $[A|B]$ *is stabilizable under the overall capacity constraint* $\mathfrak{C}$. *We can see that under the framework of channel/controller co-design, mean-square stabilization can be easily verified. A feasible allocation is given by*

$$\mathfrak{C}_1 = \frac{1}{2}\log 5, \quad \mathfrak{C}_2 = \frac{1}{2}\log 2.$$

Then, $SNR_1 = 4$, $SNR_2 = 1$, *and* $W = \begin{bmatrix} 1.25 & 1 \\ 1 & 2 \end{bmatrix}$. *Let the mean and variance of the random input gains be* $M = \begin{bmatrix} 1 & 0 \\ 0 & 1 \end{bmatrix}$ *and* $\Sigma^2 = \begin{bmatrix} \frac{1}{4} & 0 \\ 0 & 1 \end{bmatrix}$, *respectively.*

Two sets of C, D *are considered. First, let*

$$C = \begin{bmatrix} 0 & 1 & 0 \\ 0 & 0 & 1 \end{bmatrix} \quad \text{and } D = \begin{bmatrix} 1 & 0 \\ 0 & 2 \end{bmatrix}.$$

According to Theorem 4.7, to determine the solvability of the LQ optimal control with the above parameters, we need to examine the mean-square detectability of the stochastic system (4.15) or the observability of 1 if 1 is a distinguished eigenvalue of $\begin{bmatrix} \boldsymbol{A}_{-M^{-1}[W\odot(D'D)]^{-1}D'C} \\ \hline \boldsymbol{C}_{-M^{-1}[W\odot(D'D)]^{-1}D'C} \end{bmatrix}$. *We have*

$$\hat{F} \triangleq -M^{-1}[W \odot (D'D)]^{-1}D'C = \begin{bmatrix} 0 & -0.8 & 0 \\ 0 & 0 & -0.25 \end{bmatrix}.$$

Then, the stochastic system in (4.15) is equal to

$$\begin{aligned} x(k+1) &= \begin{bmatrix} 2 & -0.8\kappa_1(k) & 0 \\ 0 & 1-0.8\kappa_1(k) & -0.25\kappa_2(k) \\ 0 & 0 & 1-0.25\kappa_2(k) \end{bmatrix} x(k), \\ z(k) &= \begin{bmatrix} 0 & 1-0.8\kappa_1(k) & 0 \\ 0 & 0 & 1-0.5\kappa_2(k) \end{bmatrix} x(k). \end{aligned} \tag{4.18}$$

The set of distinguished eigenvalues of $\boldsymbol{A}_{\hat{F}}$ *is given by*

$$\sigma_{\mathcal{P}}(\boldsymbol{A}_{\hat{F}}) = \{4\}$$

by taking the approach described in Section 4.4. A basis of the eigenspace corresponding to $\lambda_1 = 4$ *is* $Y = \begin{bmatrix} 1 & 0 & 0 \\ 0 & 0 & 0 \\ 0 & 0 & 0 \end{bmatrix}$. *The operator* $\boldsymbol{C}_{\hat{F}}$ *is equal to*

$$\boldsymbol{C}_{\hat{F}} : X \in \mathcal{S}_3 \mapsto \begin{bmatrix} 0 & 0.2 & 0 \\ 0 & 0 & 0.5 \end{bmatrix} X \begin{bmatrix} 0 & 0 \\ 0.2 & 0 \\ 0 & 0.5 \end{bmatrix} + \begin{bmatrix} 1 & 0 \\ 0 & 2 \end{bmatrix}$$

$$\times \left\{ \begin{bmatrix} \frac{1}{4} & 0 \\ 0 & 1 \end{bmatrix} \odot \left(\begin{bmatrix} 0 & -0.8 & 0 \\ 0 & 0 & -0.25 \end{bmatrix} X \begin{bmatrix} 0 & 0 \\ -0.8 & 0 \\ 0 & -0.25 \end{bmatrix} \right) \right\} \begin{bmatrix} 1 & 0 \\ 0 & 2 \end{bmatrix} \in \mathcal{S}_2.$$

Evidently, $\mathbf{C}_{\hat{F}}(Y) = 0$, i.e. $\lambda_1 = 4$ is unobservable, or it is equivalent to say that the stochastic system (4.18) is not mean-square detectable. If we only get the sufficient condition in terms of the observability or detectability of certain stochastic or cone-invariant system, we cannot conclude whether or not the optimal state feedback controller exists. Fortunately, we have an explicit necessary and sufficient condition shown in Theorem 4.7, which tells us that the optimal controller does exist, as the system parameters satisfy the necessary and sufficient condition. By solving the convex optimization problem (4.13), we get the mean-square stabilizing solution

$$X = \begin{bmatrix} 75.9593 & -13.0253 & 7.6301 \\ -13.0253 & 5.2438 & -2.7106 \\ 7.6301 & -2.7106 & 3.6209 \end{bmatrix}$$

with the associated state feedback gain

$$F(X) = \begin{bmatrix} -1.7776 & 0.0904 & -0.0787 \\ 0.3831 & -0.1528 & -0.2106 \end{bmatrix}.$$

By examining the sufficient condition for the mean-square stabilization stated in Section 4.4, $F(X)$ is indeed mean-square stabilizing. Note that there are some other solutions to this MARE (4.6). For example,

$$Y = \begin{bmatrix} 0 & 0 & 0 \\ 0 & 0.2498 & -0.0174 \\ 0 & -0.0174 & 1.3435 \end{bmatrix}$$

is also a solution with

$$F(Y) = \begin{bmatrix} 0 & -0.7994 & 0.0558 \\ 0 & -0.0042 & -0.3004 \end{bmatrix}.$$

However, $F(Y)$ is not mean-square stabilizing, as $\rho(\mathbf{A}_{F(Y)}) > 1$.

Now, consider another set of parameters. Let

$$C = \begin{bmatrix} 1 & 1 & 0 \\ 0 & 0 & 0 \end{bmatrix} \quad \textit{and} \quad D = \begin{bmatrix} 1 & 0 \\ 0 & 2 \end{bmatrix}.$$

Then we have

$$\hat{F} \triangleq -M^{-1}[W \odot (D'D)]^{-1}D'C = \begin{bmatrix} -0.8 & -0.8 & 0 \\ 0 & 0 & 0 \end{bmatrix}$$

and the stochastic system (4.15) is equal to

$$\begin{aligned} x(k+1) &= \begin{bmatrix} 2 - 0.8\kappa_1(k) & -0.8\kappa_1(k) & 0 \\ -0.8\kappa_1(k) & 1 - 0.8\kappa_1(k) & 0 \\ 0 & 0 & 1 \end{bmatrix} x(k), \\ z(k) &= \begin{bmatrix} 1 - 0.8\kappa_1(k) & 1 - 0.8\kappa_1(k) & 0 \\ 0 & 0 & 0 \end{bmatrix} x(k). \end{aligned} \tag{4.19}$$

The set of distinguished eigenvalues of $\boldsymbol{A}_{\hat{F}}$ *is*

$$\sigma_{\mathcal{P}}(\boldsymbol{A}_{\hat{F}}) = \{1, 2.7056\}.$$

The basis of the eigenspace corresponding to $\lambda_1 = 1$ *is* $Y = \begin{bmatrix} 0 & 0 & 0 \\ 0 & 0 & 0 \\ 0 & 0 & 1 \end{bmatrix}$. *Clearly,* $\boldsymbol{C}_{\hat{F}}(Y) = 0$, *where* $\boldsymbol{C}_{\hat{F}}$ *is equal to*

$$\boldsymbol{C}_{\hat{F}} : X \in \mathcal{S}_3 \mapsto \begin{bmatrix} 0.2 & 0.2 & 0 \\ 0 & 0 & 0 \end{bmatrix} X \begin{bmatrix} 0.2 & 0 \\ 0.2 & 0 \\ 0 & 0 \end{bmatrix} + \begin{bmatrix} 1 & 0 \\ 0 & 2 \end{bmatrix}$$

$$\times \left\{ \begin{bmatrix} \frac{1}{4} & 0 \\ 0 & 1 \end{bmatrix} \odot \left(\begin{bmatrix} -0.8 & -0.8 & 0 \\ 0 & 0 & 0 \end{bmatrix} X \begin{bmatrix} -0.8 & 0 \\ -0.8 & 0 \\ 0 & 0 \end{bmatrix} \right) \right\} \begin{bmatrix} 1 & 0 \\ 0 & 2 \end{bmatrix} \in \mathcal{S}_2.$$

Therefore, the distinguished eigenvalue $\lambda_1 = 1$ *is unobservable. By Theorem 4.7, the optimal controller does not exist because condition (ii) is not satisfied. Indeed, the maximal solution to the associated MARE*

$$X = \begin{bmatrix} 37.9752 & -7.2136 & 0 \\ -7.2136 & 3.8077 & 0 \\ 0 & 0 & 0 \end{bmatrix}$$

is just a strong solution, as $\rho(\boldsymbol{A}_{F(X)}) = 1$, *where* $F(X) = \begin{bmatrix} -1.7111 & 0.0454 & 0 \\ 0.5507 & -0.2339 & 0 \end{bmatrix}$.

4.6 Conclusion

In this chapter, we investigate the LQ optimal control for discrete-time LTI systems with random input gains and focus on the solvability of the associated MARE. By employing the tool of cone-invariant operators, an explicit necessary and sufficient condition ensuring the existence of the mean-square stabilizing solution to the MARE is obtained. Such a condition is compatible with the one ensuring the stabilizing solution to the standard ARE and it indicates that the common condition of observability or detectability of certain stochastic systems is unnecessary. This result can be easily extended to other MAREs arising in the stochastic optimal control problems with different settings [Tan and Zhang, 2016], as well as the stochastic optimal filtering problems [Zhou et al., 2018, Feng et al., 2020], and finds applications in other topics [Yang et al., 2015]. Another important feature of this stochastic LQ optimal control problem is that we can put the problem under the framework of channel/controller codesign and analytically solve the mean-square stabilization problem, which also makes our work more than a special case of general stochastic LQ optimal control problems.

Bibliography

R. Bowen. Entropy for group endomorphisms and homogeneous spaces. *Transactions of the American Mathematical Society*, 153:401–414, 1971.

V. Dragan, T. Morozan, and A. M. Stoica. *Mathematical methods in robust control of discrete-time linear stochastic systems*. Springer, New York, 2010.

N. Elia. Remote stabilization over fading channels. *Systems & Control Letters*, 54(3):237–249, 2005.

L. Farina and S. Rinaldi. *Positive linear systems: theory and applications*. John Wiley ç Sons, 2011.

Y. Feng, Z. Chen, and X. Chen. Distributed h_∞ gaussian consensus filtering for discrete-time systems over lossy sensor networks. *SIAM Journal on Control and Optimization*, 58(1):34–58, 2020.

M. G. Krein and M. A. Rutman. Linear operators leaving invariant a cone in a Banach space. *American Mathematical Society*, 1:199–325, 1950.

P. Lancaster and L. Rodman. *Algebraic Riccati equations*. Oxford University Press, 1995.

K. Mahler. An application of Jensen's formula to polynomials. *Mathematika*, 7(02):98–100, 1960.

L. Qiu. On the generalized eigenspace approach for solving Riccati equations. In *Reprints of the 14th IFAC World Congress*, volume D, pages 281–286, 1999.

B. S. Tam and S. F. Wu. On the Collatz-Wielandt sets associated with a cone-preserving map. *Linear Algebra and its Applications*, 125:77–95, 1989.

C. Tan and H. Zhang. Necessary and sufficient stabilizing conditions for networked control systems with simultaneous transmission delay and packet dropout. *IEEE Transactions on Automatic Control*, 62(8):4011–4016, 2016.

V. M. Ungureanu. Stabilizing solution for a discrete-time modified algebraic Riccati equation in infinite dimensions. *Discrete Dynamics in Nature and Society*, 2015:293930, 2015.

W. M. Wonham. On pole assignment in multi-input controllable linear systems. *IEEE Transactions on Automatic Control*, 12(6):660–665, 1967.

N. Xiao, L. Xie, and L. Qiu. Feedback stabilization of discrete-time networked systems over fading channels. *IEEE Transactions on Automatic Control*, 57(9):2176–2189, 2012.

C. Yang, J. Wu, X. Ren, W. Yang, H. Shi, and L. Shi. Deterministic sensor selection for centralized state estimation under limited communication resource. *IEEE Transactions on Signal Processing*, 63(9):2336–2348, 2015.

J. Zheng and L. Qiu. Existence of a mean-square stabilizing solution to a modified algebraic riccati equation. *SIAM Journal on Control and Optimization*, 56(1):367–387, 2018.

K. Zhou, J. C. Doyle, and K. Glover. *Robust and optimal control*, volume 40. Prentice Hall Upper Saddle River, NJ, 1996.

J. Zhou, G. Gu, and X. Chen. Distributed kalman filtering over wireless sensor networks in the presence of data packet drops. *IEEE Transactions on Automatic Control*, 64(4):1603–1610, 2018.

5

Multisensor Kalman Filtering with Intermittent Measurements

5.1 Introduction

In this chapter, we extend the stability theory on Kalman filtering with intermittent measurements from a single-sensor scenario to a multisensor scenario. Consider that a group of sensors take measurement of the states of a process and then send the data to a remote estimator (see Figure 5.1). The estimator receives the measurements intermittently, which may be caused by that the channels have packet dropouts or that the sensors schedule the data transmission stochastically. Based on the received measurements, the estimator computes the estimates of the process states by multisensor Kalman filtering.

Because of the intermittent measurements, the estimator may be unstable. To investigate this stability issue, a notion of transmission capacity, which is related to the communication rates of sensors, is proposed. It is shown that the expected estimation error covariance diverges for all feasible communication rates collections of the sensors when the transmission capacity is below a certain value; meanwhile, when the transmission capacity is above another certain value, there exists a feasible communication rates collection such that the expected estimation error covariance is bounded.

The remainder of this chapter is organized as follows: the problem formulation is given in Section 5.2 and the main analysis is given in Section 5.3. Examples and conclusions are given in Sections 5.4 and 5.5, respectively.

Control over Communication Networks: Modeling, Analysis, and Design of Networked Control Systems and Multi-Agent Systems over Imperfect Communication Channels, First Edition.
Jianying Zheng, Liang Xu, Qinglei Hu, and Lihua Xie.

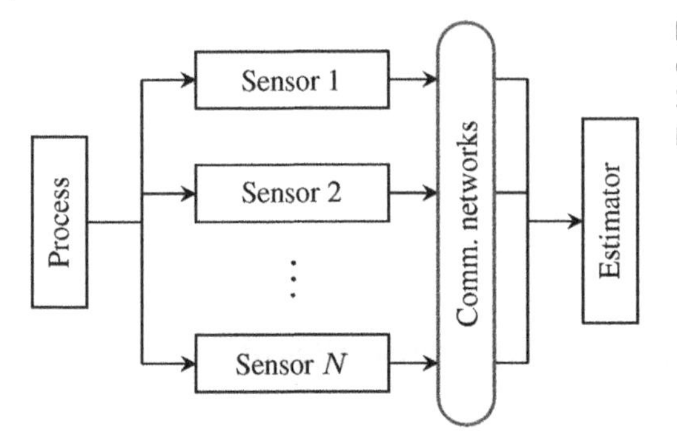

Figure 5.1 The structure of the centralized sensor network. Source: Yang et al. (2018)/with permission of IEEE.

5.2 Problem Formulation

The system studied in this chapter is shown in Figure 5.1. In this section, the models of the system components are provided, and the stability problem is proposed.

Consider a single process whose states are observed by a total number of N sensors:

$$x_{k+1} = Ax_k + w_k, \tag{5.1}$$

$$y_k^i = C_i x_k + v_k^i, \quad i = 1, 2, \dots, N. \tag{5.2}$$

In the equations above, $x_k \in \mathbb{R}^n$ is the process state at time k and $y_k^i \in \mathbb{R}^{p_i}$ is the measurement taken by sensor i, where n and p_i are the corresponding dimensions. The sequences $\{w_k\}$ and $\{v_k^i\}$ are zero-mean white Gaussian noise processes with $\mathbb{E}\left\{w_k w_j'\right\} = \delta_{kj} Q$ $(Q \geq 0)$ and $\mathbb{E}\left\{v_k^i (v_j^i)'\right\} = \delta_{kj} R_i$ $(R_i > 0)$, respectively. Meanwhile, $\{w_k\}$ and $\{v_k^i\}$ are mutually independent processes, i.e. $\mathbb{E}\left\{w_k (v_j^i)'\right\} = 0$, $\forall j, k$. The initial state x_0 is a Gaussian random vector with distribution $\mathcal{N}(0, \Pi)$ and is uncorrelated with w_k and v_k^i, $\forall k, i$. Assume that C_i has full row rank.

The sensors send their measurements to a remote estimator via wireless communication channels, while the estimator receives the measurements intermittently. The intermittent measurements may result from various scenarios: the sensors do the transmission constantly, but the channels are unreliable, or the channels are reliable, but the sensors do the transmission only at scheduled time instants, or so on. Eventually, these scenarios can be modeled as intermittent communication.

Define the *communication variable* γ_k^i as

$$\gamma_k^i = \begin{cases} 1, & y_k^i \text{ is received,} \\ 0, & y_k^i \text{ is not received.} \end{cases}$$

We consider the scenario that $\{\gamma_k^i\}$ is a Bernoulli process with mean $\mathbb{E}\left\{\gamma_k^i\right\} = \lambda_i$, which is the communication rate of sensor i. Define the *communication rates collection* for the group of sensors as $\lambda \triangleq \{\lambda_1, \lambda_2, \dots, \lambda_N\}$.

The remote estimator uses multisensor Kalman filtering to calculate the minimum mean-square error (MMSE) estimate of the state x_k based on the received measurements. Define

$$\tilde{y}_k \triangleq \left(\gamma_k^1 (y_k^1)', \gamma_k^2 (y_k^2)', \dots, \gamma_k^N (y_k^N)'\right)' \tag{5.3}$$

and let $\tilde{\mathbf{Y}}_k \triangleq \{\tilde{y}_1, \tilde{y}_2, \dots, \tilde{y}_k\}$. Define $\hat{x}_{k|k-1}$ and $\hat{x}_{k|k}$:

$$\hat{x}_{k|k-1} \triangleq \mathbb{E}\left\{x_k | \tilde{\mathbf{Y}}_{k-1}\right\},$$
$$\hat{x}_{k|k} \triangleq \mathbb{E}\left\{x_k | \tilde{\mathbf{Y}}_k\right\}.$$

Let $P_{k|k-1}$ and $P_{k|k}$ be the estimation error covariance matrices associated with $\hat{x}_{k|k-1}$ and $\hat{x}_{k|k}$, respectively:

$$P_{k|k-1} \triangleq \mathbb{E}\left\{(x_k - \hat{x}_{k|k-1})(x_k - \hat{x}_{k|k-1})' | \tilde{\mathbf{Y}}_{k-1}\right\},$$
$$P_{k|k} \triangleq \mathbb{E}\left\{(x_k - \hat{x}_{k|k})(x_k - \hat{x}_{k|k})' | \tilde{\mathbf{Y}}_k\right\}.$$

At time k, the estimator first calculates $\hat{x}_{k|k-1}$ and $P_{k|k-1}$:

$$\hat{x}_{k|k-1} = A\hat{x}_{k-1|k-1}, \tag{5.4}$$

$$P_{k|k-1} = AP_{k-1|k-1}A' + Q, \tag{5.5}$$

where the recursion starts from $\hat{x}_{0|0} = 0$ and $P_{0|0} = \Pi$. After receiving available measurements of time k, the estimator fuses them to obtain $\tilde{y}_k$ and computes the following quantities:

$$\tilde{C}_k \triangleq (\gamma_k^1 C_1', \gamma_k^2 C_2', \dots, \gamma_k^N C_N')', \tag{5.6}$$

$$\tilde{R}_k \triangleq \mathrm{diag}\{\gamma_k^1 R_1, \gamma_k^2 R_2, \dots, \gamma_k^N R_N\}. \tag{5.7}$$

Then it computes $\hat{x}_{k|k}$ and $P_{k|k}$ as follows:

$$P_{k|k} = \left(P_{k|k-1}^{-1} + \sum_{i=1}^{N} \gamma_k^i C_i' R_i^{-1} C_i\right)^{-1}, \tag{5.8}$$

$$K_k = P_{k|k}\tilde{C}_k'\tilde{R}_k^{\dagger}, \tag{5.9}$$

$$\hat{x}_{k|k} = \hat{x}_{k|k-1} + K_k(\tilde{y}_k - \tilde{C}_k\hat{x}_{k|k-1}), \tag{5.10}$$

where $\dagger$ represents the Moore–Penrose pseudoinverse.

Remark 5.1 The Eqs. (5.4)–(5.10) are a set of feasible formulas of Kalman filter and mainly facilitate the theoretical analysis. One may seek formulas that are more computationally efficient.

When $\lambda_1, \lambda_2, \dots, \lambda_N$ are all close to zero, the estimator will receive few measurements, and the estimation error is likely to be unbounded. Namely, the communication rates influence the stability of the estimator. The main objective of this chapter is to find the relationship between λ and the performance of the estimation error.

5.3 Stability Analysis

In this section, we investigate the stability problem of the multisensor Kalman filtering with intermittent measurements.

5.3.1 Transmission Capacity

When involving a single sensor, the stability is related to the communication rate of this sensor alone. For multiple channels, we introduce the concept of *transmission capacity* to represent the overall effect of the communication rates of multiple sensors.

For sensor i, denote the variance of the random process $\{\gamma_k^i\}$ as σ_i^2. Define

$$q_i \triangleq \frac{\lambda_i^2}{\sigma_i^2}$$

and the *individual transmission capacity* as

$$\mathfrak{C}_i \triangleq \frac{1}{2}\ln(1+q_i).$$

Moreover, define the *overall transmission capacity* as

$$\mathfrak{C} \triangleq \sum_{i=1}^{N}\mathfrak{C}_i.$$

Remark 5.2 We have $q_i = \frac{\lambda_i}{1-\lambda_i}$. It can be calculated that

$$\mathfrak{C} = -\sum_{i=1}^{N}\frac{1}{2}\ln(1-\lambda_i),$$

which shows the connection between the transmission capacity and the multiple communication rates of the Bernoulli processes. Moreover, the transmission capacity also reflects the communication resources distributed or assigned in the corresponding channels.

5.3.2 Preliminaries

In this subsection, we present necessary preliminaries. We study the *a priori* error covariance $P_{k|k-1}$ for stability analysis. In the following part, P_k^- is used for short of $P_{k|k-1}$.

We introduce an alternative formulation for estimation error covariances. As defined in Yang et al. [2013], the *sensing precision matrix* S_i for sensor i is given as $S_i \triangleq C_i' R_i^{-1} C_i$. Let $H_i = \sqrt{R_i^{-1}} C_i$. Then S_i can be factorized as $S_i = H_i' H_i$. Define $H \triangleq (H_1', H_2', \ldots, H_N')'$. Let

$$\Gamma_k \triangleq \mathrm{diag}\{\gamma_k^1 I_{p_1}, \gamma_k^2 I_{p_2}, \ldots, \gamma_k^N I_{p_N}\},$$

where I_{p_i} is the identity matrix with order p_i, i.e. the order of y_k^i. By combining Eqs. (5.5) and (5.8), the recursive update equation of P_k^- is as follows:

$$P_{k+1}^- = A\left((P_k^-)^{-1} + H'\Gamma_k H\right)^{-1} A' + Q. \tag{5.11}$$

Since P_k^- depends on Γ_k and is thus stochastic, only statistical properties can be deduced. Alternatively, we study $\mathbb{E}\left\{P_k^-\right\}$ to eliminate the uncertainty caused by Γ_k.

Sinopoli et al. [2004] studied the one single-sensor case. They proved the existence of a critical arrival rate below which the estimation error may diverge. They also proposed lower and upper bounds of the estimation error covariance, and based on them, they provided lower and upper bounds of the critical arrival rate.

The analysis for multiple sensors follows a parallel way. In Section 5.3.3, we study the lower bound of $\mathbb{E}\left\{P_k^-\right\}$. We first present a lower bound (Proposition 5.1). Then we give a necessary and sufficient condition on the convergence and divergence of this lower bound (Theorem 5.1). Furthermore, we present a critical overall transmission capacity $\underline{\mathfrak{C}}$: when $\mathfrak{C} < \underline{\mathfrak{C}}$, the lower bound diverges for any feasible communication rates collection λ, which leads to that $\liminf_{k\to\infty} \mathbb{E}\left\{P_k^-\right\}$ diverges (Theorem 5.2). In Section 5.3.4, we study the upper bound of $\mathbb{E}\left\{P_k^-\right\}$. We first give an upper bound (Theorem 5.3). We further present a necessary and sufficient condition on the convergence of the upper bound (Theorem 5.4) and a sufficient one on that in terms of transmission capacities (Theorem 5.5). Then we give a critical overall transmission capacity $\overline{\mathfrak{C}}$: above $\overline{\mathfrak{C}}$, there exists a feasible communication rates collection λ which guarantees the convergence of the upper bound and hence that of $\limsup_{k\to\infty} \mathbb{E}\left\{P_k^-\right\}$ (Theorem 5.6). Section 5.3.5 presents the results for some special cases.

5.3.3 Lower Bound

The proposition below gives a lower bound of $\{\mathbb{E}\left\{P_k^-\right\}\}$.

Proposition 5.1 Define

$$m(X) \triangleq \prod_{i=1}^{N}(1 - \lambda_i) AXA' + Q. \tag{5.12}$$

Let the sequence $\{G_k\}$ be constructed as follows: $G_1 = \mathbf{0}$ and $G_{k+1} = m(G_k)$. Then $G_k \leq \mathbb{E}\left\{P_k^-\right\}$, $\forall k$.

Proof: Liu and Goldsmith [2004] provides the proof when $N = 2$, and the general case is stated by Corollary 4.3.1 in Rong [2012]. □

The next theorem gives a necessary and sufficient condition on the convergence and divergence of the lower bound $\{G_k\}$.

Theorem 5.1 *If A is unstable and $(A, \sqrt{Q})$ is controllable, a necessary and sufficient condition for $\{G_k\}$ to converge is*

$$\mathfrak{C} > \ln \rho(A), \tag{5.13}$$

where $\rho(A)$ denotes the spectrum radius of A. Furthermore, $\{G_k\}$ diverges if and only if $\mathfrak{C} \leq \ln \rho(A)$.

Proof: Define $\tilde{A} \triangleq \left(\prod_{i=1}^{N}(1-\lambda_i)\right)^{\frac{1}{2}} A$. Then

$$m(X) = \tilde{A}X\tilde{A}' + Q.$$

Since $(A, \sqrt{Q})$ is controllable, $(\tilde{A}, \sqrt{Q})$ is also controllable. Then $X = m(X)$ has a unique strictly positive definite solution if and only if $\rho(\tilde{A}) < 1$, which leads to

$$\left(\prod_{i=1}^{N}(1-\lambda_i)\right)^{\frac{1}{2}} \rho(A) < 1. \tag{5.14}$$

Equation (5.14) yields

$$\ln \rho(A) < -\sum_{i=1}^{N} \frac{1}{2} \ln(1-\lambda_i),$$

i.e. $\mathfrak{C} > \ln \rho(A)$.

Assume that (5.13) holds. Then $\tilde{A}$ is stable, and $X = m(X)$ has a unique positive definite solution $\overline{G}$, which is given by $\overline{G} = \sum_{k=0}^{\infty} \tilde{A}^k Q(\tilde{A}')^k$. Hence,

$$\lim_{k\to\infty} G_k = \tilde{A}^{k+1} G_1 (\tilde{A}')^{k+1} + \sum_{k=0}^{\infty} \tilde{A}^k Q(\tilde{A}')^k = \sum_{k=0}^{\infty} \tilde{A}^k Q(\tilde{A}')^k = \overline{G}.$$

Therefore, it is proved that $\{G_k\}$ converges. On the other hand, if $\{G_k\}$ converges, since

$$\lim_{k\to\infty} G_k = \sum_{k=0}^{\infty} \tilde{A}^k Q(\tilde{A}')^k,$$

the series of the right-hand side converges. We denote the series by $\overline{G}$. Notice that $\overline{G}$ is the solution to $X = m(X)$. Since $(\tilde{A}, \sqrt{Q})$ is controllable, according to the property of controllability, $\overline{G}$ has full rank. Hence, it is positive definite. The fact that

$\overline{G} = m(\overline{G})$ and $\overline{G} > 0$ leads to that $\tilde{A}$ is stable, i.e. $\mathfrak{C} > \ln \rho(A)$. Consequently, it is proved that $\{G_k\}$ converges if and only if $\mathfrak{C} > \ln \rho(A)$.

Furthermore, $\mathfrak{C} \leq \ln \rho(A)$ is equivalent to that $\{G_k\}$ does not converge. Notice that $\{G_k\}$ is monotonically increasing. Then the case that $\{G_k\}$ does not converge is equivalent to that $\{G_k\}$ is unbounded. Hence, $\{G_k\}$ diverges if and only if $\mathfrak{C} \leq \ln \rho(A)$. □

The following theorem gives a condition on the critical transmission capacity regarding the divergence of $\{\mathbb{E}\left\{P_k^-\right\}\}$.

Theorem 5.2 *Assume that* $(A, \sqrt{Q})$ *is controllable. Let* $\underline{\mathfrak{C}} \triangleq \ln \rho(A)$. *When the overall transmission capacity* $\mathfrak{C} \leq \underline{\mathfrak{C}}$, $\{\mathbb{E}\left\{P_k^-\right\}\}$ *diverges for all initial values. Moreover, when* $\mathfrak{C} > \underline{\mathfrak{C}}$, *there exists a unique positive definite* $\overline{G}$ *satisfying* $\overline{G} = m(\overline{G})$, *and* $\liminf_{k\to\infty} \mathbb{E}\left\{P_k^-\right\} \geq \overline{G}$.

Proof: From Proposition 5.1, we have $G_k \leq \mathbb{E}\left\{P_k^-\right\}, \forall k$. When $\mathfrak{C} \leq \underline{\mathfrak{C}}$, $\lim_{k\to\infty} G_k = \infty$ according to Theorem 5.1. Then

$$\lim_{k\to\infty} \mathbb{E}\left\{P_k^-\right\} \geq \lim_{k\to\infty} G_k = \infty,$$

i.e. $\left\{\mathbb{E}\left\{P_k^-\right\}\right\}$ diverges. When $\mathfrak{C} > \underline{\mathfrak{C}}$, according to the proof of Theorem 5.1, $X = m(X)$ has a unique positive definite solution $\overline{G}$. Since $G_k \leq \mathbb{E}\left\{P_k^-\right\}$, we have $\liminf_{k\to\infty} \mathbb{E}\left\{P_k^-\right\} \geq \lim_{k\to\infty} G_k = \overline{G}$, which completes the proof. □

Remark 5.3 When A is stable, we have

$$\overline{G} = \begin{cases} Q, & \lambda_i \to 1, \quad \exists i, \\ \Sigma, & \lambda_i \to 0, \quad \forall i, \end{cases}$$

where Σ satisfies $\Sigma - A\Sigma A' = Q$.

Remark 5.4 It is worth to mention that (5.8) implies another lower bound for $\mathbb{E}\left\{P_k^-\right\}$. Define an operator $\overline{g} : \mathcal{P}_n \times \mathcal{P}_n \to \mathcal{P}_n$:

$$\overline{g}(X; S) \triangleq A(X^{-1} + S)^{-1}A' + Q.$$

Construct a sequence $\{F_k\}$, where $F_1 = P_1^-$, and $F_{k+1} = \overline{g}\left(F_k; \sum_{i=1}^N \lambda_i S_i\right)$. Then

$$F_k \leq \mathbb{E}\left\{P_k^-\right\}, \quad \forall k,$$

which results from the convexity of $\overline{g}(X; S)$ in S [Yang et al., 2013]. This bound is shown by Mo and Sinopoli [2008]. Since $\{F_k\}$ is a loose lower bound and always converges under the condition that $(A, \sqrt{Q})$ is controllable, it is not able to provide a certain critical transmission capacity.

5.3.4 Upper Bound

We study the upper bound of $\mathbb{E}\left\{P_k^-\right\}$ in this part. Define the modified algebraic Riccati equation associated with the multisensor Kalman filter with intermittent measurements as follows:

$$g_\lambda(X) \triangleq AXA' + Q - AXH'\left(W \odot (HXH' + I)\right)^{-1} HXA', \tag{5.15}$$

where $\odot$ is the Hadamard product denoting the elementwise matrix multiplication, $W = \mathbf{1}\mathbf{1}' + D_{\text{SNR}}^{-1}\ \mathcal{I}$, $\mathbf{1}$ is the column vector with all components equal to 1, and

$$D_{\text{SNR}} \triangleq \text{diag}\{q_1 I_{p_1}, q_2 I_{p_2}, \dots, q_N I_{p_N}\},$$
$$\mathcal{I} \triangleq \text{diag}\{\mathbf{1}_{p_1}\mathbf{1}'_{p_1}, \mathbf{1}_{p_2}\mathbf{1}'_{p_2}, \dots, \mathbf{1}_{p_N}\mathbf{1}'_{p_N}\}.$$

Since $q_i = \frac{\lambda_i}{1-\lambda_i}$, W is the function of λ_i's. The subscript λ in $g_\lambda(X)$ indicates that $g_\lambda(X)$ depends on λ.

Some properties of $g_\lambda(X)$ are presented in Lemma 5.1.

Lemma 5.1 *The following results hold:*

1. *If $Y \geq X > 0$, then $g_\lambda(Y) \geq g_\lambda(X)$;*
2. *$g_\lambda(X)$ is concave with respect to X.*

Proof: Let

$$\psi_\lambda(L, X) = AXA' + Q + L\Lambda\left[W \odot (HXH' + I)\right]\Lambda L' - AXH'\Lambda L' - L\Lambda HXA',$$

where Λ is defined in (5.16), and $L(X) = AXH'\left[W \odot (HXH' + I)\right]^{-1}\Lambda^{-1}$. Notice that

$$g_\lambda(X) = \psi_\lambda\left(L(X), X\right) = \min_L \psi_\lambda(L, X).$$

We first prove the monotonicity.

$$\begin{aligned}\psi_\lambda(L, X) &= AXA' + L\Lambda HXH'\Lambda L' - AXH'\Lambda L' - L\Lambda HXA' \\ &\quad + L\Lambda\left((D_{\text{SNR}}^{-1}\ \mathcal{I}) \odot (HXH')\right)\Lambda L' + L\Lambda(W \odot I)\Lambda L' + Q \\ &= (A - L\Lambda H)X(A - L\Lambda H)' + L\Lambda(W \odot I)\Lambda L' \\ &\quad + Q + \sum_{i=1}^{N} L\Lambda\left(\frac{1}{q_i}\overline{H}_i X \overline{H}_i'\right)\Lambda L',\end{aligned}$$

where $\overline{H}_i = [\mathbf{0}, H_i', \mathbf{0}]'$ which is obtained by replacing in H all the blocks by $\mathbf{0}$ matrices except H_i. Hence, $\psi_\lambda(L, X)$ is quadratic in X. For X, Y satisfying $Y \geq X > 0$, it is easy to show that $\psi_\lambda(L, X) \leq \psi_\lambda(L, Y)$. Then

$$g_\lambda(X) = \psi_\lambda\left(L(X), X\right) \leq \psi_\lambda\left(L(Y), X\right) \leq \psi_\lambda\left(L(Y), Y\right) = g_\lambda(Y),$$

which shows the monotonicity. To prove the concavity, let $Z = \alpha X + \beta Y$, where X, Y are positive definite matrices, and α, β are positive real numbers with $\alpha + \beta = 1$. Then

$$\begin{aligned}
g_\lambda(Z) &= g_\lambda(\alpha X + \beta Y) = \psi_\lambda(L(Z), Z) \\
&= AZA' + Q + L(Z)\Lambda\left(W \odot (HZH' + I)\right)\Lambda L(Z)' \\
&\quad - AZH'\Lambda L(Z)' - L(Z)\Lambda HZA' \\
&= \alpha\left[AXA' + Q + L(Z)\Lambda\left(W \odot (HXH' + I)\right)\Lambda L(Z)'\right. \\
&\quad \left.-AXH'\Lambda L(Z)' - L(Z)\Lambda HXA'\right] \\
&\quad + \beta\left[AYA' + Q + L(Z)\Lambda\left(W \odot (HYH' + I)\right)\Lambda L(Z)'\right. \\
&\quad \left.-AYH'\Lambda L(Z)' - L(Z)\Lambda HYA'\right] \\
&= \alpha\psi_\lambda(L(Z), X) + \beta\psi_\lambda(L(Z), Y) \\
&\geq \alpha\psi_\lambda(L(X), X) + \beta\psi_\lambda(L(Y), Y) \\
&= \alpha g_\lambda(X) + \beta g_\lambda(Y),
\end{aligned}$$

which shows the concavity. □

Next, theorem presents an upper bound of $\{\boldsymbol{E}[P_k^-]\}$.

Theorem 5.3 *Construct a sequence* $\{V_k\}$ *as follows:* $V_1 = P_1^-$ *and* $V_{k+1} = g_\lambda(V_k)$. *Then* $\mathbb{E}\left\{P_k^-\right\} \leq V_k$, $\forall k$.

Proof: Define an operator

$$\Phi(L, X, \Gamma) \triangleq (A - L\Gamma H)X(A - L\Gamma H)' + Q + L\Gamma I \Gamma L'.$$

From (5.11), we have

$$\begin{aligned}
P_{k+1}^- &= A\left((P_k^-)^{-1} + H'\Gamma_k I \Gamma_k H\right)^{-1} A' + Q \\
&= AP_k^- A' + Q - AP_k^- H'\Gamma_k(\Gamma_k HP_k^- H'\Gamma_k + I)^{-1}\Gamma_k HP_k^- A' \\
&= AP_k^- A' + Q - AP_k^- H'\Gamma_k(\Gamma_k HP_k^- H'\Gamma_k + \Gamma_k)^{\dagger}\Gamma_k HP_k^- A',
\end{aligned}$$

where the second equality follows by Matrix Inversion Lemma, and the third one can be verified by simple calculation. This is the conventional expression for the *a priori* error covariance P_k^-. Then P_{k+1}^- satisfies

$$P_{k+1}^- = \Phi(\overline{K}_k, P_k^-, \Gamma_k),$$

where $\overline{K}_k = AP_k^- H'\Gamma_k(\Gamma_k HP_k^- H'\Gamma_k + \Gamma_k)^{\dagger}$. According to the theory of Kalman filtering, $\overline{K}_k$ is the gain which minimizes $\Phi(L, P_k^-, \Gamma_k)$. Hence, we have

$$P_{k+1}^- = \min_L \Phi(L, P_k^-, \Gamma_k).$$

Define

$$\Lambda \triangleq \text{diag}\{\lambda_1 I_{p_1}, \lambda_2 Ip_2, \dots, \lambda_N I_{p_N}\} \tag{5.16}$$

and

$$L_k \triangleq AP_k^- H' \left(W \odot (HP_k^- H' + I)\right)^{-1} \Lambda^{-1}.$$

We use mathematical induction to prove the argument. When $k = 0$, the argument holds. Assume when $k = l$, $\mathbb{E}\{P_l^-\} \leq V_l$ holds. For time $k = l + 1$, we have

$$P_{l+1}^- \leq \Phi(L_l, P_l^-, \Gamma_l).$$

Take expectation of both sides with respect to $\{\Gamma_k\}$:

$$\begin{aligned}
\mathbb{E}\left\{P_{l+1}^-\right\} &\leq \mathbb{E}\left\{\Phi(L_l, P_l^-, \Gamma_l)\right\} \\
&= \mathbb{E}\left[(A - L_l\Gamma_l H)P_l^-(A - L_l\Gamma_l H)' + Q + L_l\Gamma_l I\Gamma_l L_l'\right] \\
&= \mathbb{E}\left[AP_l^- A' + Q + L_l\Gamma_l HP_l^- H'\Gamma_l L_l' + L_l\Gamma_l I\Gamma_l L_l' \right. \\
&\quad \left. -AP_l^- H'\Gamma_l L_l' - L_l\Gamma_l HP_l^- A'\right].
\end{aligned}$$

Since P_l^- depends on $\Gamma_1, \Gamma_2, \dots, \Gamma_{l-1}$, it is independent from Γ_l. Hence, we first take expectation of the right-hand side with respect to Γ_l. We have $\mathbb{E}\left\{\Gamma_l\right\} = \Lambda$ and $\mathbb{E}\left\{\Gamma_l X\Gamma_l\right\} = \Lambda\,(W \odot X)\,\Lambda$ for a constant matrix X. Then

$$\begin{aligned}
\mathbb{E}\left\{P_{l+1}^-\right\} &\leq \mathbb{E}\left[AP_l^- A' + Q + L_l\Lambda\left(W \odot (HP_l^- H' + I)\right)\Lambda L_l' \right. \\
&\quad \left. -AP_l^- H'\Lambda L_l' - L_l\Lambda HP_l^- A'\right].
\end{aligned}$$

The expectation on the right-hand side is with respect to $\Gamma_1, \Gamma_2, \dots, \Gamma_{l-1}$. Take the expression of L_l into the right-hand side and it turns to be $\mathbb{E}\left\{g_\lambda(P_l^-)\right\}$. Hence,

$$\mathbb{E}\left\{P_{l+1}^-\right\} \leq \mathbb{E}\left\{g_\lambda(P_l^-)\right\} \leq g_\lambda(\mathbf{E}[P_l^-]) \leq g_\lambda(V_l) = V_{l+1},$$

where the second and third inequalities follow from the concavity and the increasing monotonicity of $g_\lambda(X)$, respectively. From mathematical induction, $\mathbb{E}\left\{P_k^-\right\} \leq V_k$ holds for all $k \geq 0$. □

The theorem below provides a necessary and sufficient condition for the convergence of the upper bound $\{V_k\}$.

Theorem 5.4 *A necessary and sufficient condition for $\{V_k\}$ to converge for an arbitrary initial value, namely $X = g_\lambda(X)$ has one unique positive definite solution, is given as follows:*

1. *$\exists\, P > 0$, L, and λ, such that*

$$P > (A - L\Lambda H)P(A - L\Lambda H)' + L\left((\Sigma^2 \mathcal{I}) \odot (HPH')\right)L', \tag{5.17}$$

where Λ is defined in (5.16), *and*

$$\Sigma \triangleq \text{diag}\{\sigma_1 I_{p_1}, \sigma_2 I_{p_2}, \dots, \sigma_N I_{p_N}\}.$$

2. $\begin{bmatrix} A - e^{j\omega} I & B \end{bmatrix}$ *has full row rank for all* $\omega \in \mathbb{R}$, *where* $BB' = Q$.

Proof: According to Corollary 4.1 in Chapter 4, the modified algebraic Riccati equation (MARE) (5.15) has one unique positive semidefinite solution if and only if

1. for the stochastic system

$$e_{k+1} = (A - L\Gamma_k H)' e_k, \tag{5.18}$$

   there exists a static L such that $\lim_{k\to\infty} \mathbb{E}\left\{e_k e_k'\right\} = 0$,
2. the system

$$\begin{aligned} x_{k+1} &= A' x_k, \\ y_k &= B' x_k, \end{aligned} \tag{5.19}$$

   has no unobservable eigenvalues on the unit circle.

For system (5.18), let $\Pi_k \triangleq \mathbb{E}\left\{e_k e_k'\right\}$. We have

$$\Pi_k = (A - L\Lambda H)' \Pi_{k-1} (A - L\Lambda H) + H' \left((\Sigma^2 I) \odot (L' \Pi_{k-1} L) \right) H.$$

Similar to the result in Chapter 9 of Boyd et al. [1994], $\lim_{k\to\infty} \Pi_k = 0$ is equivalent to that $\exists\, P > 0$, L and λ, such that (5.17) is satisfied.

It is simple to verify that condition (2) is equivalent to that system (5.19) has no unobservable eigenvalues on the unit circle. □

Remark 5.5 If the system model is generalized to that $\{w_k\}$ and $\{v_k\}$ are correlated with $\mathbb{E}\left\{w_k v_k'\right\} = S$, the corresponding MARE is

$$g_\lambda(X) \triangleq AXA' + Q - (AXC' + S) \cdot \left(W \odot (CXC' + R)\right)^{-1} (CXA' + S'),$$

which is well studied in Chapter 4.

In Theorem 5.4, condition (2) is simple to check, while condition (1) is still not straightforward. In the following part, we relate the convergence of $\{V_k\}$ to the conditions on scheduling capacities.

Define the Mahler measure [Mahler, 1960] of A:

$$\mathcal{M}(A) \triangleq \prod_{i=1}^{n} \max\{1, |\lambda_i(A)|\},$$

where $\lambda_i(A)$ are the eigenvalues of A. Define the topological entropy [Bowen, 1971] of A:

$$\mathfrak{h}(A) = \ln \mathcal{M}(A).$$

Assume that the pair (H, A) is observable and H has row rank m. Then one can choose from $H_i, i = 1, 2, \dots, N$, a subset of rows, denoted as $h_{i1}, h_{i2}, \dots, h_{it_i}, 0 \le t_i \le p_i$, and let $\mathcal{H}_i = (h'_{i1}, h'_{i2}, \dots, h'_{it_i})'$, $\mathcal{H} = (\mathcal{H}'_1, \mathcal{H}'_2, \dots, \mathcal{H}'_N)'$, such that $\mathcal{H}$ has m rows with full row rank and $(\mathcal{H}, A)$ is observable. According to Wonham decomposition [Wonham, 1967], A and $\mathcal{H}$ can be transformed to $\tilde{A}$ and $\tilde{\mathcal{H}}$ by some similarity transformation $\mathfrak{T}$ with the following form:

$$\tilde{A} = \begin{bmatrix} \tilde{A}_1 & 0 & \dots & 0 \\ \star & \tilde{A}_2 & \dots & 0 \\ \vdots & \vdots & \ddots & \vdots \\ \star & \star & \dots & \tilde{A}_N \end{bmatrix}, \quad \tilde{\mathcal{H}} = \begin{bmatrix} \tilde{\mathcal{H}}_1 & 0 & \dots & 0 \\ 0 & \tilde{\mathcal{H}}_2 & \dots & 0 \\ \vdots & \vdots & \ddots & \vdots \\ 0 & 0 & \dots & \tilde{\mathcal{H}}_N \end{bmatrix}, \tag{5.20}$$

in which

$$\tilde{A}_i = \begin{bmatrix} \tilde{A}_{i1} & 0 & \dots & 0 \\ \star & \tilde{A}_{i2} & \dots & 0 \\ \vdots & \vdots & \ddots & \vdots \\ \star & \star & \dots & \tilde{A}_{it_i} \end{bmatrix}, \quad \tilde{\mathcal{H}}_i = \begin{bmatrix} \tilde{h}_{i1} & 0 & \dots & 0 \\ 0 & \tilde{h}_{i2} & \dots & 0 \\ \vdots & \vdots & \ddots & \vdots \\ 0 & 0 & \dots & \tilde{h}_{it_i} \end{bmatrix},$$

where $\tilde{A}_{ij} \in \mathbb{R}^{n_{ij} \times n_{ij}}$, $\tilde{h}_{ij} \in \mathbb{R}^{n_{ij} \times 1}$, $\sum n_{ij} = n$, and pair $(\tilde{h}_{ij}, \tilde{A}_{ij})$ is observable. The following theorem provides sufficient conditions on the convergence of $\{V_k\}$ in terms of scheduling capacities.

Theorem 5.5 *Assume that the pair (H, A) is observable and condition (2) in Theorem 5.4 is satisfied. If*

$$\mathfrak{C} > \sum_{i=1}^{N} \max_j \{\mathfrak{h}(\tilde{A}_{ij})\}, \tag{5.21}$$

where $\tilde{A}_{ij}$ is given in (5.20), there exists a communication rates collection λ, such that $\{V_k\}$ converges for all initial values. In particular, if

$$\mathfrak{C} > \mathfrak{h}(A), \tag{5.22}$$

$\{V_k\}$ converges.

Proof: We only need to show that condition (1) in Theorem 5.4 holds. To prove this, we use the similar method in the proof of Theorem 3.1 in Xiao et al. [2012]. Since (5.21) holds, one can find a communication rates collection λ with associated $\mathfrak{C}_i$ satisfying $\mathfrak{C}_i > \max_j\{\mathfrak{h}(\tilde{A}_{ij})\}$, i.e. $1 + \frac{\lambda_i^2}{\sigma_i^2} > \left[\mathcal{M}(\tilde{A}_{ij})\right]^2, \forall i, j$. According to Theorem 6.4 and Corollary 8.4 in Elia [2005], for all i, j, there exists $\tilde{P}_{ij} > 0$ and $\tilde{l}_{ij} \in \mathbb{R}^{n_{ij} \times 1}$, such that

$$\tilde{P}_{ij} > (\tilde{A}_{ij} - \tilde{l}_{ij}\lambda_i\tilde{h}_{ij})\tilde{P}_{ij}(\tilde{A}_{ij} - \tilde{l}_{ij}\lambda_i\tilde{h}_{ij})' + \sigma_i^2\tilde{l}_{ij}\tilde{h}_{ij}\tilde{P}_{ij}\tilde{h}'_{ij}\tilde{l}'_{ij}.$$

First, we prove that for all i, there exist $\tilde{P}_i$ and $\tilde{\mathcal{L}}_i$, such that

$$\tilde{P}_i > (\tilde{A}_i - \tilde{\mathcal{L}}_i\lambda_i\tilde{\mathcal{H}}_i)\tilde{P}_i(\tilde{A}_i - \tilde{\mathcal{L}}_i\lambda_i\tilde{\mathcal{H}}_i)' + \sigma_i^2\tilde{\mathcal{L}}_i\tilde{\mathcal{H}}_i\tilde{P}_i\tilde{\mathcal{H}}_i'\tilde{\mathcal{L}}_i' \tag{5.23}$$

holds, where $\tilde{A}_i$ and $\tilde{\mathcal{H}}_i$ are given in (5.20). Denote

$$\tilde{A}_i^s = \begin{bmatrix} \tilde{A}_{i1} & & 0 \\ & \ddots & \\ \star & & \tilde{A}_{is} \end{bmatrix}, \quad \tilde{\mathcal{H}}_i^s = \begin{bmatrix} \tilde{h}_{i1} & & 0 \\ & \ddots & \\ 0 & & \tilde{h}_{is} \end{bmatrix},$$

where $s = 1, 2, \dots, t_i$. We show that there exist $\tilde{P}_i^s$ and $\tilde{\mathcal{L}}_i^s$, such that

$$\tilde{P}_i^s > (\tilde{A}_i^s - \tilde{\mathcal{L}}_i^s\lambda_i\tilde{\mathcal{H}}_i^s)\tilde{P}_i^s(\tilde{A}_i^s - \tilde{\mathcal{L}}_i^s\lambda_i\tilde{\mathcal{H}}_i^s)' + \sigma_i^2\tilde{\mathcal{L}}_i^s\tilde{\mathcal{H}}_i^s\tilde{P}_i^s(\tilde{\mathcal{H}}_i^s)'(\tilde{\mathcal{L}}_i^s)' \tag{5.24}$$

holds for $s = 1, 2, \dots, t_i$. We prove this by mathematical induction. When $s = 1$, the case is proved, where $\tilde{P}_i^1 = \tilde{P}_{i1}$, $\tilde{\mathcal{L}}_i^1 = \tilde{l}_{i1}$. Assume the case $s = m - 1$ is verified. When $s = m$, let

$$\tilde{P}_i^m = \begin{bmatrix} \tilde{P}_i^{m-1} & 0 \\ & \alpha\tilde{P}_{im} \end{bmatrix}, \quad \tilde{\mathcal{L}}_i^m = \begin{bmatrix} \tilde{\mathcal{L}}_i^{m-1} & 0 \\ 0 & \tilde{l}_{im} \end{bmatrix}, \tag{5.25}$$

where α is to be given later. Denote

$$\delta_{ij} = \tilde{P}_{ij} - (\tilde{A}_{ij} - \tilde{l}_{ij}\lambda_i\tilde{h}_{ij})\tilde{P}_{ij}(\tilde{A}_{ij} - \tilde{l}_{ij}\lambda_i\tilde{h}_{ij})' - \sigma_i^2\tilde{l}_{ij}\tilde{h}_{ij}\tilde{P}_{ij}\tilde{h}_{ij}'\tilde{l}_{ij}',$$
$$\Delta_i^s = \tilde{P}_i^s - (\tilde{A}_i^s - \tilde{\mathcal{L}}_i^s\lambda_i\tilde{\mathcal{H}}_i^s)\tilde{P}_i^s(\tilde{A}_i^s - \tilde{\mathcal{L}}_i^s\lambda_i\tilde{\mathcal{H}}_i^s)' - \sigma_i^2\tilde{\mathcal{L}}_i^s\tilde{\mathcal{H}}_i^s\tilde{P}_i^s(\tilde{\mathcal{H}}_i^s)'(\tilde{\mathcal{L}}_i^s)',$$

and

$$\tilde{A}_i^m = \begin{bmatrix} \tilde{A}_i^{m-1} & 0 \\ A_L & \tilde{A}_{im} \end{bmatrix}.$$

Notice that there exists $\beta > 0$ such that $\delta_{im} > \beta I$. Choose sufficiently large α such that

$$\alpha\beta I > \mathcal{K}\Delta_i^{m-1}\mathcal{K}' + A_L\tilde{P}_i^{m-1}A_L'$$

holds, where $\mathcal{K} = A_L\tilde{P}_i^{m-1}(\tilde{A}_i^{m-1} - \tilde{\mathcal{L}}_i^{m-1}\lambda_i\tilde{\mathcal{H}}_i^{m-1})'$. Straightforward calculation shows that $\tilde{P}_i^m$ and $\tilde{\mathcal{L}}_i^m$ given by (5.25) satisfy (5.24). By mathematical induction, there exist $\tilde{P}_i^s$ and $\tilde{\mathcal{L}}_i^s$ such that (5.24) holds for $s = 1, 2, \dots, t_i$. When $s = t_i$, (5.24) is exactly (5.23), with $\tilde{P}_i = \tilde{P}_i^{t_i}$ and $\tilde{\mathcal{L}}_i = \tilde{\mathcal{L}}_i^{t_i}$. From (5.23), in a similar way one can further show that $\tilde{\mathcal{L}} = \text{diag}\{\tilde{\mathcal{L}}_1, \tilde{\mathcal{L}}_2, \dots, \tilde{\mathcal{L}}_N\}$ and $\tilde{P} = \text{diag}\{\tilde{P}_1, \alpha_1\tilde{P}_2, \dots, \alpha_{N-1}\tilde{P}_N\}$ with sufficiently large $\alpha_1, \alpha_2, \dots, \alpha_N$ satisfying

$$\tilde{P} > (\tilde{A} - \tilde{\mathcal{L}}\Lambda\tilde{\mathcal{H}})\tilde{P}(\tilde{A} - \tilde{\mathcal{L}}\Lambda\tilde{\mathcal{H}})' + \tilde{\mathcal{L}}\left((\overline{\Sigma}^2\overline{\mathcal{I}}) \odot (\tilde{\mathcal{H}}\tilde{P}\tilde{\mathcal{H}}')\right)\tilde{\mathcal{L}}',$$

where $\overline{\Sigma}$ and $\overline{\mathcal{I}}$ have proper dimensions. By the inverse transformation $\mathfrak{T}^{-1}$ ($\mathfrak{T}$ is the one in Wonham decomposition), we have

$$P > (A - \mathcal{L}\Lambda\mathcal{H})P(A - \mathcal{L}\Lambda\mathcal{H})' + \mathcal{L}\left((\overline{\Sigma}^2\overline{\mathcal{I}}) \odot (\mathcal{H}P\mathcal{H}')\right)\mathcal{L}'.$$

Without loss of generality, we assume that H has the form of

$$H = (\mathcal{H}_1', \overline{\mathcal{H}}_1', \mathcal{H}_2', \overline{\mathcal{H}}_2', \ldots, \mathcal{H}_N', \overline{\mathcal{H}}_N')'.$$

Correspondingly, let $L = (\mathcal{L}_1, 0, \mathcal{L}_2, 0, \ldots, \mathcal{L}_N, 0)$. It is easy to see that P and L satisfies

$$P > (A - L\Lambda H)P(A - L\Lambda H)' + L\left((\Sigma^2 \mathcal{I}) \odot (HPH')\right)L'.$$

By Theorem 5.4, the convergence of $\{V_k\}$ is guaranteed.

Since

$$\mathfrak{h}(A) = \sum_{i=1}^{N}\sum_{j=1}^{t_i}\{\mathfrak{h}(\tilde{A}_{ij})\} > \sum_{i=1}^{N}\max_j\{\mathfrak{h}(\tilde{A}_{ij})\},$$

(5.22) also leads to the convergence of $\{V_k\}$. □

The following theorem gives a condition on the critical overall capacity related to the convergence of $\{\mathbb{E}\{P_k^-\}\}$.

Theorem 5.6 *Assume that condition (2) in Theorem 5.4 is satisfied. If* $\mathfrak{C} > \overline{\mathfrak{C}}$, *where* $\overline{\mathfrak{C}} \triangleq \mathfrak{h}(A)$, *there exists a communication rates collection* λ *such that* $\limsup_{k\to\infty}\mathbb{E}\{P_k^-\}$ *converges for an arbitrary initial value. Moreover, let* $\overline{V}$ *satisfy* $\overline{V} = g_\lambda(\overline{V})$. *Then* $\overline{V}$ *exists and* $\limsup_{k\to\infty}\mathbb{E}\{P_k^-\} \leq \overline{V}$.

Proof: The arguments are verified simply by combining the result of Theorems 5.3 and 5.5. □

5.3.5 Special Cases

In this subsection, we present further results on some special cases. Theorem 5.5 only provides sufficient conditions on the convergence of $\{V_k\}$. For some special systems, they become necessary and sufficient conditions.

Theorem 5.7 *Assume that* A *has no zero eigenvalues and the pair* (H, A) *is observable. If* $\{V_k\}$ *converges for an arbitrary initial value, then*

$$\sum_{i=1}^{N} p_i \mathfrak{C}_i > \ln\det(A). \tag{5.26}$$

Proof: The proof is similar to the one of Theorem 1 in Xiao et al. [2012].

Corollary 5.1 *Assume that condition (2) in Theorem 5.4 is satisfied. When all the eigenvalues of* A *are unstable and the measurements of all sensors are scalar valued,* *there exists a communication rates collection λ such that $\{V_k\}$ converges if and only if $\mathfrak{C} > \mathfrak{h}(A)$.*

Proof: Notice that when all the eigenvalues of A are unstable, $\det(A) = \mathcal{M}(A)$. Meanwhile, $p_i = 1, \forall i$. Hence, (5.26) coincides with (5.22). □

When $N = 1$, the system is reduced to the one considered in Sinopoli et al. [2004]. Denote the mean of the packet arrival rate of the single channel by λ. The MARE (5.15) is reduced to

$$g_\lambda(X) = AXA' + Q - \lambda AXH'(HXH' + I)^{-1}HXA'. \tag{5.27}$$

Notice that Sinopoli et al. [2004] only gives a sufficient condition on the unique existence of a positive definite solution to (5.27) in terms of an linear matrix inequality (LMI). Theorems 5.4 and 5.5 improve the results. Sinopoli et al. [2004] proves the existence of a critical packet arrival rate λ_c upon which the expected estimated error covariance of the filter converges. It also gives the lower and upper bounds of λ_c, where the lower bound λ is stated by a closed form, while the upper bound $\overline{\lambda}$ is provided by solving an LMI. According to Theorem 5.5, the upper bound can be improved as a closed form, i.e. $\overline{\lambda} = 1 - \frac{1}{\mathcal{M}^2(A)}$. This is also obtained by Rong [2012].

In some special scenarios, the two capacity bounds $\overline{\mathfrak{C}}$ and $\mathfrak{C}$ meet.

Lemma 5.2 *When A has only one unstable eigenvalue, we have $\mathfrak{C} = \overline{\mathfrak{C}}$.*

Proof: Since A has only one unstable eigenvalue, the Mahler measure coincides with the spectrum radius, i.e. $\mathcal{M}(A) = \rho(A)$. Thus, the argument follows. □

Corollary 5.2 *For a first-order system, when A is unstable, we have $\mathfrak{C} = \overline{\mathfrak{C}}$.*

5.4 Examples

In this section, we provide an example to demonstrate the lower and upper bounds of $\mathbb{E}\left\{P_k^-\right\}$.

Example 5.1 *Consider a system with $n = 2$, $N = 8$. We plot the lower bound G_k, upper bound V_k, and a realization of $\mathbb{E}\left\{P_k^-\right\}$ by Monte Carlo method. We randomly choose $\lambda = (0.2443, 0.4572, 0.7435, 0.6174, 0.1747, 0.6024, 0.8031, 0.3663)$. The system has the following matrices which are generated randomly:*

$$A = \begin{bmatrix} -0.82 & 0.53 \\ 0.34 & 0.78 \end{bmatrix}, \qquad Q = \begin{bmatrix} 0.53 & 0.21 \\ 0.21 & 0.50 \end{bmatrix},$$

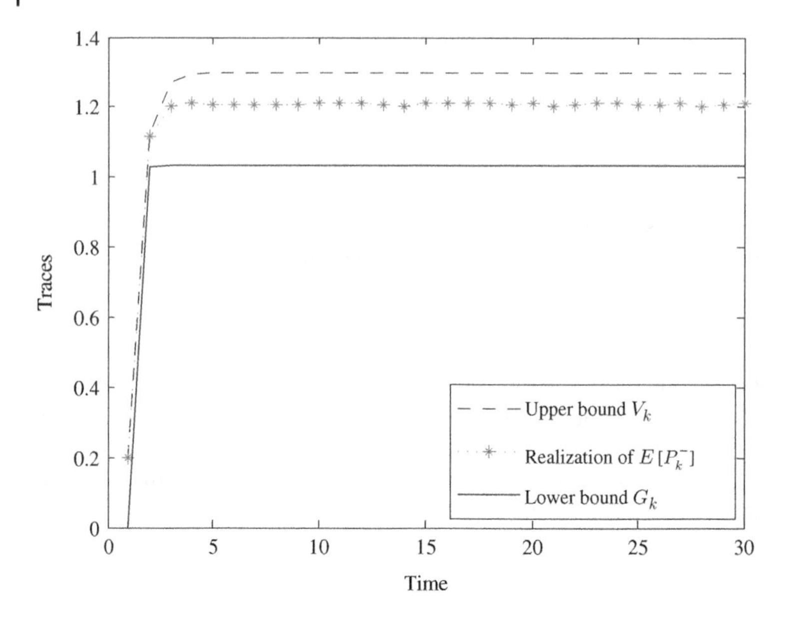

Figure 5.2 The lower and upper bounds of $\mathbb{E}\left\{P_k^-\right\}$. Source: Yang et al. (2018)/with permission of IEEE.

$$
\begin{aligned}
H_1 &= [-0.69, 0.67], & H_2 &= \begin{bmatrix} 1.18 & -0.36 \\ 0.70 & -0.38 \end{bmatrix}, \\
H_3 &= [-1.58, 1.43], & H_4 &= [-1.71, -2.12], \\
H_5 &= \begin{bmatrix} -0.65 & 0.49 \\ -2.01 & 0.46 \end{bmatrix}, & H_6 &= \begin{bmatrix} 1.79 & 2.2 \\ 0.59 & 0.31 \end{bmatrix}, \\
H_7 &= \begin{bmatrix} 1.28 & 2.07 \\ -1.00 & -0.92 \end{bmatrix}, & H_8 &= \begin{bmatrix} 0.18 & 0.21 \\ 2.04 & 0.91 \end{bmatrix}.
\end{aligned}
$$

The plot is presented in Figure 5.2. The plot shows that the two bounds are not tight.

5.5 Conclusions

This chapter studies the stability issue of the multisensor Kalman filtering with intermittent measurements. The lower and upper bounds of the expected estimation error covariance are provided, and the conditions on the divergence and convergence of them are given. Based on those results, it is shown that the expected estimation error covariance diverges for all feasible communication rates collections of the sensors when the transmission capacity is below a certain value; meanwhile, when the transmission capacity is above another certain value, there exists a feasible communication rates collection such that the expected estimation error covariance is bounded.

Bibliography

R. Bowen. Entropy for group endomorphisms and homogeneous spaces. *Transactions of the American Mathematical Society*, 153:401–414, 1971.

S. Boyd, L. E. Ghaoui, E. Feron, and V. Balakrishnan. *Linear Matrix Inequalities in System and Control Theory*. Society for Industrial and Applied Mathematics, 1994.

N. Elia. Remote stabilization over fading channels. *Systems & Control Letters*, 54(3):237–249, 2005.

X. Liu and A. Goldsmith. Kalman filtering with partial observation losses. In *Proceedings of the 43rd IEEE Conference on Decision and Control*, volume 4, pages 4180–4186. IEEE, 2004.

K. Mahler. An application of Jensen's formula to polynomials. *Mathematika*, 7(02):98–100, 1960.

Y. Mo and B. Sinopoli. A characterization of the critical value for Kalman filtering with intermittent observations. In *Proceedings of the 47th IEEE Conference on Decision and Control*, pages 2692–2697, 2008.

B. Rong. State estimation over packet-dropping channels. Master's thesis, Hong Kong University of Science and Technology, 2012.

B. Sinopoli, L. Schenato, M. Franceschetti, K. Poolla, M. I. Jordan, and S. S. Sastry. Kalman filtering with intermittent observations. *IEEE Transactions on Automatic Control*, 49(9):1453–1464, 2004.

W. M. Wonham. On pole assignment in multi-input controllable linear systems. *IEEE Transactions on Automatic Control*, 12(6):660–665, 1967.

N. Xiao, L. Xie, and L. Qiu. Feedback stabilization of discrete-time networked systems over fading channels. *IEEE Transactions on Automatic Control*, 57(9):2176–2189, 2012.

C. Yang, J. Wu, W. Zhang, and L. Shi. Schedule communication for decentralized state estimation. *IEEE Transactions on Signal Processing*, 61(10):2525–2535, 2013.

6

Remote State Estimation with Stochastic Event-Triggered Sensor Schedule and Packet Drops

6.1 Introduction

Sensor scheduling algorithms, which use scheduled transmissions to reduce the communication frequency to prolong the service time of sensor devices, are widely used in sensor networks [Yang and Shi, 2011, Han et al., 2015]. Wireless communications are mostly utilized in sensor networks, and packet drops are inevitable in wireless communications. However, for the stochastic event-triggered sensor scheduling algorithms proposed in Han et al. [2015], no packet drops are considered. This chapter studies the remote state estimation problem of linear time-invariant systems with stochastic event-triggered sensor schedules in the presence of packet drops between the sensor and the estimator. We aim to provide closed-form expressions for the optimal estimator under this setting. We show that the system state conditioned on the available information at the estimator side is Gaussian mixture distributed. The minimum mean-square error (MMSE) estimator can be obtained from a bank of Kalman filters.

This chapter is organized as follows: the problem formulation is given in Section 6.2. The optimal estimator is studied in Section 6.3. Two suboptimal estimators with fixed computation requirements are proposed in Section 6.4. Simulation comparisons are provided in Section 6.5. This chapter ends with some concluding remarks in Section 6.6.

6.2 Problem Formulation

In this chapter, we are interested in the remote state estimation problem with stochastic event-triggered sensor schedule in the presence of packet drops depicted in Figure 6.1. The linear process is

$$\begin{aligned} x_{k+1} &= Ax_k + w_k, \\ y_k &= Cx_k + v_k, \end{aligned}$$

Control over Communication Networks: Modeling, Analysis, and Design of Networked Control Systems and Multi-Agent Systems over Imperfect Communication Channels, First Edition.
Jianying Zheng, Liang Xu, Qinglei Hu, and Lihua Xie.

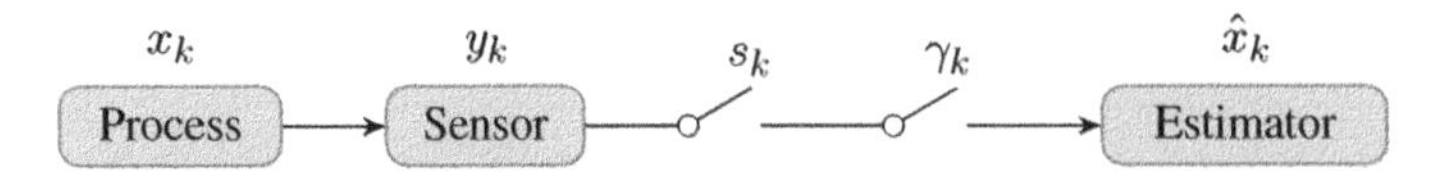

Figure 6.1 Remote state estimation with stochastic event-triggered sensor schedule in the presence of packet drops. Source: Xu et al. (2020)/with permission of IEEE.

where $x_k \in \mathbb{R}^n$ is the process state, $y_k \in \mathbb{R}^m$ is the measurement output, and w_k and v_k are the process and measurement noises. We assume that $\{w_k\}_{k\geq 0}$ and $\{v_k\}_{k\geq 0}$ are white Gaussian processes with covariance matrices Q and R, respectively. Moreover, the initial system state satisfies $x_0 \sim \mathcal{N}_{x_0}(0, \Sigma_0)$ and is independent with w_k and v_k.

After receiving y_k, the sensor follows the stochastic event-triggered schedule [Han et al., 2015] to decide whether to transmit y_k to the estimator or not. Let s_k denote the decision variable by the sensor. When $s_k = 1$, the sensor transmits y_k to the estimator and $s_k = 0$, otherwise. The stochastic event-triggered sensor schedule operates as follows. At time k, the sensor randomly generates a variable ζ_k from the uniform distribution on $[0, 1]$. Then ζ_k is compared with $e^{-y_k' Y y_k}$, where $Y > 0$. The sensor schedules transmissions based on the following rule:

$$s_k = \begin{cases} 0, & \text{if } \zeta_k \leq e^{-y_k' Y y_k}, \\ 1, & \text{if } \zeta_k > e^{-y_k' Y y_k}. \end{cases} \tag{6.1}$$

It is shown in Han et al. [2015] that the design (6.1) together with the random variable ζ_k can avoid the nonlinearity introduced by the truncated Gaussian prior conditional distribution of the system state in deterministic event-triggered schedule in Wu et al. [2013].

Remark 6.1 There are several reasons for not implementing the Kalman filter at the sensor side. Firstly, the reason is that the sensor might be primitive [Han et al., 2015], so it does not have a sufficient computation capability to run a local Kalman filter. Secondly, the system parameters might not be available to the sensor. Thirdly, in decentralized settings where there are multiple sensors measuring the same process, only the fusion center which has access to all the sensor measurements can run the Kalman filter. In the end, the state dimension might be larger than the output dimension. Therefore, it reduces the communication cost to transmit the sensor output and perform the Kalman filter at the estimator side.

Remark 6.2 In this chapter, we only consider the open-loop stochastic event-triggered sensor schedule. The results for the closed-loop stochastic event-triggered sensor schedule [Han et al., 2015] can be obtained in a similar way.

The communication channel between the sensor and the estimator suffers from i.i.d. packet drops, which are described by the i.i.d. stochastic process $\{\gamma_k\}_{k\geq 0}$. When $\gamma_k = 1$, the transmission is successful, and $\gamma_k = 0$, otherwise. Moreover, we assume $\gamma_k \in \{0,1\}$ has a Bernoulli distribution with $\Pr(\gamma_k = 0) = p$. Therefore, the following information is available to the estimator at time k

$$\mathcal{I}_k = \{s_0\gamma_0, \dots, s_k\gamma_k, s_0\gamma_0 y_0, \dots, s_k\gamma_k y_k\} \tag{6.2}$$

with $\mathcal{I}_{-1} = \emptyset$.

When there are no transmission packet drops, the posterior distribution $f(x_k|\mathcal{I}_k)$ is shown to be Gaussian in Han et al. [2015]. However, in the presence of packet drops, the Gaussian property no longer holds [Kung et al., 2017]. In subsequent sections, we will show that in the presence of packet drops, the posterior distribution $f(x_k|\mathcal{I}_k)$ is mixture Gaussian with exponentially increasing components with time. Moreover, the MMSE estimator is derived.

6.3 Optimal Estimator

In this section, we try to derive the MMSE estimator in the presence of packet drops between the sensor and the estimator. First of all, the following notions are defined. For any given $k \in \mathbb{N}$ and $i \in \mathbb{N}$ with $0 \leq i \leq 2^{k+1} - 1$, define the event

$$\Gamma_k^i = \{\gamma_k = b_k, \dots, \gamma_0 = b_0\},$$

where b_k is the $(k+1)$th element of the binary expansion of i, i.e. $i = b_k 2^k + b_{k-1}2^{k-1} + \cdots + b_0 2^0$. Therefore, Γ_k^i denotes a packet drop sequence $\{\gamma_k, \dots, \gamma_0\}$ specified by the index i. For any given $i \in \mathbb{N}$ and $k \in \mathbb{N}$, let

$$i_k^- = \begin{cases} i, & \text{if } i < 2^k, \\ i - 2^k, & \text{if } i \geq 2^k. \end{cases}$$

Therefore, i_k^- is the index of the subsequence $\{\gamma_{k-1}, \dots, \gamma_0\}$ extracted from the sequence $\{\gamma_k, \dots, \gamma_0\}$ specified by the index i. From the law of total probability, we have that

$$f(x_k|\mathcal{I}_k) = \sum_{i=0}^{2^{k+1}-1} f(x_k|\Gamma_k^i, \mathcal{I}_k)\Pr(\Gamma_k^i|\mathcal{I}_k), \tag{6.3}$$

$$f(x_k|\mathcal{I}_{k-1}) = \sum_{i=0}^{2^k-1} f(x_k|\Gamma_{k-1}^i, \mathcal{I}_{k-1})\Pr(\Gamma_{k-1}^i|\mathcal{I}_{k-1}). \tag{6.4}$$

We call Γ_k^i the hypothesis and $\Pr(\Gamma_k^i|\mathcal{I}_k)$ the hypothesis probability. The conditional distribution $f(x_k|\Gamma_k^i, \mathcal{I}_k)$ can be shown to be Gaussian. Therefore, $f(x_k|\mathcal{I}_k)$ is mixture Gaussian.

Remark 6.3 The Gaussian mixture phenomenon also appears in other control and estimation problems. For example, the estimation of Markov jump linear systems with unknown jump mode [Costa et al., 2005] and the estimation of linear systems with unknown control inputs [Lin et al., 2016]. In general, if there are hidden variables in the control and estimation problem, the resulting conditional distribution is a mixture distribution from the law of total probability.

It should be noted that in (6.3), there are 2^{k+1} hypotheses at time k. However, certain hypotheses might be impossible. For example, when $\gamma_k s_k = 1$, we know that $\gamma_k = 1$. Therefore, the hypotheses with $\gamma_k = 0$ should be excluded. Nevertheless, for such case, we can show that $\Pr(\Gamma_k^i | \mathcal{I}_k) = 0$. In principle, the set of hypotheses at time k is determined by the received signals $\{\gamma_k s_k, \dots, \gamma_0 s_0\}$, which however is stochastic and cannot be determined in a prior. Therefore, in the sequel, we always express $f(x_k | \mathcal{I}_k)$ as a summation of 2^{k+1} hypotheses as in (6.3) to simplify the description.

In the sequel, we will recursively derive $f(x_k | \Gamma_k^i, \mathcal{I}_k)$ and $\Pr(\Gamma_k^i | \mathcal{I}_k)$. Then in view of (6.3) and (6.4), the MMSE estimator can be obtained. The following results are required and presented first.

$$f\left(x_k | \Gamma_k^i, \mathcal{I}_{k-1}\right) = f\left(x_k | \Gamma_{k-1}^{i_k^-}, \mathcal{I}_{k-1}\right), \tag{6.5}$$

since $\{\gamma_k\}$ is i.i.d. and knowing γ_k only cannot help to improve the knowledge about x_k. We then have the following result.

Lemma 6.1

$$f\left(x_k | \Gamma_k^i, \mathcal{I}_k\right) = \mathcal{N}_{x_k}\left(\hat{x}_{k|k}^i, P_{k|k}^i\right), \quad 0 \le i \le 2^{k+1} - 1,$$
$$f\left(x_k | \Gamma_{k-1}^i, \mathcal{I}_{k-1}\right) = \mathcal{N}_{x_k}\left(\hat{x}_{k|k-1}^i, P_{k|k-1}^i\right), \quad 0 \le i \le 2^k - 1,$$

where $\hat{x}_{k|k}^i, P_{k|k}^i, \hat{x}_{k|k-1}^i, P_{k|k-1}^i$ *satisfy the following recursion.*

Time update:

$$\hat{x}_{k|k-1}^i = A\hat{x}_{k-1|k-1}^i,$$
$$P_{k|k-1}^i = AP_{k-1|k-1}^i A' + Q.$$

Measurement update:

- *For* $i < 2^k$,

$$\hat{x}_{k|k}^i = \hat{x}_{k|k-1}^i, P_{k|k}^i = P_{k|k-1}^i. \tag{6.6}$$

- *For* $i \ge 2^k$,

$$\hat{x}_{k|k}^i = \left(I - K_k^{i_k^-} C\right)\hat{x}_{k|k-1}^{i_k^-} + K_k^{i_k^-} s_k \gamma_k y_k, \tag{6.7}$$

$$P^{i}_{k|k} = P^{i_k^-}_{k|k-1} - K^{i_k^-}_{k} CP^{i_k^-}_{k|k-1}, \tag{6.8}$$

$$K^{i_k^-}_{k} = P^{i_k^-}_{k|k-1} C'[CP^{i_k^-}_{k|k-1} C' + R + (1 - s_k\gamma_k)Y^{-1}]^{-1} \tag{6.9}$$

with initial conditions

$$\hat{x}^{0}_{0|-1} = 0, \quad P^{0}_{0|-1} = \Sigma_0.$$

Proof: The proofs of the initialization and the time update are straightforward. The measurement update (6.6) follows from the fact that for $i < 2^k$, we have $\gamma_k = 0$. Therefore, no new information is available, and the measurement update is not needed.

The measurement update (6.7), (6.8), and (6.9) follow from the fact that for $i \geq 2^k$, we have $\gamma_k = 1$. Therefore, the measurement update is the same as Han et al. [2015]. □

Next, we calculate the probabilities of $\alpha^{i}_{k|k-1} = \Pr(\Gamma^{i}_{k}|\mathcal{I}_{k-1}), \alpha^{i}_{k|k} = \Pr(\Gamma^{i}_{k}|\mathcal{I}_{k})$ and have the following result.

Lemma 6.2 $\alpha^{i}_{k|k-1}$ *and* $\alpha^{i}_{k|k}$ *with* $0 \leq i \leq 2^{k+1} - 1$ *satisfy the following recursion:*
Time update:

- *For* $i < 2^k$,

$$\alpha^{i}_{k|k-1} = p\alpha^{i}_{k-1|k-1}. \tag{6.10}$$

- *For* $i \geq 2^k$,

$$\alpha^{i}_{k|k-1} = (1 - p)\alpha^{i_k^-}_{k-1|k-1}. \tag{6.11}$$

Measurement update:

$$\alpha^{i}_{k|k} = \frac{\Pr\left(s_k\gamma_k|\Gamma^{i}_{k}, \mathcal{I}_{k-1}\right)\alpha^{i}_{k|k-1}}{\sum_{j=0}^{2^{k+1}-1}\Pr\left(s_k\gamma_k|\Gamma^{j}_{k}, \mathcal{I}_{k-1}\right)\alpha^{j}_{k|k-1}},$$

where

- *for* $j < 2^k$,

$$\Pr\left(s_k\gamma_k|\Gamma^{j}_{k}, \mathcal{I}_{k-1}\right) = 1 - s_k\gamma_k,$$

- *for* $j \geq 2^k$,

$$\Pr\left(s_k\gamma_k|\Gamma_k^j, \mathcal{I}_{k-1}\right) = s_k\gamma_k + \frac{1 - 2s_k\gamma_k}{\sqrt{\left|\left(CP_{k|k-1}^{j_k^-}C' + R\right)Y + I\right|}} \times e^{-\frac{1}{2}\left(C\hat{x}_{k|k-1}^{j_k^-}\right)'\left[Y^{-1}+\left(CP_{k|k-1}^{j_k^-}C'+R\right)\right]^{-1}C\hat{x}_{k|k-1}^{j_k^-}}$$

with the initial condition

$$\alpha_{0|-1}^0 = p, \quad \alpha_{0|-1}^1 = 1 - p.$$

Proof: Time update: (6.10) follows from the fact that for $i < 2^k$, we have $\gamma_k = 0$. Therefore,

$$\begin{aligned}
\alpha_{k|k-1}^i &= \Pr\left(\Gamma_{k-1}^i, \gamma_k = 0|\mathcal{I}_{k-1}\right) \\
&= \Pr(\gamma_k = 0)\Pr\left(\Gamma_{k-1}^i|\mathcal{I}_{k-1}\right) \\
&= p\alpha_{k-1|k-1}^i.
\end{aligned}$$

On the other hand, (6.11) follows from the fact that for $i \geq 2^k$, we have $\gamma_k = 1$. Therefore,

$$\begin{aligned}
\alpha_{k|k-1}^i &= \Pr(\Gamma_{k-1}^{i_k^-}, \gamma_k = 1|\mathcal{I}_{k-1}) \\
&= \Pr(\gamma_k = 1)\Pr(\Gamma_{k-1}^{i_k^-}|\mathcal{I}_{k-1}) \\
&= (1-p)\alpha_{k-1|k-1}^{i_k^-},
\end{aligned}$$

which is (6.11).

Measurement update: Since

$$\begin{aligned}
\alpha_{k|k}^i &= \Pr\left(\Gamma_k^i|\mathcal{I}_k\right) \\
&= \Pr\left(\Gamma_k^i|s_k\gamma_k, s_k\gamma_k y_k, \mathcal{I}_{k-1}\right) \\
&= \Pr\left(\Gamma_k^i|s_k\gamma_k, \mathcal{I}_{k-1}\right) \\
&= \frac{\Pr\left(s_k\gamma_k|\Gamma_k^i, \mathcal{I}_{k-1}\right)\Pr\left(\Gamma_k^i|\mathcal{I}_{k-1}\right)}{\Pr\left(s_k\gamma_k|\mathcal{I}_{k-1}\right)} \\
&= \frac{\Pr\left(s_k\gamma_k|\Gamma_k^i, \mathcal{I}_{k-1}\right)\Pr\left(\Gamma_k^i|\mathcal{I}_{k-1}\right)}{\sum_{j=0}^{2^{k+1}-1}\Pr\left(s_k\gamma_k|\Gamma_k^j, \mathcal{I}_{k-1}\right)\Pr\left(\Gamma_k^j|\mathcal{I}_{k-1}\right)} \\
&= \frac{\Pr\left(s_k\gamma_k|\Gamma_k^i, \mathcal{I}_{k-1}\right)\alpha_{k|k-1}^i}{\sum_{j=0}^{2^{k+1}-1}\Pr\left(s_k\gamma_k|\Gamma_k^j, \mathcal{I}_{k-1}\right)\alpha_{k|k-1}^j},
\end{aligned}$$

where the third equality follows from the fact that when $s_k\gamma_k = 0$, $s_k\gamma_k y_k = 0$, it is useless to know $s_k\gamma_k y_k$; when $s_k\gamma_k = 1$, knowing $s_k\gamma_k y_k$ is equivalent to know y_k, which is also useless in predicting Γ^i_k. Next, we will show how to calculate $\Pr\left(s_k\gamma_k|\Gamma^i_k, \mathcal{I}_{k-1}\right)$.

When $i < 2^k$, since $\gamma_k = 0$, we have that $s_k\gamma_k \equiv 0$. Therefore,

$$\Pr\left(s_k\gamma_k|\Gamma^i_k, \mathcal{I}_{k-1}\right) = 1 - s_k\gamma_k. \tag{6.12}$$

When $i \geq 2^k$, we have $\gamma_k = 1$. Let $M_k^{i_k^-} = CP_{k|k-1}^{i_k^-}C' + R$, we then have

$$\begin{aligned}
&\Pr\left(s_k\gamma_k|\Gamma^i_k, \mathcal{I}_{k-1}\right)\\
&= \Pr\left(s_k|\Gamma^i_k, \mathcal{I}_{k-1}\right)\\
&= \int_{\mathbb{R}^m} \Pr\left(s_k|y_k, \Gamma^i_k, \mathcal{I}_{k-1}\right) f\left(y_k|\Gamma^i_k, \mathcal{I}_{k-1}\right) \mathrm{d}y_k\\
&= \int_{\mathbb{R}^m} \Pr(s_k|y_k) f\left(Cx_k + v_k|\Gamma^i_k, \mathcal{I}_{k-1}\right) \mathrm{d}y_k\\
&\overset{(a)}{=} \int_{\mathbb{R}^m} \left(s_k\left(1 - 2e^{-\frac{1}{2}y_k'Yy_k}\right) + e^{-\frac{1}{2}y_k'Yy_k}\right) f\left(Cx_k + v_k|\Gamma^{i_k^-}_{k-1}, \mathcal{I}_{k-1}\right) \mathrm{d}y_k\\
&= \int_{\mathbb{R}^m} \left(s_k\left(1 - 2e^{-\frac{1}{2}y_k'Yy_k}\right) + e^{-\frac{1}{2}y_k'Yy_k}\right) \mathcal{N}_{y_k}\left(C\hat{x}_{k|k-1}^{i_k^-}, M_k^{i_k^-}\right) \mathrm{d}y_k\\
&= s_k + (1 - 2s_k)\int_{\mathbb{R}^m} e^{-\frac{1}{2}y_k'Yy_k} \mathcal{N}_{y_k}\left(C\hat{x}_{k|k-1}^{i_k^-}, M_k^{i_k^-}\right) \mathrm{d}y_k\\
&= s_k + \frac{1 - 2s_k}{\sqrt{(2\pi)^m|M_k^{i_k^-}|}}\int_{\mathbb{R}^m} e^{-\frac{1}{2}y_k'Yy_k - \frac{1}{2}\left(y_k - C\hat{x}_{k|k-1}^{i_k^-}\right)'\left(M_k^{i_k^-}\right)^{-1}\left(y_k - C\hat{x}_{k|k-1}^{i_k^-}\right)} \mathrm{d}y_k\\
&= s_k + \frac{1 - 2s_k}{\sqrt{(2\pi)^m|M_k^{i_k^-}|}} e^{-\frac{1}{2}\left(C\hat{x}_{k|k-1}^{i_k^-}\right)'\left(M_k^{i_k^-}\right)^{-1}C\hat{x}_{k|k-1}^{i_k^-}}\\
&\quad \times \int_{\mathbb{R}^m} e^{-\frac{1}{2}y_k'\left[Y + \left(M_k^{i_k^-}\right)^{-1}\right]y_k + \left(C\hat{x}_{k|k-1}^{i_k^-}\right)'\left(M_k^{i_k^-}\right)^{-1}y_k} \mathrm{d}y_k\\
&\overset{(b)}{=} s_k + \frac{1 - 2s_k}{\sqrt{(2\pi)^m|M_k^{i_k^-}|}} e^{-\frac{1}{2}\left(C\hat{x}_{k|k-1}^{i_k^-}\right)'\left(M_k^{i_k^-}\right)^{-1}C\hat{x}_{k|k-1}^{i_k^-}} \sqrt{\frac{(2\pi)^m}{\left|Y + \left(M_k^{i_k^-}\right)^{-1}\right|}}\\
&\quad \times e^{\frac{1}{2}\left(C\hat{x}_{k|k-1}^{i_k^-}\right)'\left(M_k^{i_k^-}\right)^{-1}\left[Y + \left(M_k^{i_k^-}\right)^{-1}\right]^{-1}\left(M_k^{i_k^-}\right)^{-1}C\hat{x}_{k|k-1}^{i_k^-}}\\
&\overset{(c)}{=} s_k + \frac{1 - 2s_k}{\sqrt{\left|\left(M_k^{i_k^-}\right)Y + I\right|}} e^{-\frac{1}{2}\left(C\hat{x}_{k|k-1}^{i_k^-}\right)'\left[Y^{-1} + \left(M_k^{i_k^-}\right)\right]^{-1}C\hat{x}_{k|k-1}^{i_k^-}},
\end{aligned} \tag{6.13}$$

where (a) follows from (6.5); (b) follows from the Gaussian integral and (c) follows from the matrix inversion lemma. □

The following notions are defined:

$$\hat{x}_{k|k} = \mathcal{E}x_k|\mathcal{I}_k, \qquad \hat{x}_{k|k-1} = \mathcal{E}x_k|\mathcal{I}_{k-1},$$
$$e_{k|k} = x_k - \hat{x}_{k|k}, \qquad e_{k|k-1} = x_k - \hat{x}_{k|k-1},$$
$$P_{k|k} = \mathcal{E}e_{k|k}e'_{k|k}, \qquad P_{k|k-1} = \mathcal{E}e_{k|k-1}e'_{k|k-1}.$$

In view of Lemma 6.1 and Lemma 6.2, the following is straightforward from (6.3) and (6.4).

Theorem 6.1 *With the stochastic event-trigger sensor schedule and in the presence of packet drops, the conditional probability density functions of x_k are Gaussian mixture, i.e.*

$$f(x_k|\mathcal{I}_k) = \sum_{i=0}^{2^{k+1}-1} \alpha^i_{k|k}\mathcal{N}_{x_k}\left(\hat{x}^i_{k|k}, P^i_{k|k}\right),$$
$$f(x_k|\mathcal{I}_{k-1}) = \sum_{i=0}^{2^{k}-1} \alpha^i_{k-1|k-1}\mathcal{N}_{x_k}\left(\hat{x}^i_{k|k-1}, P^i_{k|k-1}\right),$$

where $\hat{x}^i_{k|k}, P^i_{k|k}, \hat{x}^i_{k|k-1}, P^i_{k|k-1}, \alpha^i_{k|k}, \alpha^i_{k|k-1}$ are computed in Lemma 6.1 and Lemma 6.2. Moreover, the optimal estimate with the corresponding estimation error covariance can be calculated by the Gaussian sum filter [Anderson and Moore, 1979] and are given by

$$\hat{x}_{k|k} = \sum_{i=0}^{2^{k+1}-1} \alpha^i_{k|k}\hat{x}^i_{k|k},$$
$$P_{k|k} = \sum_{i=0}^{2^{k+1}-1} \alpha^i_{k|k}\left(P^i_{k|k} + \left(\hat{x}^i_{k|k} - \hat{x}_{k|k}\right)\left(\hat{x}^i_{k|k} - \hat{x}_{k|k}\right)'\right),$$
$$\hat{x}_{k|k-1} = \sum_{i=0}^{2^{k}-1} \alpha^i_{k-1|k-1}\hat{x}^i_{k|k-1},$$
$$P_{k|k-1} = \sum_{i=0}^{2^{k}-1} \alpha^i_{k-1|k-1}\left(P^i_{k|k-1} + \left(\hat{x}^i_{k|k-1} - \hat{x}_{k|k-1}\right)\left(\hat{x}^i_{k|k-1} - \hat{x}_{k|k-1}\right)'\right).$$

We can verify that when there are no packet drops, the optimal estimator degenerates to the one given in Han et al. [2015]. Besides, the time update of the MMSE estimator satisfies the following simple recursion:

$$\hat{x}_{k|k-1} = \sum_{i=0}^{2^{k}-1} \alpha^i_{k-1|k-1}A\hat{x}^i_{k-1|k-1} = A\hat{x}_{k-1|k-1},$$

$$
\begin{aligned}
P_{k|k-1} = \sum_{i=0}^{2^k-1} \alpha^i_{k-1|k-1} \Big[& AP^i_{k-1|k-1}A' + Q \\
& + A(\hat{x}^i_{k-1|k-1} - \hat{x}_{k-1|k-1})(\hat{x}^i_{k-1|k-1} - \hat{x}_{k-1|k-1})'A' \Big] \\
= AP_{k-1|k-1}A' + Q. &
\end{aligned}
$$

However, there are no such simple relations for the measurement update.

It is immediate from Theorem 6.1 that the optimal estimator requires exponentially increasing computation and memory with time, which cannot be applied to practical applications. Therefore, in the following, two suboptimal estimators with constant resource requirements are proposed.

6.4 Suboptimal Estimators

In this section, we propose two suboptimal estimators with constant resource requirements: the fixed memory estimator and the particle filter. The two suboptimal estimators are obtained by limiting the hypothesis length and numbers, respectively. In the following, we will describe the two suboptimal estimators in detail.

6.4.1 Fixed Memory Estimator

The problem considered in this chapter is similar to the state estimation problem of linear systems with multiplicative and additive noises in Jaffer and Gupta [1971], where it is shown that the optimal nonlinear filter is obtained from a bank of Kalman filters, which requires ever-increasing memory and computation with time. Fixed memory suboptimal estimators are therefore proposed in Jaffer and Gupta [1971] to overcome the computational complexity. The approximations consist of restricting the probability density $f(x_k|\mathcal{I}_k)$ to depend on at most the last N random variables $\gamma_k, \dots, \gamma_{k-N+1}$ and approximate each conditional probability density $f(x_k|\gamma_k, \dots, \gamma_{k-N+1}, \mathcal{I}_k)$ with a Gaussian distribution. Moreover, a hypothesis merging operation is introduced at every step to prevent the increase of hypothesis numbers with time. The same principle is utilized to derive a suboptimal estimator for the problem considered in this chapter. The detailed derivations are given below.

Let Υ_k^N denote the sequence $\{\gamma_k, \dots, \gamma_{k-N+1}\}$. At time k, instead of being conditioned on all the past history $\gamma_k, \dots, \gamma_0$, we are only conditioned on the past N steps Υ_k^N, where $N \geq 2$ have the following relation:

$$f(x_k|\mathcal{I}_k) = \sum_{\Upsilon_k^N} \Pr(\Upsilon_k^N|\mathcal{I}_k) f(x_k|\Upsilon_k^N, \mathcal{I}_k). \tag{6.14}$$

It is clear from Section 6.3 that $f(x_k|\Upsilon_k^N, \mathcal{I}_k)$ is mixture Gaussian with 2^{k+1-N} components. To obtain an approximate estimator, we make the approximation that

$$f\left(x_k|\Upsilon_k^N, \mathcal{I}_k\right) \approx \mathcal{N}_{x_k}\left(\hat{x}_k\left(\Upsilon_k^N\right), P_k\left(\Upsilon_k^N\right)\right), \tag{6.15}$$

where equality holds exactly when $k = N - 1$.

Therefore, an approximate estimator from (6.14) and (6.15) is given by

$$\hat{x}_{k|k} = \sum_{\Upsilon_k^N} \Pr\left(\Upsilon_k^N|\mathcal{I}_k\right)\hat{x}_k\left(\Upsilon_k^N\right), \tag{6.16}$$

$$P_{k|k} = \sum_{\Upsilon_k^N} \Pr\left(\Upsilon_k^N|\mathcal{I}_k\right)\left(P_k\left(\Upsilon_k^N\right) + \left(\hat{x}_k\left(\Upsilon_k^N\right) - \hat{x}_{k|k}\right)(*)'\right). \tag{6.17}$$

In the following, we show the need to introduce a hypothesis merging step and how to recursively calculate $\hat{x}_k\left(\Upsilon_k^N\right)$, $P_k\left(\Upsilon_k^N\right)$, and $\Pr\left(\Upsilon_k^N|\mathcal{I}_k\right)$. Suppose at time $k-1$, (6.15) holds, we have

$$\begin{aligned}
f\left(x_{k-1}|\Upsilon_{k-1}^{N-1}, \mathcal{I}_{k-1}\right) &= \sum_{\gamma_{k-N}} f\left(x_{k-1}, \gamma_{k-N}|\Upsilon_{k-1}^{N-1}, \mathcal{I}_{k-1}\right)\\
&= \sum_{\gamma_{k-N}} \frac{f\left(x_{k-1}|\Upsilon_{k-1}^{N}, \mathcal{I}_{k-1}\right)\Pr\left(\Upsilon_{k-1}^{N}|\mathcal{I}_{k-1}\right)}{\Pr\left(\Upsilon_{k-1}^{N-1}|\mathcal{I}_{k-1}\right)}\\
&= \sum_{\gamma_{k-N}} \frac{f\left(x_{k-1}|\Upsilon_{k-1}^{N}, \mathcal{I}_{k-1}\right)\Pr\left(\Upsilon_{k-1}^{N}|\mathcal{I}_{k-1}\right)}{\sum_{\gamma_{k-N}}\Pr\left(\Upsilon_{k-1}^{N}|\mathcal{I}_{k-1}\right)}.
\end{aligned}$$

Therefore, if $f\left(x_{k-1}|\Upsilon_{k-1}^{N}, \mathcal{I}_{k-1}\right)$ is Gaussian, $f\left(x_{k-1}|\Upsilon_{k-1}^{N-1}, \mathcal{I}_{k-1}\right)$ is mixture Gaussian with two components. As a result, $f\left(x_k|\Upsilon_k^N, \mathcal{I}_k\right)$ is rarely Gaussian, which makes (6.15) invalid. We therefore introduce a hypothesis merging step by applying a Gaussian mixture reduction to $f\left(x_{k-1}|\Upsilon_{k-1}^{N-1}, \mathcal{I}_{k-1}\right)$ with moment match and make the approximation that

$$f\left(x_{k-1}|\Upsilon_{k-1}^{N-1}, \mathcal{I}_{k-1}\right) \approx \mathcal{N}_{x_{k-1}}\left(\hat{x}_{k-1}\left(\Upsilon_{k-1}^{N-1}\right), P_{k-1}\left(\Upsilon_{k-1}^{N-1}\right)\right), \tag{6.18}$$

where

$$\hat{x}_{k-1}\left(\Upsilon_{k-1}^{N-1}\right) = \sum_{\gamma_{k-N}} \frac{\hat{x}_{k-1}\left(\Upsilon_{k-1}^{N}\right)\Pr\left(\Upsilon_{k-1}^{N}|\mathcal{I}_{k-1}\right)}{\sum_{\gamma_{k-N}}\Pr\left(\Upsilon_{k-1}^{N}|\mathcal{I}_{k-1}\right)}, \tag{6.19}$$

$$\begin{aligned}
P_{k-1}\left(\Upsilon_{k-1}^{N-1}\right) = \sum_{\gamma_{k-N}} &\frac{\Pr\left(\Upsilon_{k-1}^{N}|\mathcal{I}_{k-1}\right)}{\sum_{\gamma_{k-N}}\Pr\left(\Upsilon_{k-1}^{N}|\mathcal{I}_{k-1}\right)}\left(P_{k-1}\left(\Upsilon_{k-1}^{N}\right)\right.\\
&\left.+\left(\hat{x}_{k-1}\left(\Upsilon_{k-1}^{N}\right) - \hat{x}_{k-1}\left(\Upsilon_{k-1}^{N-1}\right)\right)(*)'\right).
\end{aligned} \tag{6.20}$$

Under the approximation (6.18), $f\left(x_k|\Upsilon_k^N, \mathcal{I}_k\right)$ is Gaussian and its mean $\hat{x}_k\left(\Upsilon_k^N\right)$ and covariance $P_k\left(\Upsilon_k^N\right)$ can be obtained from $\hat{x}_{k-1}\left(\Upsilon_{k-1}^{N-1}\right)$ and $P_{k-1}\left(\Upsilon_{k-1}^{N-1}\right)$ via the Kalman filter with the new information $\{\gamma_k, \gamma_k s_k, \gamma_k s_k y_k\}$ in a similar way to Lemma 6.1.

For the hypothesis probability recursion, we first have that

$$\Pr\left(\Upsilon_{k-1}^{N-1}|\mathcal{I}_{k-1}\right) = \sum_{\gamma_{k-N}} \Pr\left(\Upsilon_{k-1}^{N}|\mathcal{I}_{k-1}\right). \tag{6.21}$$

Then $\Pr\left(\Upsilon_k^N|\mathcal{I}_k\right)$ can be obtained from $\Pr\left(\Upsilon_{k-1}^{N-1}|\mathcal{I}_{k-1}\right)$ with the new information $\{\gamma_k s_k, \gamma_k s_k y_k\}$ in a similar way to Lemma 6.2. The fixed memory suboptimal estimator is described in Algorithm 6.1.

6.4.2 Particle Filter

Particle filter is a well-established numerical method to approximate nonlinear and non-Gaussian probability distributions by using a set of samples [Arulampalam et al., 2002]. We can utilize the particle filter to approximate the posterior distribution $f(x_k, \Gamma_k|\mathcal{I}_k)$ and then to obtain a suboptimal estimator. However, the structure of the considered problem allows us to use the Rao–Blackwellization method [Doucet et al., 2000] and work more efficiently by sampling only from a conditional distribution to reduce the computational burden. Specifically, since $f(x_k|\Gamma_k, \mathcal{I}_k)$ is Gaussian, it can be analytically calculated from the Kalman filter. We only need to use samples to approximate $\Pr(\Gamma_k|\mathcal{I}_k)$ and then merge the two parts together to obtain an approximation to the desired posterior distribution $f(x_k|\mathcal{I}_k)$. The detailed derivation is given below.

In our estimation problem, it is clear from (6.3) that the difficulty is caused by the ever-expanding probability space for the hypothesis Γ_k. The particle filter motivates us to use finite samples to approximate the entire probability space of Γ_k, i.e.

$$\Pr(\Gamma_k|\mathcal{I}_k) \approx \sum_{i=1}^{N} \delta\left(\Gamma - \Gamma_k^i\right) \Pr\left(\Gamma_k^i|\mathcal{I}_k\right),$$

where δ is the Dirac delta measure and N is the number of samples. With a slight abuse of notion, we use Γ_k^i here to denote the ith sample drawn from $\Pr(\Gamma_k|\mathcal{I}_k)$. Therefore, an approximate to the posterior is given by

$$f(x_k|\mathcal{I}_k) \approx \sum_{i=1}^{N} f\left(x_k|\Gamma_k^i, \mathcal{I}_k\right) \Pr\left(\Gamma_k^i|\mathcal{I}_k\right), \tag{6.22}$$

based on which we can obtain a suboptimal estimator.

However, since $\Pr(\Gamma_k|\mathcal{I}_k)$ is unknown, we cannot directly sample from this probability distribution. The commonly used method to overcome the circumstance is to sample particles $\{\Gamma_k^i\}_{i=1,\ldots,N}$ from an importance density $q(\Gamma_k|\mathcal{I}_k)$. Then we can approximate $\Pr(\Gamma_k|\mathcal{I}_k)$ with

$$\Pr(\Gamma_k|\mathcal{I}_k) \approx \sum_{i=1}^{N} w_k^i \delta(\Gamma_k - \Gamma_k^i), \tag{6.23}$$

Algorithm 6.1: Fixed Memory Estimator

For $k < N$, run the optimal estimator to obtain $\hat{x}_{k|k}$ and $P_{k|k}$.
For $k \geq N$,

1. **Hypothesis merging**: compute $\hat{x}_{k-1}(\Upsilon_{k-1}^{N-1}), P_{k-1}(\Upsilon_{k-1}^{N-1})$ from (6.19) and (6.20).
2. **Hypothesis time and measurement update**:
   - If $\gamma_k = 0$,

$$\hat{x}_k(\Upsilon_k^N) = A\hat{x}_{k-1}(\Upsilon_{k-1}^{N-1}),$$
$$P_k(\Upsilon_k^N) = AP_{k-1}(\Upsilon_{k-1}^{N-1})A' + Q.$$

   - If $\gamma_k = 1$,

$$\hat{x}_k^- = A\hat{x}_{k-1}(\Upsilon_{k-1}^{N-1}),$$
$$P_k^- = AP_{k-1}(\Upsilon_{k-1}^{N-1})A' + Q,$$
$$K_k = P_k^- C'[CP_k^- C' + R + (1 - s_k\gamma_k)Y^{-1}]^{-1},$$
$$\hat{x}_k(\Upsilon_k^N) = (I - K_k C)\hat{x}_k^- + K_k s_k \gamma_k y_k,$$
$$P_k(\Upsilon_k^N) = (I - K_k C)P_k^-.$$

3. **Hypothesis probability merging**: compute $\Pr(\Upsilon_{k-1}^{N-1}|\mathcal{I}_{k-1})$ via (6.21).
4. **Hypothesis probability time and measurement update**: compute

$$\Pr(\Upsilon_k^N|\mathcal{I}_k) \propto \Pr(\Upsilon_k^N|\mathcal{I}_{k-1})\Pr(s_k\gamma_k|\Upsilon_k^N, \mathcal{I}_{k-1}),$$

   where
   - if $\gamma_k = 0$,

$$\Pr(\Upsilon_k^N|\mathcal{I}_{k-1}) = p\Pr(\Upsilon_{k-1}^{N-1}|\mathcal{I}_{k-1}),$$
$$\Pr(s_k\gamma_k|\Upsilon_k^N, \mathcal{I}_{k-1}) = 1 - s_k\gamma_k.$$

   - if $\gamma_k = 1$,

$$\Pr(\Upsilon_k^N|\mathcal{I}_{k-1}) = (1-p)\Pr(\Upsilon_{k-1}^{N-1}|\mathcal{I}_{k-1}),$$
$$\Pr(s_k\gamma_k|\Upsilon_k^N, \mathcal{I}_{k-1}) = s_k\gamma_k + \frac{1 - 2s_k\gamma_k}{\sqrt{|M_k Y + I|}}e^{-\frac{1}{2}(C\hat{x}_k^-)'[Y^{-1}+M_k]^{-1}C\hat{x}_k^-},$$

   where $M_k = CP_k^- C' + R$.
5. **State estimate**: compute $\hat{x}_{k|k}$ and $P_{k|k}$ from (6.16) and (6.17).

---

where w_k^i is the normalized importance weight and $w_k^i \propto \Pr(\Gamma_k^i|\mathcal{I}_k)/q(\Gamma_k^i|\mathcal{I}_k)$. Moreover, if the importance density is chosen to factorize such that

$$q(\Gamma_k|\mathcal{I}_k) = q(\gamma_k|\Gamma_{k-1}, \mathcal{I}_k)q(\Gamma_{k-1}|\mathcal{I}_{k-1}),$$

one can obtain new particle $\Gamma_k^i \sim q(\Gamma_k|\mathcal{I}_k)$ by augmenting each existing particle $\Gamma_{k-1}^i \sim q(\Gamma_{k-1}|\mathcal{I}_{k-1})$ with the new state $\gamma_k^i \sim q(\gamma_k|\Gamma_{k-1}, \mathcal{I}_k)$.

It has been proved that the degeneracy problem is inevitable with the above sequential importance sampling [Arulampalam et al., 2002]. That is, after a few iterations, all but one particle will have weights that are very close to zero. Therefore, a large computation is devoted to updating particles that have negligible contribution to the final estimate. One way to alleviate this problem is to select a good importance density. It is shown in Doucet et al. [2000] that the optimal importance density to minimize some degeneracy measure is given by

$$q(\gamma_k|\Gamma_{k-1}^i, \mathcal{I}_k) = \Pr(\gamma_k|\Gamma_{k-1}^i, \mathcal{I}_k).$$

For the considered problem in this chapter, we can analytically calculate $\Pr(\gamma_k|\Gamma_{k-1}^i, \mathcal{I}_k)$, which is given as follows. If $\gamma_k s_k = 1$, $\Pr(\gamma_k|\Gamma_{k-1}^i, \mathcal{I}_k) = \gamma_k$. If $\gamma_k s_k = 0$,

$$\begin{aligned}
\Pr(\gamma_k|\Gamma_{k-1}^i, \mathcal{I}_k) &= \Pr(\gamma_k|\Gamma_{k-1}^i, \gamma_k s_k = 0, \mathcal{I}_{k-1}) \\
&\propto \Pr(\gamma_k, \gamma_k s_k = 0|\Gamma_{k-1}^i, \mathcal{I}_{k-1}) \\
&= \Pr(\gamma_k s_k = 0|\Gamma_k^i, \mathcal{I}_{k-1}) \Pr(\gamma_k).
\end{aligned}$$

Therefore, we have if $\gamma_k s_k = 0$,

$$\Pr(\gamma_k = 0|\Gamma_{k-1}^i, \mathcal{I}_k) \propto p \Pr(\gamma_k s_k = 0|\gamma_k^i = 0, \Gamma_{k-1}^i, \mathcal{I}_{k-1}) = p, \tag{6.24}$$

and

$$\begin{aligned}
\Pr(\gamma_k = 1|\Gamma_{k-1}^i, \mathcal{I}_k) &\propto (1-p) \Pr(\gamma_k s_k = 0|\gamma_k^i = 1, \Gamma_{k-1}^i, \mathcal{I}_{k-1}) \\
&\overset{(a)}{=} \frac{1-p}{\sqrt{|M_k^i Y + I|}} e^{-\frac{1}{2}(C\hat{x}_{k|k-1}^i)'[Y^{-1}+M_k^i]^{-1}(*)},
\end{aligned} \tag{6.25}$$

where $M_k^i = CP_{k|k-1}^i C' + R$, and (a) can be calculated similarly as (6.13).

As $\Pr(\gamma_k|\Gamma_{k-1}^i, \mathcal{I}_k)$ is known, we can select it as the importance density to alleviate the degeneracy problem. Moreover, we can verify that the particle weight becomes $w_k^i = 1/N$ with this optimal importance density. Based on the approximations (6.22) and (6.23), we have the following state estimate:

$$\hat{x}_{k|k} = \frac{1}{N}\sum_{i=1}^{N} \hat{x}_{k|k}^i, \tag{6.26}$$

$$P_{k|k} = \frac{1}{N}\sum_{i=1}^{N} \left(P_{k|k}^i + (\hat{x}_{k|k}^i - \hat{x}_{k|k})(\hat{x}_{k|k}^i - \hat{x}_{k|k})'\right), \tag{6.27}$$

where $\hat{x}^i_{k|k}$ and $P^i_{k|k}$ are the mean and covariance of $f(x_k|\Gamma^i_k, \mathcal{I}_k)$, respectively. The detailed particle filter is described in Algorithm 6.2.

Algorithm 6.2: Particle Filter

---

1. **Time update for each particle**: For $i = 1, \dots, N$,
   - If $k = 0$, then
   $$x^i_{0|-1} = 0, P^i_{0|-1} = \Sigma_0.$$
   - If $k > 0$, then
   $$\hat{x}^i_{k|k-1} = A\hat{x}^i_{k-1|k-1},$$
   $$P^i_{k|k-1} = AP^i_{k-1|k-1}A' + Q.$$
2. **Sampling new particles**: If $\gamma_k s_k = 1$, let $\gamma^i_k = 1$. If $\gamma_k s_k = 0$, generate γ^i_k from the distribution described by (6.24), (6.25).
3. **Measurement update for each particle**:
   - If $\gamma^i_k = 0$,
   $$\hat{x}^i_{k|k} = \hat{x}^i_{k|k-1}, P^i_{k|k} = P^i_{k|k-1}.$$
   - If $\gamma^i_k = 1$,
   $$K^i_k = P^i_{k|k-1}C'[CP^i_{k|k-1}C' + R + (1 - s_k\gamma_k)Y^{-1}]^{-1},$$
   $$\hat{x}^i_{k|k} = (I - K^i_k C)\hat{x}^i_{k|k-1} + s_k\gamma_k K^i_k y_k,$$
   $$P^i_{k|k} = P^i_{k|k-1} - K^i_k CP^i_{k|k-1}.$$
4. **State estimate**: compute $\hat{x}_{k|k}$ and $P_{k|k}$ from (6.26) and (6.27).

Remark 6.4 The variational Bayesian (VB) method can also be adopted here to design suboptimal estimators with constant resource requirements. The VB method approximates the complex posterior distribution with a proposal distribution, which is parameterized in certain forms to represent necessary statistics. These parameters will be determined by optimizing the statistical distance between the implicit posterior distribution and the proposal distribution. The VB method has been utilized to approximate Gaussian mixture distributions in Ma et al. [2019]. Interested readers can refer to Ma et al. [2019] and references therein for details.

Remark 6.5 The proposed estimators can be extended to the case with Markovian packet drops. Since $f(x_k|\Gamma^i_k, \mathcal{I}_k)$ is Gaussian if packet drops are independent with the system state, the conditional posterior distribution $f(x_k|\mathcal{I}_k)$ is still mixture Gaussian even for Markovian packet drops. We can use similar approach to derive the MMSE estimator and sub-optimal estimators. The main difference compared with the i.i.d. case is that the iteration of $\Pr(\Gamma_k^i|\mathcal{I}_k)$ is different and more complex for correlated packet drops.

6.5 Simulations

In simulations, we adopt the same system parameters as in Han et al. [2015], which are

$$A = \begin{bmatrix} 0.8 & \\ & 0.95 \end{bmatrix}, \quad C = \begin{bmatrix} 1 & 1 \end{bmatrix}, \quad \Sigma_0 = Q = \begin{bmatrix} 1 & \\ & 1 \end{bmatrix}, \quad R = 1.$$

We conduct simulations with the optimal estimator, the oracle estimator, the OLSET-KF estimator in Han et al. [2015], the fixed memory estimator, and the particle filter. The OLSET-KF estimator does not consider packet drops. When the estimator fails to receive a packet, it always assumes that this is caused by the hold of transmission from the scheduler. The oracle estimator is the optimal estimator under the assumption that the estimator knows the value of γ_k at each step, which is given in Algorithm 6.3. Clearly, the oracle estimator has the

---

Algorithm 6.3: Oracle Estimator

1. **Initialization**:

$$\hat{x}_{0|-1} = 0, P_{0|-1} = \Sigma_0.$$

2. **Measurement update**:
   - If $\gamma_k = 0$,

$$\hat{x}_{k|k} = \hat{x}_{k|k-1}, P_{k|k} = P_{k|k-1}.$$

   - If $\gamma_k = 1$,

$$\begin{aligned} \hat{x}_{k|k} &= (I - K_k C)\hat{x}_{k|k-1} + K_k s_k \gamma_k y_k, \\ P_{k|k} &= P_{k|k-1} - K_k C P_{k|k-1}, \\ K_k &= P_{k|k-1} C' [C P_{k|k-1} C' + R + (1 - s_k \gamma_k) Y^{-1}]^{-1}. \end{aligned}$$

3. **Time update**:

$$\begin{aligned} \hat{x}_{k+1|k} &= A\hat{x}_{k|k}, \\ P_{k+1|k} &= A P_{k|k} A' + Q. \end{aligned}$$

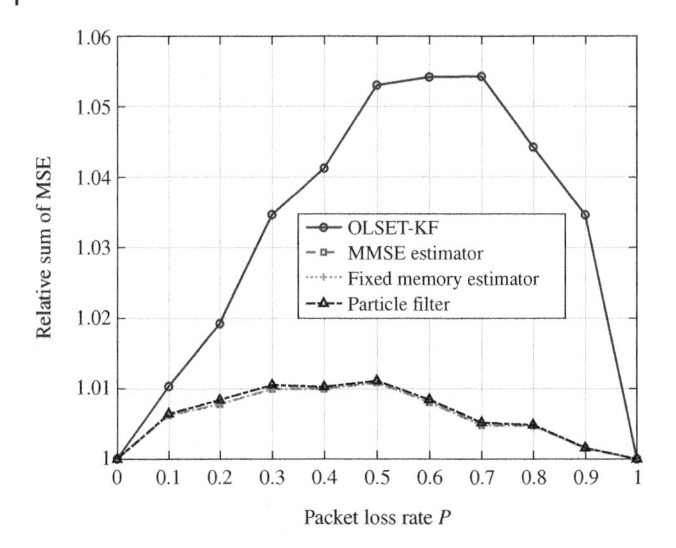

Figure 6.2 Relative sum of MSE of different estimator under different packet drop rate. Source: Xu et al. (2020)/with permission of IEEE.

smallest mean-square error (MSE) and can be used as a benchmark to evaluate the performance of other estimators. In simulations, the schedule parameter Y is selected as $Y = 1$, $N = 2$ is selected for the fixed memory estimator and the particle numbers is set to 20 in the particle filter. The source code for all the simulations in this section is available at Gaussian-mixture simulation [2019].

Firstly, we compare the performance of different estimators, and we adopt Monte Carlo methods with 1500 independent experiments to evaluate the sum of MSE $\sum_{k=0}^{9} \mathcal{E}||x_k - \hat{x}_{k|k}||^2$ under different packet drop rates. The simulation results are illustrated in Figure 6.2, where the relative sum of MSE is plotted. The relative sum of MSE is defined as the sum of MSE of an estimator divided by the sum of MSE of the oracle estimator. It is clear from Figure 6.2 that the estimation error of the fixed memory estimator and the particle filter are close to that of the optimal estimator and is much smaller than the OLSET-KF, which shows the superior performance of the proposed suboptimal estimators and also indicates the advantage of considering packet drops in the remote state estimation problem.

Moreover, in the case of $p = 0$ and $p = 1$, the sum of MSE of all the estimators are equal. This is because when $p = 0$ ($p = 1$), the optimal estimator (suboptimal estimators) assigns zero probability to all the hypotheses with a $\gamma_k = 0$ ($\gamma_k = 1$). Therefore, only the hypothesis with $\gamma_k = 1$ ($\gamma_k = 0$) for all k is preserved. As a result, the estimate of the optimal estimator (suboptimal estimators) is the same with the oracle estimator. Therefore, for the case that $p = 0$ ($p = 1$), the optimal estimator (suboptimal estimators) and the oracle estimator have the same sum of MSE. The recursions of the OLSET-KF and the oracle estimator are the same for the

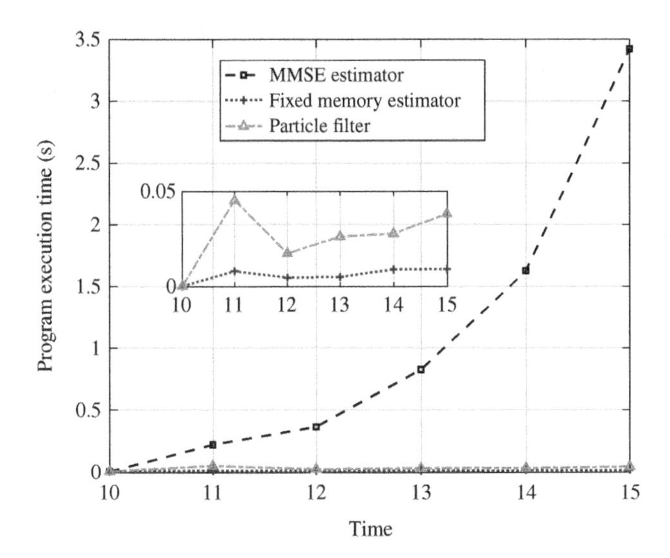

Figure 6.3 Execution time of the optimal estimator and the suboptimal estimators with respect to time. Source: Xu et al. (2020)/with permission of IEEE.

case $p = 0$, where there are no packet drops. For the case that $p = 1$, even though the recursions of the OLSET-KF and the oracle estimator are different, since they both start with $\hat{x}_{0|-1} = 0$, their estimates would always be $\hat{x}_{k|k} = 0$. Therefore, for the case that $p = 0$ and $p = 1$, the OLSET-KF and the oracle estimator have the same sum of MSE.

In the second simulation, we evaluate how the execution time of the MMSE estimator and the suboptimal estimators grow with time. We let $p = 0.5$ and generate a sequence of noisy observations. Then we use the optimal estimator and the suboptimal estimators to estimate the system state. The simulation is conducted with python/numpy and evaluated on Macbook Pro with 2.3 GHz Intel Core i5 processor. The program execution time with respect to time is illustrated in Figure 6.3. It is clear that the program execution time of the optimal estimator is increasing exponentially with time. However, the program execution time of the two sub-optimal estimators does not change much when time increases, which demonstrates the effectiveness of the proposed suboptimal estimators in reducing the computational complexity.

6.6 Conclusions

This chapter studies the remote state estimation problem of linear systems with stochastic event-triggered sensor schedulers in the presence of packet drops.

The posterior distributions at the estimator side are computed. Recursive MMSE estimators are derived. Suboptimal estimators to reduce the computational complexity are proposed.

Bibliography

B. D. O. Anderson and J. B. Moore. *Optimal filtering*. Prentice Hall, 1979.

M. S. Arulampalam, S. Maskell, N. Gordon, and T. Clapp. A tutorial on particle filters for online nonlinear/non-gaussian Bayesian tracking. *IEEE Transactions on Signal Processing*, 50(2):174–188, 2002.

O. L. V. Costa, M. D. Fragoso, and R. P. Marques. *Discrete-time Markov jump linear systems*. Probability and its applications. Springer, London, 2005.

A. Doucet, S. Godsill, and C. Andrieu. On sequential Monte Carlo sampling methods for Bayesian filtering. *Statistics and Computing*, 10(3):197–208, 2000.

GitHub reopsitory Gaussian-mixture simulation, 2019.

D. Han, Y. Mo, J. Wu, S. Weerakkody, B. Sinopoli, and L. Shi. Stochastic event-triggered sensor schedule for remote state estimation. *IEEE Transactions on Automatic Control*, 60(10):2661–2675, 2015.

A. G. Jaffer and S. C. Gupta. On estimation of discrete processes under multiplicative and additive noise conditions. *Information Sciences*, 3(3):267–276, 1971.

E. Kung, J. Wu, D. Shi, and L. Shi. On the nonexistence of event-based triggers that preserve Gaussian state in presence of package-drop. In *Proceedings of the 2017 American Control Conference*, pages 1233–1237, 2017.

H. Lin, H. Su, Z. Shu, Z. G. Wu, and Y. Xu. Optimal estimation in UDP-like networked control systems with intermittent inputs: Stability analysis and suboptimal filter design. *IEEE Transactions on Automatic Control*, 61(7):1794–1809, 2016.

Y. Ma, S. Zhao, and B. Huang. Multiple-model state estimation based on variational Bayesian inference. *IEEE Transactions on Automatic Control*, 64(4):1679–1685, 2019.

J. Wu, Q. Jia, K. H. Johansson, and L. Shi. Event-based sensor data scheduling: Trade-off between communication rate and estimation quality. *IEEE Transactions on Automatic Control*, 58(4):1041–1046, 2013.

C. Yang and L. Shi. Deterministic sensor data scheduling under limited communication resource. *IEEE Transactions on Signal Processing*, 59(10):5050–5056, 2011.

7

Distributed Consensus over Undirected Fading Networks

7.1 Introduction

The earlier chapters studied the networked control problem of single-agent systems over fading channels. It is still unknown how the channel fading affects the consensus problem of multi-agent systems (MASs). Previously, the consensus problem of MASs had been studied under perfect communication assumptions in Ma and Zhang [2010], You and Xie [2011], Li et al. [2010], and Trentelman et al. [2013]. However, since fading is unavoidable in wireless networks, which are commonly used by most MASs nowadays, it is necessary to consider its impact on the consensusability of MASs. In this chapter, we consider a distributed consensus problem, in which there are multiple agents that are connected through faded communication channels. Each agent can only receive corrupted information about its neighborhoods' states. The MAS wants to reach an agreement about all the agents' states. We aim to characterize the requirement on the fading parameters and the communication topology that can ensure the existence of a linear distributed consensus controller. The derived results would shed light on how the fading communication networks affect distributed control systems.

The rest of the chapter is organized as follows: Section 7.2 provides background materials and the problem formulation. Section 7.3 deals with the case of identical fading networks, and the mean-square consensus problem over nonidentical fading networks is discussed in Section 7.4. Section 7.5 provides the numerical simulations, and this chapter ends with a summary in Section 7.6.

Control over Communication Networks: Modeling, Analysis, and Design of Networked Control Systems and Multi-Agent Systems over Imperfect Communication Channels, First Edition.
Jianying Zheng, Liang Xu, Qinglei Hu, and Lihua Xie.

7.2 Problem Formulation

Consider an MAS with N agents. An undirected graph $\mathcal{G} = (\mathcal{V}, \mathcal{E})$ is used to characterize the interaction among agents, as in Section 1.4.1. The discrete-time dynamics of agent i has the following form:

$$x_i(t+1) = Ax_i(t) + Bu_i(t), \quad y_i(t) = Cx_i(t), \tag{7.1}$$

where $i = 1, 2, \ldots, N$, and $x_i \in \mathbb{R}^n, y_i \in \mathbb{R}^p, u_i \in \mathbb{R}^m$ represent the agent state, output, and control input, respectively. Without loss of generality, we assume that B has full-column rank and C has full-row rank.

The agents communicate information to their neighbors through fading channels (see Figure 7.1). Specifically, in this chapter, we let the jth agent send the information $Cq_j(t) - y_j(t)$ to the ith agent at time t with $q_j \in \mathbb{R}^n$ representing the jth agent's controller state as specified later. At the channel output side, the ith agent receives the deteriorated information:

$$r_{ij}(t) = \gamma_{ij}(t)(Cq_j(t) - y_j(t)) + \omega_{ij}(t)$$

with γ_{ij} modeling the channel fading and ω_{ij} denoting a zero-mean white communication noise with bounded variance. For simplicity, we assume that all components of $Cq_j(t) - y_j(t)$ are transmitted together over the same fading channel and do not consider the channel input power constraint in this chapter. Combining all the received information from its neighbors, agent i generates the control input by using the following controller:

$$\begin{aligned} q_i(t+1) &= (A + BK)\, q_i(t) + F \sum_{j \in \mathcal{N}_i} \left[\gamma_{ij}(t)\left(Cq_i(t) - y_i(t)\right) - r_{ij}(t)\right], \\ u_i(t) &= Kq_i(t), \end{aligned} \tag{7.2}$$

where $i = 1, 2, \ldots, N$, and F and K are controller parameters to be designed.

Remark 7.1 The fading factors of MASs appear in the consensus protocol in a similar way as the coupling terms c_{ij} in Li et al. [2013a], Li et al. [2013b], and Li and Chen [2017], which design adaptive updating laws for c_{ij} to achieve a fully distributed consensus control. However, they are different in the following aspects: Firstly, γ_{ij} in our formulation arises from the channel fading, which is part of the model and is stochastic, while c_{ij} is a design parameter, which is part of the

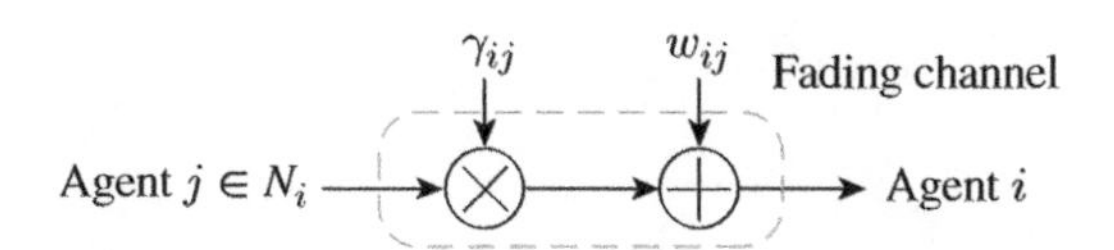

Figure 7.1 Information transmission from agent j to agent i. Source: Xu et al. (2020)/with permission of Elsevier.

controller in Li et al. [2013a], Li et al. [2013b], and Li and Chen [2017]. Secondly, we try to determine the relations of the agent dynamics, the network topology, and the fading statistics to ensure the existence of a consensus control law, while they aim at designing one admissible consensus protocol to achieve a fully distributed consensus control.

Let $\varepsilon_i = \left[x_i', q_i'\right]'$, $\varepsilon = [\varepsilon_1', \varepsilon_2', \dots, \varepsilon_N']'$, and define the consensus error as $\delta = \varepsilon - \frac{1}{N}((\mathbf{11}') \otimes I_{2n})\varepsilon$. The mean-square consensus is defined as the mean-square boundedness of the consensus error, i.e. $\lim_{t\to\infty}\mathbb{E}\{\delta(t)\delta(t)'\} \leq M$, where $M > 0$ is a constant matrix. We aim to derive conditions on the fading statistics, the agent dynamics, and the communication topology under which there exist F and K in the controller (7.2) such that the multi-agent system (7.1) can achieve mean-square consensus. To avoid triviality, we make the following assumption as in Section II.B of You and Xie [2011].

Assumption 7.1 All the eigenvalues of A are either on or outside the unit disk.

In this chapter, the mean-square consensus problem is studied under undirected graphs with identical fading networks and nonidentical fading networks, respectively. The case with directed graphs is studied in Chapter 8.

7.3 Identical Fading Networks

In this section, we consider the scenario where all the fading channels are identical, which is a reasonable assumption for MASs operating in a small area with similar physical configurations.

Assumption 7.2 The channel fading is identical and i.i.d., i.e. $\gamma_{ij}(t) = \gamma(t)$ for all $t \geq 0$, $i, j = 1, 2, \dots, N$, and the sequence $\{\gamma(t)\}_{t\geq 0}$ is i.i.d. with mean μ and variance σ^2.

Throughout this chapter, if the state of a stochastic dynamical system converges to zero in mean-square sense, we say the dynamical system is mean-square stable. Let $\mathcal{L}$ be the Laplacian matrix of the communication graph $\mathcal{G}$. Then the error dynamics of δ under Assumption 7.2 is

$$\delta(t+1) = (I \otimes \mathcal{A} + \gamma(t)\mathcal{L} \otimes \mathcal{H})\delta(t) + C(t),$$

with $\mathcal{A} = \begin{bmatrix} A & BK \\ 0 & A+BK \end{bmatrix}$, $\mathcal{H} = \begin{bmatrix} 0 & 0 \\ -FC & FC \end{bmatrix}$ and

$$C(t) = \left\{I - \frac{1}{N}\left[(\mathbf{11}') \otimes I_{2n}\right]\right\}\left\{\sum_{j=1}^{N}[0', -\omega_{1j}(t)'F'], \dots, \sum_{j=1}^{N}[0', -\omega_{Nj}(t)'F']\right\}'.$$

Since $\omega_{ij}(t)$ is with bounded variance, so is $C(t)$. Because the consensusability is defined as the mean-square boundedness of δ, if the following dynamics is mean-square stable

$$\delta(t+1) = \left(I_N \otimes \mathcal{A} + \gamma(t)\mathcal{L} \otimes \mathcal{H}\right)\delta(t), \tag{7.3}$$

i.e. $\lim_{t\to\infty} \mathbb{E}\{\delta(t)\delta(t)'\} = 0$, mean-square consensus of the MAS can be achieved. Thus, we focus on studying the requirement under which system (7.3) is mean-square stable. The following lemma, which describes the solvability of a modified Riccati inequality, is critical in networked control over fading channels of a single-agent system. The extension of networked control over fading channels from single-agent systems to MASs relies closely on Lemma 7.1.

Lemma 7.1 ***(Schenato et al. [2007])*** *Under Assumption 7.1 and assuming that (C, A) is observable, there exists a solution $P > 0$ to the following modified Riccati inequality:*

$$P > APA' - \theta APC'(CPC')^{-1}CPA', \tag{7.4}$$

if and only if θ is greater than a critical value $\theta_c \in [0, 1)$.

Remark 7.2 The value θ_c is of great importance for determining the critical erasure probability in Kalman filtering over intermittent channels [Sinopoli et al., 2004, Schenato et al., 2007, Mo and Sinopoli, 2008]. It has been shown that the critical value θ_c is only determined by the pair (A, C) [Mo and Sinopoli, 2008]. However, an explicit expression of θ_c is only available for some specific situations. For example, it has been shown that when rank$(C) = 1$, $\theta_c = 1 - \frac{1}{\prod_i |\lambda_i(A)|^2}$ and when C is square and invertible, $\theta_c = 1 - \frac{1}{\max_i |\lambda_i(A)|^2}$. For other cases, the critical value θ_c can be obtained by solving a quasiconvex linear matrix inequality (LMI) optimization problem [Schenato et al., 2007].

The basic idea in this subsection is to transform the mean-square stabilization problem of (7.3) into an equivalent simultaneous mean-square stabilization problem, i.e. to determine whether there exist common control gains F and K that can simultaneously stabilize a series of subdynamics in mean-square sense. Let $h = \left(I_N \otimes \begin{bmatrix} I_n & -I_n \\ 0 & I_n \end{bmatrix}\right)\delta$, then

$$h(t+1) = \left(I_N \otimes \bar{A} + \gamma(t)\mathcal{L} \otimes \overline{\mathcal{H}}\right)h(t) \tag{7.5}$$

with $\bar{A} = \begin{bmatrix} A & 0 \\ 0 & A+BK \end{bmatrix}$, $\overline{\mathcal{H}} = \begin{bmatrix} FC & 0 \\ -FC & 0 \end{bmatrix}$. The mean-square stability of (7.3) is equivalent to that of (7.5). If the undirected graph $\mathcal{G}$ is connected, we can select

$\phi_i \in \mathbb{R}^N$ such that $\mathcal{L}\phi_i = \lambda_i \phi_i$ and form the unitary matrix $\Theta = [\mathbf{1}/\sqrt{N}, \phi_2, \phi_3, \dots, \phi_N]$ with $\mathrm{diag}(0, \lambda_2, \lambda_3, \dots, \lambda_N) = \Theta' \mathcal{L} \Theta$ and $0 < \lambda_2 \le \lambda_3 \le \dots \le \lambda_N$ [Ren and Beard, 2008]. Let $g = [g_1', g_2', \dots, g_N']' = (\Theta' \otimes I_{2n})h$, then $g_1 \equiv 0$ and

$$g_i(t+1) = (\bar{A} + \lambda_i \gamma(t) \overline{\mathcal{H}}) g_i(t) \tag{7.6}$$

for $i = 2, 3, \dots, N$. Thus, the mean-square stability of (7.5) is equivalent to the simultaneous mean-square stability of (7.6) with $i = 2, 3, \dots, N$.

In the following, we will show that the mean-square stability of (7.6) for any i can be obtained from that of a low-dimensional system, which physically implies that dynamic output feedback control has the same effect as state feedback control if the communication topology is undirected and connected.

Lemma 7.2 *Under Assumptions 7.1 and 7.2, there exist F and K, such that system (7.6) is mean-square stable if and only if (A, B) is controllable and $g_{1i}(t+1) = (A + \lambda_i \gamma(t) FC) g_{1i}(t)$ is mean-square stable.*

Proof: (Sufficiency) Suppose there exists F, such that

$$g_{1i}(t+1) = (A + \lambda_i \gamma(t) FC) g_{1i}(t)$$

is mean-square stable, then there exists $P_{1i} > 0$, such that [Xiao et al., 2012]

$$P_{1i} > (A + \lambda_i \mu FC) P_{1i} (A + \lambda_i \mu FC)' + \lambda_i^2 \sigma^2 FCP_{1i} C' F'.$$

Since (A, B) is controllable, there exist $P_{2i} > 0$ and K such that

$$P_{2i} - (A + BK) P_{2i} (A + BK)' > Q_i$$

for any $Q_i > 0$. Let

$$\begin{aligned}
Q_i &= \lambda_i^2 (\mu^2 + \sigma^2) FCP_{1i} C' F' + M_i' H_i^{-1} M_i, \\
M_i &= (A + \lambda_i \mu FC) P_{1i} (\lambda_i \mu FC)' + \lambda_i^2 \sigma^2 FCP_{1i} C' F', \\
H_i &= P_{1i} - \lambda_i^2 \sigma^2 FCP_{1i} C' F' - (A + \lambda_i \mu FC) P_{1i} (A + \lambda_i \mu FC)', \\
\overline{P}_i &= \begin{bmatrix} P_{1i} & 0 \\ 0 & P_{2i} \end{bmatrix}.
\end{aligned}$$

Based on the Schur complement lemma [Bernstein, 2009], it is trivial to show that

$$\overline{P}_i > (\bar{A} + \lambda_i \mu \overline{\mathcal{H}}) \overline{P}_i (\bar{A} + \lambda_i \mu \overline{\mathcal{H}})' + \lambda_i^2 \sigma^2 \overline{\mathcal{H}} \, \overline{P}_i \overline{\mathcal{H}}',$$

which implies the mean-square stability of (7.6) and thus proves the sufficiency.

(Necessity) Since the system (7.6) is mean-square stable, decomposing $g_i = [g_{1i}', g_{2i}']'$ as

$$g_{1i}(t+1) = \left(A + \lambda_i \gamma(t) FC\right) g_{1i}(t), \tag{7.7}$$

$$g_{2i}(t+1) = (A + BK) g_{2i}(t) - \lambda_i \gamma(t) FC g_{1i}(t). \tag{7.8}$$

Then, the subdynamics (7.7) should be mean-square stable. Besides, from Lyapunov inequality [Papoulis and Pillai, 2002] in probability theory, the mean-square stability of (7.6) implies that the first-moment dynamics $\mathbb{E}\{g_i(t+1)\} = \begin{bmatrix} A+\lambda_i\mu FC & 0 \\ -\lambda_i\mu FC & A+BK \end{bmatrix} \mathbb{E}\{g_i(t)\}$ is stable, which indicates that $A+BK$ is stable. Thus, under Assumption 7.1, (A,B) is controllable. This completes the proof of the necessity. □

In view of Lemma 7.2, we have the following result.

Theorem 7.1 *Under Assumptions 7.1 and 7.2, the MAS* (7.1) *is mean-square consensusable by the controller* (7.2) *under a connected undirected communication topology $\mathcal{G}$ if (A,B) is controllable, (C,A) is observable, and*

$$\theta_1 \triangleq \frac{\mu^2}{\mu^2+\sigma^2} \times \left[1 - \left(\frac{\lambda_N - \lambda_2}{\lambda_N + \lambda_2}\right)^2\right] > \theta_c, \tag{7.9}$$

where θ_c is given in Lemma 7.1, λ_N and λ_2 are the largest and second smallest eigenvalues of the Laplacian matrix $\mathcal{L}$, respectively. Moreover, if (7.9) *holds, there exists a solution $P_0 > 0$ to the modified Riccati inequality* (7.4) *with $\theta = \theta_1$, and a pair of control gains that ensures mean-square consensus can be given by*

$$F = -\frac{2\mu}{(\lambda_2+\lambda_N)(\mu^2+\sigma^2)} AP_0C'(CP_0C')^{-1}$$

and any K satisfying that $A+BK$ is stable.

Proof: If (7.9) is satisfied and (C,A) is observable, in view of Lemma 7.1, there exists a solution $P_0 > 0$ to the modified Riccati inequality (7.4) with $\theta = \theta_1$. It is trivial to show that $\overline{\theta}_i > \theta_1$ for all $i = 2,3,\dots,N$ with $\overline{\theta}_i = \frac{\mu^2}{\mu^2+\sigma^2} \times \frac{4(\lambda_i(\lambda_2+\lambda_N)-\lambda_i^2)}{(\lambda_N+\lambda_2)^2}$. Thus, we have

$$P_0 > AP_0A' - \overline{\theta}_i AP_0C'(CP_0C')^{-1}CP_0A',$$

which can be equivalently formulated as

$$P_0 > AP_0A' + \lambda_i\mu AP_0C'F' + \lambda_i\mu FCP_0A' + \lambda_i^2(\mu^2+\sigma^2)FCP_0C'F'$$

with $F = -\frac{2\mu}{(\lambda_2+\lambda_N)(\mu^2+\sigma^2)} AP_0C'(CP_0C')^{-1}$. This implies that

$$g_{1i}(t+1) = (A + \lambda_i\gamma(t)FC)g_{1i}(t)$$

is mean-square stable for all $i = 2,3,\dots,N$. Since (A,B) is controllable, in view of Lemma 7.2, we know that (7.6) with $i = 2,3,\dots,N$ are simultaneously mean-square stable, which indicates that mean-square consensus of MAS (7.1) is achieved and this completes the proof. □

In the following, we show that the sufficient condition in Theorem 7.1 is also necessary for single-output systems, i.e. $p = 1$. The following lemmas are needed to prove such a result and stated first.

Lemma 7.3 *Let $P > 0$. Suppose there exists a vector v, such that $v'Pv > \phi^2$ and $\phi > 0$, then there exists a vector x, such that the following inequalities hold:*

$$x'P^{-1}x < 1, \quad x'v > \phi.$$

Proof: Let us choose $x = \alpha Pv$, then

$$x'P^{-1}x = \alpha^2 v'Pv, \; x'v = \alpha v'Pv.$$

Since $v'Pv > \phi^2$, we can choose α, such that

$$\frac{\phi}{v'Pv} < \alpha < \frac{1}{\sqrt{v'Pv}}.$$

The proof is completed. □

Lemma 7.4 *Suppose that* $\det(A) \neq 0$, *then any $P > 0$ that satisfies*

$$P - APA' + APC'(CPC')^{-1}CPA' > 0, \tag{7.10}$$

must also satisfy

$$\frac{CA^{-1}P(A')^{-1}C'}{CPC'} \leq \frac{1}{\det(A)^2}. \tag{7.11}$$

Proof: We prove this lemma by contradiction. Let

$$g(P) = P - APA' + APC'(CPC')^{-1}CPA' > 0.$$

Suppose that there exists a $P > 0$ to (7.10), such that

$$\frac{CA^{-1}P(A')^{-1}C'}{CPC'} > \frac{1}{(\det A')^2},$$

then we have

$$CA^{-1}\frac{g(P)}{CPC'}(A')^{-1}C' = \frac{CA^{-1}P(A')^{-1}C'}{CPC'} > \frac{1}{(\det A')^2}.$$

Therefore, by Lemma 7.3, there exists a F, such that

$$(F - F^*)'\left(\frac{g(P)}{CPC'}\right)^{-1}(F - F^*) < 1, \tag{7.12}$$

and

$$(F - F^*)'(A')^{-1}C' > \left|\frac{1}{\det A'}\right|, \tag{7.13}$$

where $F^* = -APC'/(CPC')$.

By Schur, complement lemma, and the fact that $CPC' > 0$, (7.12) is equivalent to

$$P > APA' - APC'(CPC')^{-1}CPA' + (F - F^*)CPC'(F - F^*)',$$

or $P > (A + FC)P(A + FC)'$. Therefore, $A' + C'F'$ is strictly stable. By matrix determinant lemma, if $A' + C'F'$ is stable, then $|\det(A' + C'F')| = |\det A'||1 + F'(A')^{-1}C'| < 1$. Therefore,

$$F'(A')^{-1}C' < -1 + \left|\frac{1}{\det A'}\right|. \tag{7.14}$$

On the other hand, by the definition of $(F^*)'$, we have

$$(F^*)'(A')^{-1}C' = -\frac{CPA'(A')^{-1}C'}{CPC'} = -1.$$

Then, we have from (7.13) that

$$F'(A')^{-1}C' > -1 + \left|\frac{1}{\det A'}\right|,$$

which contradicts with (7.14). Therefore, any $P > 0$ that satisfies (7.10) must also satisfy (7.11). The proof is completed. □

The main result is stated as follows:

Theorem 7.2 *Under Assumptions 7.1 and 7.2, when $p = 1$, the MAS (7.1) is mean-square consensusable by the controller (7.2) under a connected undirected communication topology $\mathcal{G}$ if and only if (A, B) is controllable, (C, A) is observable, and*

$$\frac{\mu^2}{\mu^2 + \sigma^2}\left[1 - \left(\frac{\lambda_N - \lambda_2}{\lambda_N + \lambda_2}\right)^2\right] > 1 - \frac{1}{(\det A)^2}. \tag{7.15}$$

Proof: The sufficiency follows from Theorem 7.1. Only the necessity is proved here. It has been shown that the mean-square stability of the closed-loop system is equivalent to the simultaneous mean-square stability of

$$g_{1i}(t + 1) = (A + \lambda_i\gamma(t)FC)g_{1i}(t), \quad i = 2, \dots, N, \tag{7.16}$$

where λ_i is the nonzero positive eigenvalue of $\mathcal{L}$ with $\lambda_2 \leq \cdots \leq \lambda_N$.

In view of Lemma 1 in Xiao et al. [2012], (7.16) is mean-square stable for i.i.d. $\{\gamma(t)\}_{t\geq 0}$ if and only if there exist $P_i > 0, i = 2, \dots, N$ and F', such that

$$P_i > (A + \lambda_i\mu FC)P_i(A' + \lambda_i\mu C'F') + \lambda_i^2\sigma^2 FCP_iC'F'$$

for all $i = 2, \dots, N$. With some manipulations, we can show that

$$\begin{aligned} & P_i - AP_iA' + \frac{\mu^2}{\mu^2+\sigma^2} AP_iC'(CP_iC')^{-1}CP_iA' \\ & > \lambda_i^2(\mu^2+\sigma^2)\left(F' + \frac{\mu}{\lambda_i(\mu^2+\sigma^2)}(CP_iC')^{-1}CP_iA'\right)' \\ & \times CP_iC'\left(F' + \frac{\mu}{\lambda_i(\mu^2+\sigma^2)}(CP_iC')^{-1}CP_iA'\right). \end{aligned}$$

Left and right multiply the above inequality with CA^{-1} and $(A')^{-1}C'$, we can obtain that

$$\left(\lambda_i\sqrt{\mu^2+\sigma^2}F'(A')^{-1}C' + \frac{\mu}{\sqrt{\mu^2+\sigma^2}}\right)^2 < \frac{CA^{-1}P_i(A')^{-1}C'}{CP_iC'} + \frac{\mu^2}{\mu^2+\sigma^2} - 1. \tag{7.17}$$

Since P_i satisfies (7.10), we have from Lemma 7.4 that

$$\frac{CA^{-1}P_i(A')^{-1}C'}{CP_iC'} \le \frac{1}{\det(A')^2}.$$

Therefore, we have from (7.17) that

$$\left(\lambda_i\sqrt{\mu^2+\sigma^2}F'(A')^{-1}C' + \frac{\mu}{\sqrt{\mu^2+\sigma^2}}\right)^2 < \frac{1}{\det(A')^2} + \frac{\mu^2}{\mu^2+\sigma^2} - 1,$$

which further indicates

$$\underline{\beta}_i < |F'(A')^{-1}C'| < \overline{\beta}_i \tag{7.18}$$

with

$$\underline{\beta}_i = \frac{-\sqrt{\frac{1}{a_0^2} + \frac{\mu^2}{\mu^2+\sigma^2} - 1} + \frac{\mu}{\sqrt{\mu^2+\sigma^2}}}{\lambda_i\sqrt{\mu^2+\sigma^2}},$$

$$\overline{\beta}_i = \frac{\sqrt{\frac{1}{a_0^2} + \frac{\mu^2}{\mu^2+\sigma^2} - 1} + \frac{\mu}{\sqrt{\mu^2+\sigma^2}}}{\lambda_i\sqrt{\mu^2+\sigma^2}},$$

where $a_0 = \det(A')$.

Since there exists a common $|F'(A')^{-1}C'|$, such that (7.18) holds for all $i = 2, \dots, N$. $\cap_i\left(\underline{\beta}_i, \overline{\beta}_i\right)$ must be nonempty, which implies $\underline{\beta}_2 < \overline{\beta}_N$. Further calculation shows that

$$\frac{\mu^2}{\mu^2+\sigma^2} \times \left[1 - \left(\frac{\lambda_N - \lambda_2}{\lambda_N + \lambda_2}\right)^2\right] > 1 - \frac{1}{a_0^2}. \tag{7.19}$$

Substituting the definitions of μ, σ^2, and a_0, note that $\det(A') = \det(A)$, we can obtain (7.15). The proof is completed. □

Remark 7.3 If each agent is a single integrator system

$$x_i(t+1) = x_i(t) + u_i(t), \quad i = 1, \dots, N, \tag{7.20}$$

the necessary and sufficient condition in Theorem 7.2 becomes

$$\frac{\mu^2}{\mu^2+\sigma^2}\left[1 - \left(\frac{\lambda_N - \lambda_2}{\lambda_N + \lambda_2}\right)\right] > 0,$$

which holds naturally. This implies that as long as the undirected graph is connected, the multi-agent system can achieve mean-square consensus, irrelevant of the channel fading level. This can be easily verified. If each agent is a single integrator system as in (7.20), the mean-square consensus problem is equivalent to the simultaneous mean-square stability problem that

$$g_{1i}(t+1) = (1 + \lambda_i \gamma(t) f_0) g_{1i}(t), \quad i = 2, \dots, N.$$

To stabilize the above systems, we should find f_0 such that

$$\mathbb{E}\{(1 + \lambda_i \gamma(t) f_0)^2\} = 1 + 2\lambda_i \mu f_0 + \lambda_i^2(\mu^2 + \sigma^2) f_0^2 < 1, \tag{7.21}$$

for all $i = 2, \dots, N$. Since $\lambda_i > 0$, the condition (7.21) is equivalent to the requirement that

$$\min_{f_0} \max_i 2\mu f_0 + \lambda_i(\mu^2 + \sigma^2) f_0^2 < 0.$$

Since

$$\begin{aligned}\min_{f_0} \max_i 2\mu f_0 + \lambda_i(\mu^2 + \sigma^2) f_0^2 &= \min_{f_0} 2\mu f_0 + \lambda_N(\mu^2 + \sigma^2) f_0^2 \\ &= -\frac{\mu^2}{\lambda_N(\mu^2 + \sigma^2)} < 0,\end{aligned}$$

for any given $\mu^2, \sigma^2, \lambda_2, \dots, \lambda_N$, we can always find f_0 such that (7.21) holds simultaneously. Therefore, for single integrator systems, as long as the undirected graph is connected, mean-square consensus can be achieved.

Remark 7.4 For agent dynamics with $\text{rank}(C) > 1$, the sufficient condition might not be necessary. The following simplified model can be used to demonstrate this point. Consider the systems

$$g_{1i}(t+1) = (A + \lambda_i \gamma(t) FC) g_{1i}(t), \quad i = 2, 3$$

with $\sigma = 0$, $A = \begin{bmatrix} c_1 & 0 \\ 0 & c_2 \end{bmatrix}$, $C = \begin{bmatrix} 1 & 0 \\ 0 & 1 \end{bmatrix}$ and $0 < c_1 < c_2$. If the sufficient condition (7.9) is also necessary for mean-square stabilization, then the following optimization problem:

$$\min_F \max_i \rho(A + \lambda_i \mu FC)$$

returns an optimal value that is less than 1 if and only if

$$c_2 < \frac{\lambda_3 + \lambda_2}{\lambda_3 - \lambda_2}. \tag{7.22}$$

However, numerical evaluation shows that when choosing $c_1 = 2, c_2 = 14, \lambda_1 = 6, \lambda_2 = 7, \mu = 1$, which contradicts the condition (7.22), the optimization problem still returns an optimal value 0.6155, with the argument $F = \begin{bmatrix} -0.3022 & 1.3053 \\ -0.0039 & -2.1593 \end{bmatrix}$. Thus, the sufficient condition (7.9) is generally not necessary for the simultaneous mean-square stabilization of multiple-input systems.

7.4 Nonidentical Fading Networks

In the presence of nonidentical fading networks, the consensus error dynamics of δ is $\delta(t+1) = \left(I_N \otimes \mathcal{A} + \mathcal{L}(t) \otimes \mathcal{H}\right) \delta(t)$ with $[\mathcal{L}(t)]_{ii} = -\sum_{j \in \mathcal{N}_i} [\mathcal{L}]_{ij} \gamma_{ij}(t)$, $[\mathcal{L}(t)]_{ij} = [\mathcal{L}]_{ij} \gamma_{ij}(t)$ for $i \neq j$. Since the channel fading γ_{ij} is coupled with elements of the graph Laplacian, the analysis of the mean-square consensus is difficult. In the following, we propose to use edge Laplacian instead of graph Laplacian to model the consensus dynamics. This method allows us to separate the fading effect from the network topology by building dynamics on edges rather than on vertexes.

7.4.1 Definition of Edge Laplacian

A virtual orientation of the edge in an undirected graph is an assignment of direction to the edge (i, j) such that one vertex is chosen to be the initial node and the other to be the terminal node. The incidence matrix $E(\mathcal{G})$ for an oriented graph $\mathcal{G}$ is a $\{0, 1, -1\}$-matrix with rows and columns indexed by vertices and edges of $\mathcal{G}$, respectively, such that

$$[E(\mathcal{G})]_{ik} = \begin{cases} +1, & \text{if } i \text{ is the initial node of edge } k, \\ -1, & \text{if } i \text{ is the terminal node of edge } k, \\ 0, & \text{otherwise.} \end{cases}$$

The graph Laplacian $\mathcal{L}$ and edge Laplacian $\mathcal{L}_e$ can be constructed from the incidence matrix, respectively, as $\mathcal{L} = E(\mathcal{G})E(\mathcal{G})'$, $\mathcal{L}_e = E(\mathcal{G})'E(\mathcal{G})$ [Zelazo et al., 2007, Zelazo and Mesbahi, 2011]. In this section, the consensus problem is studied under an undirected tree topology setting, where the eigenvalues of the edge Laplacian $\mathcal{L}_e$ are the nonzero eigenvalues of the graph Laplacian $\mathcal{L}$, i.e. $\lambda_2, \lambda_3, \dots, \lambda_N$ [Dimarogonas and Johansson, 2010]. Note that for the case with general connected undirected graphs, it is sufficient to study the mean-square consensus over an arbitrary tree subgraph in the communication topology. We limit our attention to the state feedback case in this section.

Suppose agent k sends the information x_k through the fading channel to agent j, and the jth agent receives the corrupted information as $r_{jk}(t) = \gamma_{jk}(t)x_k(t) + \omega_{jk}(t)$, where γ_{jk} represents the fading effect and ω_{jk} denotes a zero-mean white communication noise with bounded variance. The controller for agent j is designed as

$$u_j(t) = K \sum_{k \in \mathcal{N}_j} \left(\gamma_{jk}(t)x_j(t) - r_{jk}(t)\right). \tag{7.23}$$

Define the state on the ith edge as $z_i = x_j - x_k$, with j, k representing the initial node and the terminal node of the ith edge, respectively. Similarly, when only mean-square consensus is considered, ω_{jk} can be neglected without loss of generality. Assume that the fading on the same edge is equal, i.e. $\gamma_{jk} = \gamma_{kj}$, which makes sense in practice [Dey et al., 2009]. Following the definition of the incidence matrix, the controller (7.23) can be alternatively represented as $u_j(t) = K \sum_{k=1}^{N-1} e_{jk}\zeta_k(t)z_k(t)$, where ζ_k denotes the fading effect on the kth edge and e_{jk} is the jkth element of $E(\mathcal{G})$. If we define $z = [z_1', z_2', \dots, z_{N-1}']'$, the closed-loop dynamics on edges can be calculated as

$$z(t+1) = \left(I_{N-1} \otimes A + \mathcal{L}_e\zeta(t) \otimes BK\right) z(t) \tag{7.24}$$

with $\zeta = \mathrm{diag}(\zeta_1, \zeta_2, \dots, \zeta_{N-1})$.

If (7.24) is mean-square stable, the mean-square consensus of the MAS (7.1) can be achieved, i.e. $\lim_{t\to\infty}\mathbb{E}\{\| x_i(t) - x_j(t)\|^2\} = 0, \forall i, j \in \mathcal{V}$. Thus in the following, we focus on studying the mean-square stability of (7.24), and the following assumption is made.

Assumption 7.3 The channel fading sequence $\{\zeta_i(t)\}_{t\geq 0}$ is i.i.d. with mean μ_i and variance σ_i^2 for all $i = 1, 2, \dots, N-1$.

7.4.2 Sufficient Consensus Conditions

Under Assumption 7.3, we can derive a necessary and sufficient condition to ensure the mean-square stability of (7.24).

Lemma 7.5 *Under Assumption 7.3, the system* (7.24) *is mean-square stable if and only if there exist K and $P > 0$, such that*

$$P > (I \otimes A + \mathcal{L}_e\Lambda \otimes BK)'P(I \otimes A + \mathcal{L}_e\Lambda \otimes BK) + (I \otimes K)'G(I \otimes K) \tag{7.25}$$

with $G = (\Sigma \otimes \mathbf{11}') \odot ((\mathcal{L}_e \otimes B)'P(\mathcal{L}_e \otimes B))$, $\Sigma = [\sigma_{ij}]_{(N-1)\times(N-1)}$, $\sigma_{ij} = \mathbb{E}\{(\zeta_i - \mu_i)(\zeta_j - \mu_j)\}$ *for* $i \neq j$, $\sigma_{ii} = \sigma_i^2$ *and* $\Lambda = \text{diag}(\mu_1, \mu_2, \dots, \mu_{N-1})$.

Proof: This result is immediate from Lemma 1 in Xiao et al. [2012] by noting that

$$\mathcal{L}_e\zeta(t) \otimes BK = (\mathcal{L}_e \otimes B)(\zeta(t) \otimes I)(I \otimes K)$$

and treating $I_{N-1} \otimes A$, $\mathcal{L}_e \otimes B$, $I \otimes K$ and $\zeta(t) \otimes I$ as the system matrix, input matrix, output matrix, and fading effects of the MIMO system studied in Xiao et al. [2012], respectively. □

However, (7.25) cannot provide any physical insights into the mean-square consensus problem. In the following, we try to obtain some analytic conditions to ensure the mean-square consensus of the MAS (7.1) under controller (7.23). Similar to Lemma 7.1, we have the following result.

Lemma 7.6 ***(Schenato et al. [2007])*** *Under Assumption 7.1 and assuming that* (A, B) *is controllable, there exists a solution* $P > 0$ *to the following modified Riccati inequality:*

$$P > A'PA - \tau A'PB(B'PB)^{-1}B'PA \tag{7.26}$$

if and only if τ *is greater than a critical value* $\tau_c \in [0, 1)$.

The consensusability result is stated in Theorem 7.3.

Theorem 7.3 *Under Assumptions 7.1 and 7.3, the multi-agent system (7.1) is mean-square consensusable by the controller (7.23) under an undirected tree topology if there exists* κ*, such that*

$$\kappa\left(\mathcal{L}_e\Lambda + \Lambda\mathcal{L}_e\right) + \kappa^2(\Lambda\mathcal{L}_e^2\Lambda + \Sigma \odot \mathcal{L}_e^2) < -\tau_c I, \tag{7.27}$$

where τ_c *is given in Lemma 7.6. Moreover, if such* κ *exists, there exists a solution* $P_0 > 0$ *to the modified Riccati inequality (7.26), with* τ *being the smallest eigenvalue of* $-\kappa\left(\mathcal{L}_e\Lambda + \Lambda\mathcal{L}_e\right) - \kappa^2(\Lambda\mathcal{L}_e^2\Lambda + \Sigma \odot \mathcal{L}_e^2)$*, and a control gain that ensures the mean-square consensus can be given by* $K = \kappa(B'P_0B)^{-1}B'P_0A$.

Proof: If (7.27) is satisfied, in view of the solvability of (7.26), one can show that there exists $P_0 > 0$ to the matrix inequality

$$\begin{aligned} I \otimes P_0 > {} & I \otimes A'P_0A \\ & + (\kappa\left(\mathcal{L}_e\Lambda + \Lambda\mathcal{L}_e\right) + \kappa^2(\Lambda\mathcal{L}_e^2\Lambda + \Sigma \odot \mathcal{L}_e^2)) \otimes A'P_0B(B'P_0B)^{-1}B'P_0A, \end{aligned}$$

which actually is (7.25) with $K = \kappa(B'P_0B)^{-1}B'P_0A$ and $P = I \otimes P_0 > 0$. In view of Lemma 7.5, the proof is completed. □

Remark 7.5 If all the channel fading is identical, i.e. $\zeta_i(t) = \zeta_0(t)$, $\forall i = 1, 2, \ldots, N-1$ and $\mathbb{E}\{\zeta_0(t)\} = \mu$, $\mathbb{E}\{(\zeta_0(t) - \mu)^2\} = \sigma^2$, (7.27) is equivalent to

$$\min_{\kappa} \max_{i} \kappa^2(\mu^2 + \sigma^2)\lambda_i^2 + 2\kappa\mu\lambda_i < -\tau_c,$$

which further implies

$$\frac{\mu^2}{\mu^2 + \sigma^2}\left[1 - \left(\frac{\lambda_N - \lambda_2}{\lambda_N + \lambda_2}\right)^2\right] > \tau_c.$$

This is consistent with Theorem 7.1.

Theorem 7.3 implies that mean-square consensusability is determined by the edge Laplacian, the fading statistics, and the agent dynamics. In the following, we will show that under specific situations, the sufficient condition (7.27) can be further simplified.

A. The case of $\Lambda = \mu I$

With the help of Theorem 5.5.1 in Horn and Johnson [1991], we can obtain a relaxed sufficient consensus condition as: there exists κ, such that $2\kappa\mu\lambda_2 + \kappa^2\lambda_N^2(\mu^2 + \rho(\Sigma)) < -\tau_c$. Since the minimum of the left-hand side of the previous inequality is achieved at $\kappa = -\frac{\mu}{\mu^2+\rho(\Sigma)}\frac{\lambda_2}{\lambda_N^2}$ with the minimal value $-\frac{\mu^2}{\mu^2+\rho(\Sigma)}\frac{\lambda_2^2}{\lambda_N^2}$, we have the following corollary.

Corollary 7.1 *Under Assumptions 7.1 and 7.3 and if* $\Lambda = \mu I$, *the MAS* (7.1) *is mean-square consensusable by the controller* (7.23) *under an undirected tree topology if*

$$\tau_1 \triangleq \frac{\mu^2}{\mu^2 + \rho(\Sigma)}\frac{\lambda_2^2}{\lambda_N^2} > \tau_c, \tag{7.28}$$

where τ_c *is given in Lemma 7.6. Moreover, if* (7.28) *holds, there exists a solution* $P_0 > 0$ *to the modified Riccati inequality* (7.26) *with* $\tau = \tau_1$, *and a control gain that ensures mean-square consensus can be given by*

$$K = -\frac{\mu}{\mu^2 + \rho(\Sigma)}\frac{\lambda_2}{\lambda_N^2}(B'P_0B)^{-1}B'P_0A.$$

Remark 7.6 If the channel fading is uncorrelated with each other, the left-hand side of (7.28) can be alternatively represented as $\lambda_2^2 / \left(\lambda_N^2 \max_i \left[1 + \frac{\sigma_i^2}{\mu^2}\right]\right)$. Since $\text{argmax}_i \left[1 + \frac{\sigma_i^2}{\mu^2}\right] = \text{argmin}_i \left[\frac{1}{2}\ln\left(1 + \frac{\mu^2}{\sigma_i^2}\right)\right]$, the condition (7.28) implies that the consensusability is constrained by the eigenratio of the graph [You and Xie, 2011] and the minimal mean-square channel capacity [Elia, 2005] among all fading channels.

B. The case of $\Lambda \neq \mu I$

If $\Lambda \neq \mu I$, it is difficult to determine the eigenvalues of $\Lambda \mathcal{L}_e + \mathcal{L}_e \Lambda$. In the following, we will show that if

$$2 \max_i \left| \mu_i - \frac{1}{2} \right| < \frac{\lambda_2}{\lambda_N}, \tag{7.29}$$

then $\Lambda \mathcal{L}_e + \mathcal{L}_e \Lambda$ is positive definite, and we can further derive a relaxed sufficient condition to ensure mean-square consensus for the scenario of $\Lambda \neq \mu I$.

Corollary 7.2 *Under Assumptions 7.1 and 7.3 and if (7.29) holds, the multi-agent system (7.1) is mean-square consensusable by the controller (7.23) under an undirected tree topology if*

$$\tau_2 \triangleq \frac{1}{\max_i [\mu_i^2] + \rho(\Sigma)} \frac{\hat{\lambda}_2^2}{4\lambda_N^2} > \tau_c, \tag{7.30}$$

where τ_c is given in Lemma 7.6, and $\hat{\lambda}_2$ is the smallest positive eigenvalue of $\Lambda \mathcal{L}_e + \mathcal{L}_e \Lambda$. Moreover, if (7.30) holds, there exists a solution $P_0 > 0$ to the modified Riccati inequality (7.26) with $\tau = \tau_2$, and a control gain that ensures mean-square consensus can be given by $K = -\frac{1}{\max_i [\mu_i^2] + \rho(\Sigma)} \frac{\hat{\lambda}_2}{2\lambda_N^2} (B' P_0 B)^{-1} B' P_0 A$.

Proof: Let $\hat{\lambda}_2 \leq \hat{\lambda}_3 \leq \cdots \leq \hat{\lambda}_N$ be the ordered eigenvalues of $\Lambda \mathcal{L}_e + \mathcal{L}_e \Lambda$. Following Exercise 2 after Corollary 6.3.4 in Horn and Johnson [1985], one can conclude that

$$\begin{aligned} |\hat{\lambda}_2 - \lambda_2| &\leq \left\| \left(\Lambda - \frac{1}{2} I \right) \mathcal{L}_e + \mathcal{L}_e \left(\Lambda - \frac{1}{2} I \right) \right\|_2 \\ &\leq 2 \left\| \Lambda - \frac{1}{2} I \right\|_2 \|\mathcal{L}_e\|_2 \\ &\leq 2 \max_i \left| \mu_i - \frac{1}{2} \right| \lambda_N. \end{aligned}$$

If (7.29) holds, then $|\hat{\lambda}_2 - \lambda_2| < \lambda_2$, which means $0 < \hat{\lambda}_2 < 2\lambda_2$. Since $\hat{\lambda}_2$ is the smallest eigenvalue of $\Lambda \mathcal{L}_e + \mathcal{L}_e \Lambda$, all the eigenvalues of $\Lambda \mathcal{L}_e + \mathcal{L}_e \Lambda$ are positive. Thus, $\Lambda \mathcal{L}_e + \mathcal{L}_e \Lambda$ is positive definite. Besides, for all $x \in \mathbb{R}^{N-1}$, we have

$$x' \Lambda \mathcal{L}_e^2 \Lambda x \leq \lambda_N^2 (\Lambda x)' (\Lambda x) = \lambda_N^2 x' \Lambda' \Lambda x \leq \lambda_N^2 \max_i [\mu_i^2] x' x.$$

Based on the positive definiteness of $\Lambda \mathcal{L}_e + \mathcal{L}_e \Lambda$ and the fact that $\Lambda \mathcal{L}_e^2 \Lambda \leq \lambda_N^2 \max_i [\mu_i^2] I$, we can obtain a sufficient condition for (7.27) as

$$\min_\kappa \; [\kappa \hat{\lambda}_2 + \kappa^2 (\max_i \; [\mu_i^2] + \rho(\Sigma)) \lambda_N^2] < -\tau_c. \tag{7.31}$$

Following a similar line of argument as in the derivation of Corollary 7.1, we can obtain (7.30) from (7.31). The proof is completed. □

Remark 7.7 For general channel fading that does not satisfy (7.29), the consensusability condition would be more complicated. However, if we adopt the controller of the form $u_j(t) = K\sum_{k\in\mathcal{N}_j}\kappa_k\left(\gamma_{jk}(t)x_j(t) - r_{jk}(t)\right)$ for each agent j, the dynamics for z would be $z(t+1) = \left(I_{N-1}\otimes A + \mathcal{L}_e\zeta(t)\mathcal{K}\otimes BK\right)z(t)$, with $\mathcal{K} = \text{diag}\left(\kappa_1,\kappa_2,\dots,\kappa_{N-1}\right)$. Then by appropriately selecting the gain matrix $\mathcal{K}$, one can equalize the first moment of the channel fading statistics, thus we can obtain a sufficient consensus condition as in the scenario of $\Lambda = \mu I$.

Remark 7.8 One can easily show the consistency among the derived results. The results derived for nonidentical fading networks always recover the results for identical fading networks, i.e. under certain situations, Corollary 7.2 implies Corollary 7.1, and Corollary 7.1 implies Theorem 7.1.

7.5 Simulations

In this section, numerical simulations are conducted to verify the derived results. The parameters for the LTI dynamics (7.1) are given by

$$A = \begin{bmatrix} 1.1830 & -0.1421 & -0.0399 \\ 0.1764 & 0.8641 & -0.0394 \\ 0.1419 & -0.1098 & 0.9689 \end{bmatrix}, \quad B = [0.2, 0.1, -0.5]',$$
$$C = [1.3, 1.4, 1.5],$$

with $\lambda(A) = \{1.0086, 1.0068, 1.0006\}$ and $\theta_c = \tau_c = 0.0314$. In the following, simulations are conducted under two cases: identical fading networks with an undirected graph, nonidentical fading networks with an undirected tree graph. In simulations, the initial system states are randomly generated from the uniform distribution on the interval $(0, 0.5)$. All the fadings are assumed to satisfy the Rayleigh distribution with probability density function $f(x;\sigma_p) = \frac{x}{\sigma_p^2}e^{-x^2/(2\sigma_p^2)}$, where $x \geq 0$ and σ_p is the parameter for the Rayleigh distribution to be specified later in each simulation. The channel additive noise is drawn from a zero-mean normal distribution with variance one. The simulation results are presented by averaging over 1000 runs.

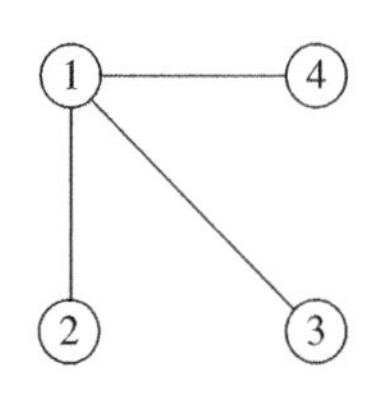

Figure 7.2 Communication topology for an undirected graph. Source: Xu et al. (2016)/ with permission of Elsevier.

Consider the consensus problem over identical fading networks with an undirected graph, where the communication topology is given in Figure 7.2, and the identical channel fading is assumed to follow Rayleigh distribution with the parameter

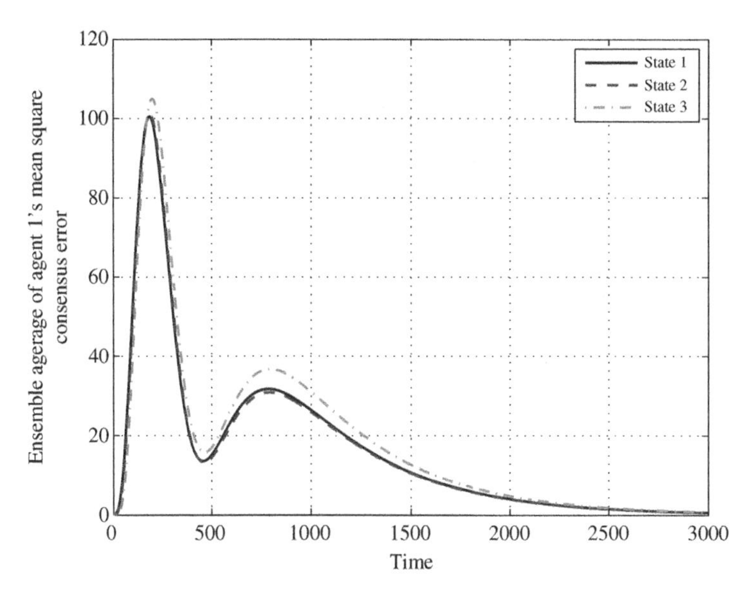

Figure 7.3 Mean-square consensus error for agent 1 under an undirected communication topology with identical fading networks. Source: Xu et al. (2016)/with permission of Elsevier.

$\sigma_p = 5$. In view of Theorem 7.1, the MAS is mean-square consensusable and one pair of control gains can be selected as

$$F = [-0.0209, 0.0014, -0.0243]', \quad K = [0.1183, -0.2153, 0.0915].$$

The mean-square consensus error for agent 1 is plotted in Figure 7.3.

For the case of consensus over nonidentical fading networks, the communication topology is assumed to be the same as in Figure 7.2, and the Rayleigh fading statistics on the communication links $1-2$, $1-3$, $1-4$ are $\sigma_{p_{12}} = 0.4980$, $\sigma_{p_{13}} = 0.4950$, $\sigma_{p_{14}} = 0.4900$, respectively. We assume that the channel fading is uncorrelated, and the sufficient condition to ensure mean-square consensus in Corollary 7.2 is satisfied. One controller gain is $K = [0.4608, -0.6829, 0.2069]$. The mean-square consensus error for agent 1 is shown in Figure 7.4.

Remark 7.9 Note that the tolerable Rayleigh fading statistics for the third simulation is smaller than the previous two simulations. This is because the fading parameters should satisfy the prerequisite (7.29), which in Corollary 7.2 is sufficient only, and is adopted to deal with the complexity caused by $\Lambda \neq \mu I$. Nevertheless, as noted in Remark 7.7, this limitation can be removed by adding more design freedom to the controller.

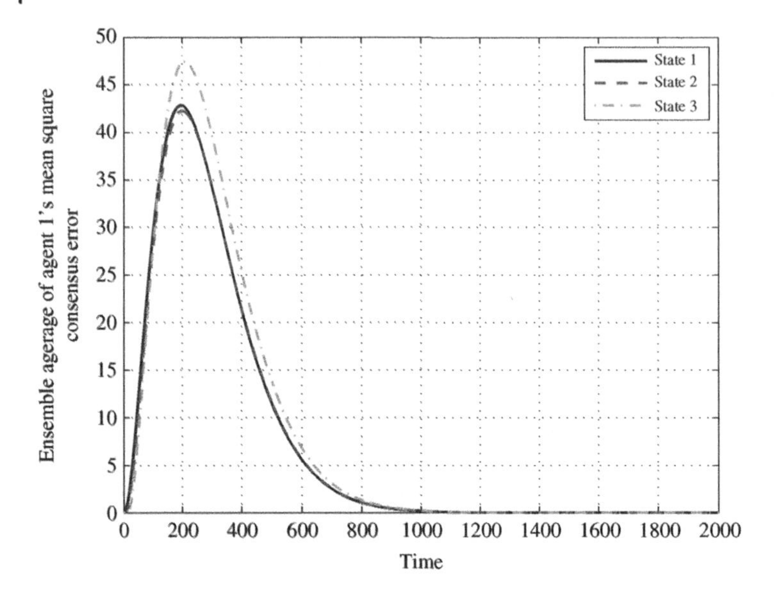

Figure 7.4 Mean-square consensus error for agent 1 under an undirected tree topology with nonidentical fading networks. Source: Xu et al. (2016)/with permission of Elsevier.

7.6 Conclusions

This chapter studies the consensusability problem of discrete-time linear MASs over undirected fading networks. It aims to decide whether there exists a distributed controller such that the underlying MAS can achieve mean-square consensus over fading channels. Conditions to ensure mean-square consensus are derived for the scenarios of undirected communication topologies with identical fading networks and undirected communications topologies with nonidentical fading networks, respectively. For scalar systems, the sufficient condition is shown to be necessary. The results indicate that the effect of fading networks on consensusability is determined by the statistics of channel fading. Finally, simulations are conducted to validate the theoretical results.

Bibliography

D. S. Bernstein. *Matrix mathematics: theory, facts, and formulas*. Princeton University Press, Princeton, NJ, 2009.

S. Dey, A. S. Leong, and J. S. Evans. Kalman filtering with faded measurements. *Automatica*, 45(10):2223–2233, 2009.

D. V. Dimarogonas and K. H. Johansson. Stability analysis for multi-agent systems using the incidence matrix: Quantized communication and formation control. *Automatica*, 46(4):695–700, 2010.

N. Elia. Remote stabilization over fading channels. *Systems & Control Letters*, 54(3):237–249, 2005.

R. A. Horn and C. R. Johnson. *Matrix analysis*. Cambridge University Press, 1985.

R. A. Horn and C. R. Johnson. *Topics in matrix analysis*. Cambridge University Press, New York, 1991.

Z. Li and J. Chen. Robust consensus of linear feedback protocols over uncertain network graphs. *IEEE Transactions on Automatic Control*, 62(8):4251–4258, 2017.

Z. Li, Z. Duan, G. Chen, and L. Huang. Consensus of multiagent systems and synchronization of complex networks: A unified viewpoint. *IEEE Transactions on Circuits and Systems I: Regular Papers*, 57(1):213–224, 2010.

Z. Li, W. Ren, X. Liu, and M. Fu. Consensus of multi-agent systems with general linear and lipschitz nonlinear dynamics using distributed adaptive protocols. *IEEE Transactions on Automatic Control*, 58(7):1786–1791, 2013a.

Z. Li, W. Ren, X. Liu, and L. Xie. Distributed consensus of linear multi-agent systems with adaptive dynamic protocols. *Automatica*, 49(7):1986–1995, 2013b.

C. Q. Ma and J. F. Zhang. Necessary and sufficient conditions for consensusability of linear multi-agent systems. *IEEE Transactions on Automatic Control*, 55(5):1263–1268, 2010.

Y. Mo and B. Sinopoli. A characterization of the critical value for Kalman filtering with intermittent observations. In *Proceedings of the 47th IEEE Conference on Decision and Control*, pages 2692–2697, 2008.

A. Papoulis and S. U. Pillai. *Probability, random variables, and stochastic processes*. McGraw-Hill, Boston, MA, 2002.

W. Ren and R. W. Beard. *Distributed consensus in multi-vehicle cooperative control: theory and applications*. Springer-Verlag London Limited, London, 2008.

L. Schenato, B. Sinopoli, M. Franceschetti, K. Poolla, and S. S. Sastry. Foundations of control and estimation over lossy networks. *Proceedings of the IEEE*, 95(1):163–187, 2007.

B. Sinopoli, L. Schenato, M. Franceschetti, K. Poolla, M. I. Jordan, and S. S. Sastry. Kalman filtering with intermittent observations. *IEEE Transactions on Automatic Control*, 49(9):1453–1464, 2004.

H. L. Trentelman, K. Takaba, and N. Monshizadeh. Robust synchronization of uncertain linear multi-agent systems. *IEEE Transactions on Automatic Control*, 58(6):1511–1523, 2013.

N. Xiao, L. Xie, and L. Qiu. Feedback stabilization of discrete-time networked systems over fading channels. *IEEE Transactions on Automatic Control*, 57(9):2176–2189, 2012.

K. You and L. Xie. Network topology and communication data rate for consensusability of discrete-time multi-agent systems. *IEEE Transactions on Automatic Control*, 56(10):2262–2275, 2011.

D. Zelazo and M. Mesbahi. Edge agreement: Graph-theoretic performance bounds and passivity analysis. *IEEE Transactions on Automatic Control*, 56(3):544–555, 2011.

D. Zelazo, A. Rahmani, and M. Mesbahi. Agreement via the edge Laplacian. In *Proceedings of the 46th IEEE Conference on Decision and Control*, pages 2309–2314, 2007.

8

Distributed Consensus over Directed Fading Networks

8.1 Introduction

In Chapter 7, we consider multi-agent systems (MASs) over fading channels with an undirected graph setting. For consensus over identical fading networks, a decomposition method is used and the mean-square consensus problem is transformed to a simultaneous mean-square stabilization problem. For consensus over nonidentical fading networks, the edge Laplacian defined for undirected graphs by Zelazo and Mesbahi [2011] is introduced to model the consensus error dynamics, and then sufficient mean-square consensus conditions are developed. However, since the graph Laplacian for directed graphs may contain complex eigenvalues and there are no well-accepted definitions of edge Laplacian for directed graphs, the method in Chapter 7 on undirected graphs for identical and nonidentical fading networks cannot be extended directly to directed graph cases. In this chapter, we distinguish bidirectional edges from nonbidirectional edges and define the compressed in-incidence matrix (CIIM), compressed incidence matrix (CIM) and compressed edge Laplacian (CEL) to study the mean-square consensus problem over fading networks with directed graphs. Sufficient conditions for the mean-square consensus are derived based on the edge state dynamics on a directed spanning tree and the role of network topology on the mean-square consensusability is discussed.

This chapter is organized as follows: the problem formulation is provided in Section 8.2. The consensus problem over identical fading networks is studied in Section 8.3. The definitions and properties of CIIM, CIM, and CEL are discussed in Section 8.4. The consensus problem over nonidentical fading networks is further studied in Section 8.5. Simulations are provided in Section 8.6 followed by some concluding remarks in Section 8.7.

Control over Communication Networks: Modeling, Analysis, and Design of Networked Control Systems and Multi-Agent Systems over Imperfect Communication Channels, First Edition.
Jianying Zheng, Liang Xu, Qinglei Hu, and Lihua Xie.

8.2 Problem Formulation

Consider an MAS with N agents. A directed graph $\mathcal{G} = (\mathcal{V}, \mathcal{E})$ is used to characterize the interaction among agents, as in Section 1.4.1. The discrete-time dynamics of agent i is given by

$$x_i(t+1) = Ax_i(t) + Bu_i(t), \quad i = 1,2,\ldots,N, \tag{8.1}$$

where $x_i \in \mathbb{R}^n$ and $u_i \in \mathbb{R}^m$ represent the agent state and control input, respectively.

Agents communicate through fading channels. Specifically, if $(j,i) \in \mathcal{E}$, we let agent j send its state to agent i at every sampling time. The agent i then receives the corrupted information $r_{ij}(t)$ as

$$r_{ij}(t) = \gamma_{ij}(t)x_j(t) + \omega_{ij}(t)$$

with γ_{ij} modeling the channel fading and ω_{ij} representing the zero-mean additive communication noise with bounded variance. Based on the received information and its own state, agent i generates the control input with the consensus protocol below

$$u_i(t) = K\sum_{j\in\mathcal{N}_i}(\gamma_{ij}(t)x_i(t) - r_{ij}(t)), \tag{8.2}$$

where K is the consensus parameter to be designed.

In this chapter, we are interested in the consensusability problem, i.e. we aim to establish conditions on the fading statistics, the agent dynamics, and the communication topology under which there exists K in the protocol (8.2) such that the MAS (8.1) can achieve mean-square consensus, i.e. $\lim_{t\to\infty}\mathbb{E}\{\|x_i(t) - x_j(t)\|_2^2\} < c$ for some positive constant c and any i,j in $\mathcal{V}$. In view of results in Ren and Beard [2008] and You and Xie [2011], the following assumption is made without loss of generality.

Assumption 8.1

1. (A, B) is controllable and all the eigenvalues of A are either on or outside the unit disk.
2. The directed graph $\mathcal{G}$ contains a directed spanning tree.

8.3 Identical Fading Networks

In this section, we consider the scenario where the channel fading on different edges is identical.

Assumption 8.2 The channel fading on different edges is identical, i.e. $\gamma_{ij}(t) = \gamma(t)$ for all $t \geq 0$ with $(j,i) \in \mathcal{E}$, and the sequence $\{\gamma(t)\}_{t\geq 0}$ is i.i.d. with mean μ and variance σ^2.

In view of the analysis in Chapter 7, when only mean-square consensus is considered, the additive noise ω_{ij} can be ignored without loss of generality. Throughout this chapter, if the state of a stochastic dynamical system converges to zero in mean-square sense, we say that the dynamical system is mean-square stable.

8.3.1 Consensus Error Dynamics

Under Assumption 8.2 and the consensus protocol (8.2), the agent dynamics is $x_i(t+1) = Ax_i(t) + \gamma(t)BK\sum_{j\in\mathcal{N}_i}(x_i(t) - x_j(t))$, where the additive noise has been ignored. Let $X = [x_1', x_2', \dots, x_N']'$, then following the definitions of the graph Laplacian, we have $X(t+1) = (I \otimes A + \gamma(t)\mathcal{L} \otimes BK)X(t)$, where $\mathcal{L}$ is the Laplacian matrix of the communication graph $\mathscr{G}$. Let h' be the left eigenvector of $\mathcal{L}$ associated with the zero eigenvalue, satisfying $h'\mathbf{1} = 1$, and define the consensus error as $\delta(t) = [I - (\mathbf{1}h') \otimes I]X(t)$. The consensus error evolves as

$$\begin{aligned}\delta(t+1) &= [I - (\mathbf{1}h') \otimes I]X(t+1) \\ &= [I - (\mathbf{1}h') \otimes I](I \otimes A + \gamma(t)\mathcal{L} \otimes BK)X(t) \\ &= (I \otimes A + \gamma(t)\mathcal{L} \otimes BK)X(t) - [(\mathbf{1}h') \otimes A]X(t) \\ &= (I \otimes A + \gamma(t)\mathcal{L} \otimes BK)\delta(t). \end{aligned} \tag{8.3}$$

If there exists K, such that (8.3) is mean-square stable, i.e. $\lim_{t\to\infty}\mathbb{E}\{\delta(t)\delta'(t)\} = 0$, mean-square consensus of the MAS (8.1) can be achieved. Since $\mathcal{G}$ contains a directed spanning tree, in view of Lemma 1.1, the graph Laplacian has the Jordan decomposition $U^{-1}\mathcal{L}U = \begin{bmatrix} 0 & 0 \\ 0 & \triangle \end{bmatrix}$ with $U^{-1} = \begin{bmatrix} h' \\ G \end{bmatrix}$, $U = \begin{bmatrix}\mathbf{1}, Y\end{bmatrix}$ for some matrices G, Y and all the diagonal elements of $\triangle$ are the nonzero eigenvalues of $\mathcal{L}$. Define the coordinate transformation

$$g(t) = [g_1'(t), g_2'(t), \dots, g_N'(t)]' = (U^{-1} \otimes I)\delta(t),$$

then it holds that

$$\begin{aligned} g_1(t) &\equiv 0, \\ [g_2'(t+1), \dots, g_N'(t+1)]' &= (I_{N-1} \otimes A + \gamma(t)\triangle \otimes BK)[g_2'(t), \dots, g_N'(t)]'. \end{aligned} \tag{8.4}$$

Let $\lambda_2, \dots, \lambda_N$ be the nonzero eigenvalues of $\mathcal{L}$ arranged as $|\lambda_2| \leq \cdots \leq |\lambda_N|$. Then we have the following result.

Lemma 8.1 *If the following dynamics*

$$g_i(t+1) = (A + \gamma(t)\lambda_i BK)g_i(t), \quad i = 2, \dots, N, \tag{8.5}$$

are simultaneously mean-square stable, i.e. $\lim_{t\to\infty} \mathbb{E}\{g_i(t)g_i^*(t)\} = 0$ *for all* $i = 2, \dots, N$, *then* $\lim_{t\to\infty} \mathbb{E}\{\delta(t)\delta'(t)\} = 0$.

Proof: In view of the above analysis, we only need to prove that if (8.5) are simultaneously mean-square stable, then (8.4) is mean-square stable.

Since $\triangle$ is in the form of $\begin{bmatrix} \triangle_1 & & & \\ & \triangle_2 & & \\ & & \ddots & \\ & & & \triangle_{N'} \end{bmatrix}$, where $\triangle_i = \begin{bmatrix} \lambda_{i'} & 1 & & \\ & \lambda_{i'+1} & 1 & \\ & & \ddots & \ddots & 0 \\ & & & & \lambda_{i'+j'} \end{bmatrix}$, $i \in \{1, \dots, N'\}$, is a $j' \times j'$ Jordan block of $\mathcal{L}$ with $\lambda_{i'} = \cdots = \lambda_{i'+j'}$. Evidently, when the algebraic multiplicity of any eigenvalue of $\mathcal{L}$ equals to its geometric multiplicity, all Jordan blocks degenerate into scalars and $\triangle$ becomes a diagonal matrix, say $\triangle = \begin{bmatrix} \lambda_2 & & & \\ & \lambda_3 & & \\ & & \ddots & \\ & & & \lambda_N \end{bmatrix}$, then systems (8.4) and (8.5) are the same.

Now consider the case when some eigenvalue's algebraic multiplicity is less than its geometric multiplicity. Without loss of generality, we only consider the case that $\triangle = \begin{bmatrix} \lambda_2 & 1 \\ 0 & \lambda_3 \end{bmatrix}$, where $\lambda_2 = \lambda_3$. Induction can then be used to prove the result for high dimensional systems. If $g_i(t+1) = (A + \lambda_i\gamma(t)BK)g_i(t)$, $i = 2,3$, are mean-square stable, then in view of Lemma 1 in Xiao et al. [2012], there exist $P_2 > 0$ and $P_3 > 0$, such that

$$P_2 > (A + \lambda_2\mu BK)^* P_2 (A + \lambda_2\mu BK) + \lambda_2^*\lambda_2\sigma^2 K'B'P_2BK,$$
$$P_3 > (A + \lambda_3\mu BK)^* P_3 (A + \lambda_3\mu BK) + \lambda_3^*\lambda_3\sigma^2 K'B'P_3BK.$$

Thus there exists $\beta > 0$ such that

$$P_3 > (A + \lambda_3\mu BK)^* P_3 (A + \lambda_3\mu BK) + \lambda_3^*\lambda_3\sigma^2 K'B'P_3BK + \beta I.$$

Select α to be sufficiently large such that

$$\alpha\beta I > (\mu^2 + \sigma^2)K'B'P_2BK + S^*(P - \mathcal{M})S,$$

where

$$S = (A + \lambda_2\mu BK)^* P_2 \mu BK + \lambda_2\sigma^2 K'B'P_2BK,$$
$$\mathcal{M} = (A + \lambda_2\mu BK)^* P_2 (A + \lambda_2\mu BK) + \lambda_2^*\lambda_2\sigma^2 K'B'P_2BK.$$

Let $P = \begin{bmatrix} P_2 & 0 \\ 0 & \alpha P_3 \end{bmatrix}$. In view of Schur complement lemma, we can show that

$$P > (I \otimes A + \mu \triangle \otimes BK)^* P (I \otimes A + \mu \triangle \otimes BK) + \sigma^2 (\triangle \otimes BK)^* P (\triangle \otimes BK).$$

Therefore, in view of Lemma 1 in Xiao et al. [2012],

$$[g_2(t+1)', g_3(t+1)']' = (I \otimes A + \gamma(t) \triangle \otimes BK)[g_2(t)', g_3(t)']'$$

is mean-square stable. The proof is completed. □

Therefore in the sequel, we shall focus on studying the simultaneous mean-square stabilizability of (8.5).

8.3.2 Consensusability Results

Theorem 8.1 *Under Assumptions 8.1 and 8.2, the MAS (8.1) is mean-square consensusable by the protocol (8.2) under a directed communication topology, if the following condition is satisfied*

$$\tau_1 := \frac{\mu^2}{\mu^2 + \sigma^2} \left[1 - \min_{k \in \mathbb{R}} \max_{i \in \{2,\ldots,N\}} |k\lambda_i + 1|^2 \right] > \tau_c, \tag{8.6}$$

where τ_c is defined in Lemma 7.6. Moreover, if (8.6) holds, there exists a solution $P_0 > 0$ to (7.26) with $\tau = \tau_1$, and a control parameter that ensures the mean-square consensus can be given by $K = \frac{\mu k_1}{\mu^2+\sigma^2}(B'P_0B)^{-1}B'P_0A$ with $k_1 = \arg\min_{k\in\mathbb{R}} \max_i |k\lambda_i + 1|^2$.

Proof: If (8.6) holds, then

$$\begin{aligned} -\tau_1 &= \max_i \frac{\mu^2}{\mu^2 + \sigma^2} \left(\left| \frac{\mu^2 + \sigma^2}{\mu} \eta \lambda_i + 1 \right|^2 - 1 \right) \\ &= \max_i \eta^2 (\mu^2 + \sigma^2) \lambda_i^* \lambda_i + 2\eta\mu \mathrm{Re}(\lambda_i) < -\tau_c \end{aligned}$$

with $\eta = \frac{\mu k_1}{\mu^2+\sigma^2}$. In view of Lemma 7.6, there exists a $P_0 > 0$ such that

$$\begin{aligned} P_0 &> A'P_0A - \tau_1 A'P_0B(B'P_0B)^{-1}B'P_0A \\ &> A'P_0A + (\eta^2(\mu^2 + \sigma^2)\lambda_i^*\lambda_i + 2\eta\mu\mathrm{Re}(\lambda_i))A'P_0B(B'P_0B)^{-1}B'P_0A \end{aligned}$$

for all $i = 2, \ldots, N$, which also implies the existence of $P_0 > 0$ and $K = \eta(B'P_0B)^{-1}B'P_0A$ such that

$$P_0 > (A + \lambda_i \mu BK)^* P_0 (A + \lambda_i \mu BK) + \lambda_i^* \lambda_i \sigma^2 K'B'P_0BK$$

for all $i = 2, \dots, N$. Thus from Lemma 1 in Xiao et al. [2012], we know that (8.5) is mean-square stable for all $i = 2, \dots, N$. This further implies that the mean-square consensus is achieved. The proof is completed. □

Remark 8.1 Suppose all the agents are with single input, i.e. $m = 1$, then from Schenato et al. [2007], $\tau_c = 1 - \frac{1}{\prod_i |\lambda_i(A)|^2}$ with $\lambda_i(A)$ being the unstable eigenvalue of A. In the following, we will show the consistency between the mean-square consensus condition (8.6) and some existing results.

1. *Networked control over fading channels*: For a single agent, the mean-square consensus problem simplifies to the mean-square stabilization problem and (8.6) implies $\prod_i |\lambda_i(A)|^2 < \frac{\mu^2}{\sigma^2} + 1$, which recovers the necessary and sufficient stabilizability condition for networked control systems over fading channels in Elia [2005].
2. *Consensus with perfect communication channels*: If the communication channel is perfect, i.e. $\sigma^2 = 0$, $\mu = 1$, then (8.6) degenerates to

$$\min_{k \in \mathbb{R}} \max_i |k\lambda_i + 1| < \frac{1}{\prod_i |\lambda_i(A)|},$$

which is the necessary and sufficient consensus condition for MASs over directed graphs [You and Xie, 2011].
3. *Consensus over identical fading networks with undirected graphs*: If the graph is undirected, then $0 < \lambda_2 \leq \cdots \leq \lambda_N$ [Ren and Beard, 2008]. Therefore

$$\min_{k \in \mathbb{R}} \max_i |k\lambda_i + 1| = \frac{\lambda_N - \lambda_2}{\lambda_N + \lambda_2}.$$

Thus a sufficient condition to ensure mean-square consensus over identical fading networks with undirected graphs from (8.6) is $\frac{\mu^2}{\mu^2+\sigma^2}\left[1 - \left(\frac{\lambda_N-\lambda_2}{\lambda_N+\lambda_2}\right)^2\right] < 1 - \frac{1}{\prod_i |\lambda_i(A)|^2}$, which has been proved in Chapter 7 and shown to be necessary when all agents are with scalar dynamics.

In the following, we prove that the sufficient condition in Theorem 8.1 is also necessary when all agents are with scalar dynamics and the graph is directed, i.e. $n = m = 1$. Without loss of generality, let $A = a_0$, $B = 1$. and $K = k_0$.

Theorem 8.2 *Under Assumptions 8.1 and 8.2 and if $n = m = 1$, the MAS (8.1) is mean-square consensusable by the protocol (8.2) under a directed communication topology, if and only if (8.6) is satisfied with $\tau_c = 1 - \frac{1}{a_0^2}$.*

Proof: The sufficiency follows from Theorem 8.1. Only the necessity is proved here. In view of the previous analysis, for scalar agent dynamics, the MAS (8.1) is mean-square consensusable if and only if $g_i(t+1) = (a_0 + \gamma(t)\lambda_i k_0)g_i(t)$ is simultaneously mean-square stabilizable for all $i = 2, \ldots, N$, which also implies that there exists $k_0 \in \mathbb{R}$, such that

$$\mathbb{E}\{|a_0 + \gamma(t)\lambda_i k_0|^2\} = |\lambda_i|^2(\mu^2 + \sigma^2)k_0^2 + 2\text{Re}(\lambda_i)\mu a_0 k_0 + a_0^2 < 1$$

for all $i = 2, \ldots, N$ or equivalently

$$\min_{k_0} \max_i |\lambda_i|^2(\mu^2 + \sigma^2)k_0^2 + 2\text{Re}(\lambda_i)\mu a_0 k_0 + a_0^2 < 1,$$

which actually is (8.6) with $\tau_c = 1 - \frac{1}{a_0^2}$ and $k = \frac{k_0(\mu^2+\sigma^2)}{\mu a_0}$. The proof is completed. □

The sufficient consensus condition (8.6) involves solving a minimax optimization problem, which cannot be explicitly derived for general directed graphs. In the following, we propose to use the Lyapunov method to derive an explicitly sufficient consensus condition for balanced directed graphs, which is directly expressed in terms of the eigenvalues of the Laplacian matrix and avoids to solve an optimization problem.

8.3.3 Balanced Directed Graph Cases

The consensusability result for the MAS (8.1) under a balanced directed graph is stated in Theorem 8.3.

Theorem 8.3 *Under Assumptions 8.1 and 8.2, the MAS (8.1) is mean-square consensusable by the protocol (8.2) under a directed communication topology, if the directed graph is balanced and*

$$\tau_2 \triangleq \frac{\mu^2}{\mu^2 + \sigma^2} \times \frac{\tilde{\lambda}_2^2}{\eta} > \tau_c, \tag{8.7}$$

where τ_c is given in Lemma 7.6, $\eta = \rho(\mathcal{L}'\mathcal{L})$ and $\tilde{\lambda}_2$ denotes the smallest positive eigenvalue of $\mathcal{L}_s = (\mathcal{L} + \mathcal{L}')/2$. Moreover, if (8.7) holds, there exists a solution $P_0 > 0$ to the modified Riccati inequality (7.26) with $\tau = \tau_2$, and a control gain that ensures mean-square consensus is given by

$$K = -\frac{\mu}{\mu^2 + \sigma^2}\frac{\tilde{\lambda}_2}{\eta}(B'P_0B)^{-1}B'P_0A.$$

Proof: Lyapunov methods will be used to show the mean-square stability of (8.3) and thus to prove the sufficiency. Define the Lyapunov function candidate $V(t) = \mathbb{E}\{\delta(t)'(I_N \otimes P)\delta(t)\}$, where $P > 0$. We can choose K as $K = -\kappa(B'PB)^{-1}B'PA$ with $\kappa > 0$. Then

$$A'PBK = K'B'PA = -\kappa A'PB(B'PB)^{-1}B'PA,$$

which implies

$$\begin{aligned} V(t+1) &= \mathbb{E}\{\delta(t+1)'(I_N \otimes P)\delta(t+1)\} \\ &\leq \mathbb{E}\{\delta(t)'(I_N \otimes A'PA + 2\mu\mathcal{L}_s \otimes K'B'PA + \eta(\mu^2+\sigma^2)I_N \otimes K'B'PBK)\delta(t)\}. \end{aligned} \tag{8.8}$$

Since the balanced directed graph $\mathcal{G}$ contains a directed spanning tree, $\mathcal{L}_s$ is a valid graph Laplacian matrix for a connected undirected graph [You et al., 2013]. Thus, we can select $\tilde{\phi}_i \in \mathbb{R}^N$ such that $\mathcal{L}_s\tilde{\phi}_i = \tilde{\lambda}_i\tilde{\phi}_i$, with $0 = \tilde{\lambda}_1 < \tilde{\lambda}_2 \leq \cdots \leq \tilde{\lambda}_N$ and form the unitary matrix $\tilde{\Theta} = [\mathbf{1}/\sqrt{N}, \tilde{\phi}_2, \tilde{\phi}_3, \ldots, \tilde{\phi}_N]$, with

$$\mathrm{diag}(\tilde{\lambda}_1, \tilde{\lambda}_2, \ldots, \tilde{\lambda}_N) = \tilde{\Theta}'\mathcal{L}_s\tilde{\Theta}.$$

Introduce the state transformation $\tilde{f} = (\tilde{\Theta}' \otimes I_N)\delta$ with

$$\tilde{f} = [\tilde{f}_1', \tilde{f}_2', \ldots, \tilde{f}_N']',$$

then $\tilde{f}_1 \equiv 0$, and (8.8) becomes

$$V(t+1) \leq \sum_{i=2}^{N} \mathbb{E}\{\tilde{f}_i(t)'(A'PA + 2\mu\tilde{\lambda}_iK'B'PA + \eta(\mu^2+\sigma^2)K'B'PBK)\tilde{f}_i(t)\}. \tag{8.9}$$

Let $\alpha_i = 2\mu\tilde{\lambda}_i\kappa - \eta(\mu^2+\sigma^2)\kappa^2$, then

$$\begin{aligned} Q_i &= A'PA + 2\mu\tilde{\lambda}_iK'B'PA + \eta(\mu^2+\sigma^2)K'B'PBK \\ &= A'PA - \alpha_iA'PB(B'PB)^{-1}B'PA. \end{aligned}$$

If (8.7) is satisfied, there exists $\kappa = \frac{\mu}{\mu^2+\sigma^2}\frac{\tilde{\lambda}_2}{\eta}$, such that $\alpha_2 > \tau_c$. Further, since (A, B) is controllable, in view of Lemma 7.6, there exist $P > 0$ and a sufficiently small $\zeta > 0$ such that $(1-\zeta)P - Q_2 > 0$. Since $\alpha_i \geq \alpha_2$, $(1-\zeta)P - Q_i > 0$ for all $i = 2,3,\ldots,N$.

Thus, there exists $P > 0$, with

$$(1-\zeta)P > A'PA + 2\mu\tilde{\lambda}_iK'B'PA + \eta(\mu^2+\sigma^2)K'B'PBK$$

for all $i = 2,3,\ldots,N$. Further from (8.9), one can obtain that

$$\begin{aligned} V(t+1) &\leq (1-\zeta)\sum_{i=1}^{N}\mathbb{E}\{\tilde{f}_i(t)'P\tilde{f}_i(t)\} \\ &= (1-\zeta)\mathbb{E}\{\delta(t)'(I_N \otimes P)\delta(t)\} = (1-\zeta)V(t). \end{aligned}$$

Thus, $V(t)$ converges to zero exponentially, and this completes the proof. □

When the fading network is nonidentical, if we still use the graph Laplacian to model the consensus error dynamics, the channel fading would be coupled with elements of the graph Laplacian. As a result, it is difficult to analyze the consensusability condition. In the following section, we propose CIIM $\bar{E}$, CIM E, and CEL $\mathcal{L}_e$ and analyze their properties. Subsequently, it will be shown that with such definitions, we can remodel the consensus error dynamics and linearly separate the channel fading from the network topology.

8.4 Definitions and Properties of CIIM, CIM, and CEL

8.4.1 Definitions of CIIM, CIM, and CEL

If two agents i and j can communicate with each other, i.e. $(i,j) \in \mathcal{E}$ and $(j,i) \in \mathcal{E}$, we call the link between them a bidirectional edge. Otherwise, we call the edge between them (if exists) a directed edge. The total number of edges in the graph is represented by $\mathcal{F}$, where a bidirectional edge is only counted once. Thus, $\mathcal{F} \leq |\mathcal{E}|$ and $\mathcal{F} = |\mathcal{E}|$ if and only if there are no bidirectional edges in $\mathcal{G}$. First, by arbitrarily applying an orientation to every bidirectional edge in $\mathcal{G}$, the CIIM and CIM are defined as follows.

Definition 8.1 The CIIM $\bar{E}$ and CIM E are $N \times \mathcal{F}$ matrices with rows and columns indexed by nodes and edges of $\mathcal{G}$, respectively, such that

- If the edge e_p connecting two nodes i, j is bidirectional and the orientated edge is with initial node j and terminal node i, then
  (a) $[\bar{E}]_{lp} = 1$ for $l = j$, $[\bar{E}]_{lp} = -1$ for $l = i$, and $[\bar{E}]_{lp} = 0$ otherwise.
  (b) $[E]_{lp} = 1$ for $l = j$, $[E]_{lp} = -1$ for $l = i$, and $[E]_{lp} = 0$ otherwise.
- If the edge e_p is a directed edge and is with initial node j and terminal node i, then
  (a) $[\bar{E}]_{lp} = -1$ for $l = i$ and $[\bar{E}]_{lp} = 0$ otherwise.
  (b) $[E]_{lp} = 1$ for $l = j$, $[E]_{lp} = -1$ for $l = i$, and $[E]_{lp} = 0$ otherwise.

With the defined CIIM and CIM, CEL is defined as follows.

Definition 8.2 The CEL of $\mathcal{G}$ is defined as

$$\mathcal{L}_e = E'\bar{E}.$$

Remark 8.2 Different from definitions of in-incidence matrix (IIM), incidence matrix (IM) and directed edge Laplacian (DEL) for directed graphs in Zeng et al. [2016a,b), the CIIM, CIM, and CEL defined in this chapter treat a bidirectional edge only as one virtually oriented edge, rather than two directed edges with opposite directions. With such consideration, the dimension of the CEL is no

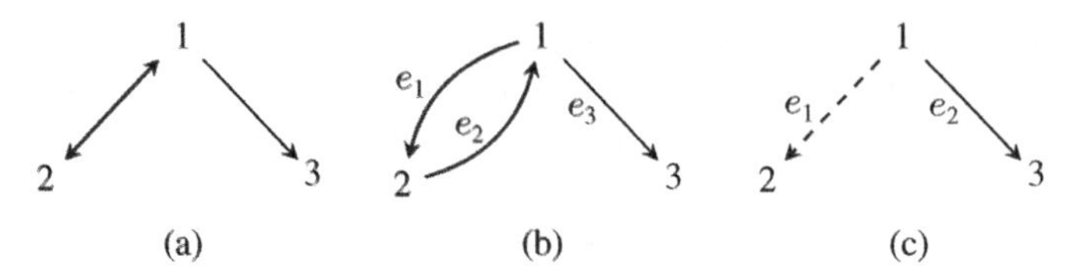

Figure 8.1 (a) A directed graph with a bidirectional edge. (b) Treat the bidirectional edges as two edges with opposite directions. (c) Apply an orientation and treat the bidirectional edge as one virtually oriented edge. Source: Xu et al. (2018)/with permission of Elsevier.

larger than that of the DEL, which would make the analysis and design of MASs simpler especially when numbers of agents and bidirectional edges are large. Moreover, CEL can degenerate to the edge Laplacian for undirected graphs in Zelazo and Mesbahi [2011], which is not possible for the DEL. Thus, the consistency of results for undirected graphs derived with CEL and undirected edge Laplacian [Zelazo and Mesbahi, 2011] can be guaranteed.

Take the directed graph in Figure 8.1a as an example. Follow the definitions in Zeng et al. [2016b], the IIM E_{IIM}, and IM E_{IM} are 3×3 matrices with rows and columns indexed by the node set $\{1,2,3\}$ and the edge set $\{e_1, e_2, e_3\}$ as illustrated in Figure 8.1b and the DEL is given by $\mathcal{L}_{\text{DEL}} = E'_{\text{IM}} E_{\text{IIM}}$. Nevertheless, the CIIM $\bar{E}$, CIM E are 3×2 matrices with rows and columns indexed by the node set $\{1,2,3\}$ and the edge set $\{e_1, e_2\}$ as illustrated in Figure 8.1c, where a dashed line is used to represent a bidirectional edge with an arbitrarily chosen direction. The expressions of these matrices are listed below.

$$E_{\text{IIM}} = \begin{bmatrix} 0 & -1 & 0 \\ -1 & 0 & 0 \\ 0 & 0 & -1 \end{bmatrix}, \bar{E} = \begin{bmatrix} 1 & 0 \\ -1 & 0 \\ 0 & -1 \end{bmatrix}, \quad \mathcal{L}_{\text{DEL}} = \begin{bmatrix} 1 & -1 & 0 \\ -1 & 1 & 0 \\ 0 & -1 & 1 \end{bmatrix},$$

$$E_{\text{IM}} = \begin{bmatrix} 1 & -1 & 1 \\ -1 & 1 & 0 \\ 0 & 0 & -1 \end{bmatrix}, \quad E = \begin{bmatrix} 1 & 1 \\ -1 & 0 \\ 0 & -1 \end{bmatrix}, \quad \mathcal{L}_e = \begin{bmatrix} 2 & 0 \\ 1 & 1 \end{bmatrix}.$$

It is immediate from the above that the dimension of $\mathcal{L}_e$ is smaller than that of $\mathcal{L}_{\text{DEL}}$. In the following, we will analyze the properties of the CIIM, CIM, and CEL and show that some desired properties are still preserved.

8.4.2 Properties of CIIM, CIM, and CEL

We have the following result about the rank of the CIM. The proof is similar to that of Theorem 8.3.1 in Godsil and Royle [2001] and is omitted here.

Proposition 8.1 When the directed graph contains a directed spanning tree, $\text{rank}(E) = N - 1$.

The graph Laplacian $\mathcal{L}$ for $\mathcal{G}$ can be reconstructed from the CIIM and CIM as follows:

Proposition 8.2 The graph Laplacian $\mathcal{L}$ has the following expression:

$$\mathcal{L} = \bar{E}E'.$$

Proof: First, consider the off-diagonal elements of $\bar{E}E'$. Suppose $i \neq j$ and there is a directed edge l connecting the node i and node j, with j being the initial node and i the terminal node of edge l. Then the lth element of $[\bar{E}]_{\text{row}i}$ is -1. The other elements of $[\bar{E}]_{\text{row}i}$ can either be 1 (i as an initial node of an oriented bidirectional edge), -1 (i as an terminal node of an edge[1]), or 0 (otherwise). Similarly, the lth element of $[E]_{\text{row}j}$ is 1. The other elements of $[E]_{\text{row}j}$ can either be 1 (j as an initial node of an edge), -1 (j as an terminal node of an edge), or 0 (otherwise). Since $[\bar{E}E']_{ij} = \sum_{p=1}^{\mathcal{F}} [\bar{E}]_{ip}[E]_{jp}$ and $[\bar{E}]_{il}[E]_{jl} = -1$. In the following, we will show that for $p \neq l$, $[\bar{E}]_{ip}[E]_{jp} = 0$. Suppose, for $p \neq l$, $[\bar{E}]_{ip}[E]_{jp} \neq 0$, then the pair $([\bar{E}]_{ip}, [E]_{jp})$ can only be of four possibilities: $[\bar{E}]_{ip} = 1$, $[E]_{jp} = 1$; $[\bar{E}]_{ip} = 1$, $[E]_{jp} = -1$; $[\bar{E}]_{ip} = -1$, $[E]_{jp} = 1$; and $[\bar{E}]_{ip} = -1$, $[E]_{jp} = -1$. The first scenario $[\bar{E}]_{ip} = 1$, $[E]_{jp} = 1$ and the fourth scenario $[\bar{E}]_{ip} = -1$, $[E]_{jp} = -1$ are not possible, since any edge p can only have one initial or terminal node. The second scenario $[\bar{E}]_{ip} = 1$, $[E]_{jp} = -1$ is also not possible since there is only a directed edge l from node j to node i. There are no other edges connecting the two nodes i and j. The third scenario $[\bar{E}]_{ip} = -1$, $[E]_{jp} = 1$ is possible only for $p = l$, which violates the assumption that $p \neq l$. Thus, when there is a directed edge from node j to node i, $[\bar{E}E]_{ij} = -1$.

Suppose there is a bidirectional edge l connecting node i and node j and a virtual orientation is assigned to this bidirectional edge. Without loss of generality, let j be assigned as the initial node and i as the terminal node. Similar to the analysis for directed edges, we can show that $[\bar{E}E']_{ij} = -1$. Now, consider the term $[\bar{E}E']_{ji}$. Since the edge l is bidirectional, the lth elements of $[\bar{E}]_{\text{row}j}$ and $[E]_{\text{row}i}$ are 1 and -1, respectively. Thus $[\bar{E}]_{jl}[E]_{il} = -1$. Using similar arguments for directed edges, we can prove that for $p \neq l$, $[\bar{E}]_{jp}[E]_{ip} = 0$. Thus, $[\bar{E}E']_{ji} = -1$. Therefore, if two nodes i and j are connected via a bidirectional edge, $[\bar{E}E']_{ij} = [\bar{E}E']_{ji} = -1$.

Similarly, when there are no edges connecting node i and node j, $[\bar{E}]_{ip}[E]_{jp} = 0$ for any edge p. Thus, $[\bar{E}E']_{ij} = 0$. Consequently, from the definition of graph Laplacian, we have $[\mathcal{L}]_{ij} = [\bar{E}E']_{ij}$ for $i \neq j$.

Now, consider the diagonal element of $\bar{E}E'$. Since $[\bar{E}E']_{ii} = \sum_{p=1}^{\mathcal{F}} [\bar{E}]_{ip}[E]_{ip}$, and $[\bar{E}]_{ip}[E]_{ip}$ can only be 1 or 0 in view of the definition of CIIM and CIM. There are two situations that may result in $[\bar{E}]_{ip}[E]_{ip} = 1$: $[\bar{E}]_{ip} = 1$, $E_{ip} = 1$ (i as the initial node of an oriented bidirectional edge), $[\bar{E}]_{ip} = -1$, $E_{ip} = -1$ (i as the terminal node

---

1 Without specifications, an edge means either a directed edge or an oriented bidirectional edge.

of an edge). Thus, the value of $[\bar{E}E']_{ii}$ equals the sum of the number of bidirectional edges that is connected to node i and the number of directed edges in which i serves as a terminal node. Thus, from the definition of the graph Laplacian, $[\bar{E}E']_{ii} = [\mathcal{L}]_{ii}$. Based on the above analysis, we have $\mathcal{L} = \bar{E}E'$. The proof is completed. □

In view of Definition 8.2 and Proposition 8.2, we further have the following result about the eigenvalue distribution of CEL.

Proposition 8.3 The CEL $\mathcal{L}_e$ and the graph Laplacian $\mathcal{L}$ share the same nonzero eigenvalues. If $\mathcal{G}$ contains a directed spanning tree, then $\mathcal{L}_e$ contains exactly $N-1$ nonzero eigenvalues which are all in the open right-half plane and zero, if exists, is a semisimple eigenvalue.[2]

Proof: Suppose λ is a nonzero eigenvalue of $\mathcal{L}$ with the associated nonzero right eigenvector q. In view of Proposition 8.2, we have $\bar{E}E'q = \lambda q$. Since $\lambda q \neq 0$, $\overline{q} = E'q \neq 0$. Left multiply $\bar{E}E'q = \lambda q$ with E', we can obtain that $E'\bar{E}E'q = \lambda E'q$, which implies $\mathcal{L}_e\overline{q} = \lambda\overline{q}$. Thus, λ is also a nonzero eigenvalue of $\mathcal{L}_e$. Similarly, we can prove that any nonzero eigenvalue of $\mathcal{L}_e$ is also a nonzero eigenvalue of $\mathcal{L}$. Thus, the graph Laplacian $\mathcal{L}$ and the CEL $\mathcal{L}_e$ share the same nonzero eigenvalues.

If the directed graph contains a directed spanning tree, from Lemma 1.1, we can further draw the conclusion that the CEL contains exactly $N-1$ nonzero eigenvalues, which are all in the open right-half plane. Thus, $\text{rank}(\mathcal{L}_e) \geq N-1$. Since $\mathcal{L}_e = E'\bar{E}$, we have $\text{rank}(\mathcal{L}_e) \leq \text{rank}(E') = N-1$ from Proposition 8.1. Thus, $\text{rank}(\mathcal{L}_e) = N-1$. In view of the rank-nullity theorem, we have that $\text{null}(\mathcal{L}_e) = \mathcal{F} - N + 1$. Thus, the geometric multiplicity of the zero eigenvalue of $\mathcal{L}_e$ is $\mathcal{F} - N + 1$. Since the algebraic multiplicity of the zero eigenvalue of $\mathcal{L}_e$ is $\mathcal{F} - N + 1$, we know that the geometric multiplicity of the zero eigenvalue of $\mathcal{L}_e$ equals to its algebraic multiplicity. The proof is completed. □

With appropriate indexing of edges, we can write the CIIM $\bar{E}$ and CIM E, respectively, as $\bar{E} = [\bar{E}_\tau, \bar{E}_c]$ and $E = [E_\tau, E_c]$, where edges in $\bar{E}_\tau, E_\tau$ are on a directed spanning tree and the remaining edges are in $\bar{E}_c, E_c$. Analogous to the property of the incidence matrix for undirected graphs in Zelazo and Mesbahi [2011], we can reconstruct E_c with E_τ from the following proposition.

Proposition 8.4 When $\mathcal{G}$ contains a directed spanning tree, there exists a matrix S, such that $E_c = E_\tau S$.

Define the matrix $R = [I, S]$, then we can decompose $\mathcal{L}_e$ as in the following proposition:

---

2 The geometric multiplicity of a semisimple eigenvalue equals to its algebraic multiplicity.

Proposition 8.5 If $\mathcal{G}$ contains a directed spanning tree, then $\mathcal{L}_e$ is similar to the following matrix:

$$\begin{bmatrix} MR' & M\theta \\ \mathbf{0} & \mathbf{0}_{(\mathcal{F}-N+1)\times(\mathcal{F}-N+1)} \end{bmatrix},$$

where $M = E_\tau' \bar{E}$ and θ is the orthonormal basis of the null space of E. The nonzero eigenvalues of $\mathcal{L}_e$ are equal to those of MR'.

Proof: Since the directed graph contains a directed spanning tree, rank$(E) = N-1$ from Proposition 8.1. We thus have $\dim(\theta) = \dim(\text{null}(E)) = \mathcal{F} - N + 1$ and $\theta'\theta = I_{\mathcal{F}-N+1}$. In view of the definition of R, we know that $E\theta = E_\tau R\theta = 0$. Since E_τ is the CIM of a directed spanning tree, in view of Proposition 8.1, we have that rank$(E_\tau) = N-1$. Thus, there exists a transformation matrix O, such that $E_\tau = O[\tilde{E}'_{\tau(N-1)\times(N-1)}, \mathbf{0}'_{1\times(N-1)}]'$ with rank$(\tilde{E}_\tau) = N-1$. Then we have that $\tilde{E}_\tau R\theta = \mathbf{0}$. Since $\tilde{E}_\tau$ is invertible, we further have $R\theta = \mathbf{0}$.

Define the matrix $T = [R', \theta]$, $Q = [R'(RR')^{-1}, \theta]'$. Since $R\theta = \mathbf{0}$, every column in R' is orthogonal to the columns of θ. Thus, the columns in R' are independent of the columns in θ. Then rank$(T) = \mathcal{F}$ and T is invertible. Since $R\theta = 0$, the direct multiplication shows that $QT = I$, thus $T^{-1} = Q$.

Applying the similarity transformation to $\mathcal{L}_e$ with Q, T, we obtain that

$$Q\mathcal{L}_e T = \begin{bmatrix} MR' & M\theta \\ \mathbf{0} & \mathbf{0} \end{bmatrix}.$$

Since the dimension of MR' is $(N-1)\times(N-1)$, and when the directed graph contains a directed spanning tree, $\mathcal{L}_e$ has $N-1$ nonzero eigenvalues, we know that the $N-1$ nonzero eigenvalues of $\mathcal{L}_e$ equals to those of MR'. The proof is completed. □

8.5 Nonidentical Fading Networks

With the aid of CIIM, CIM, and CEL, we can remodel the consensus error dynamics in terms of edge states and linearly separate the channel fading from the network topology. Since fading is mostly caused by path loss and shadowing from obstacles, for simplicity, we can assume that the fadings on the bidirectional edge are equal, i.e. $\gamma_{ij}(t) = \gamma_{ji}(t)$ if j and i are connected via a bidirectional edge, which makes sense in practical applications [Dey et al., 2009]. For general channel fading models, where $\gamma_{ij} \neq \gamma_{ji}$, the DEL can be used to formulate the consensus dynamics and similar analysis methods proposed in this section can be applicable to the study of the consensusability problem. Therefore, we can use a single-letter ζ_p to characterize the fading noise on the pth edge, i.e. $\zeta_p = \gamma_{ij}$ if the edge p

is with initial node j and terminal node i. First, apply an orientation to every bidirectional edge in the graph and define the state on the lth edge as $z_l = x_j - x_i$, with j and i being the initial and terminal node of the lth edge, respectively. Then the dynamics of z_l based on (8.1) and (8.2) is

$$\begin{aligned} z_l(t+1) &= Az_l(t) + B[u_j(t) - u_i(t)] \\ &\stackrel{(a)}{=} Az_l(t) + BK\sum_{p=1}^{\mathcal{F}} \zeta_p(t)([\bar{E}]_{jp} - [\bar{E}]_{ip})z_p(t) \\ &\stackrel{(b)}{=} Az_l(t) + BK\sum_{p=1}^{\mathcal{F}} \zeta_p(t)[E'\bar{E}]_{lp}z_p(t), \end{aligned}$$

where the additive noise has been ignored; (a) follows from

$$\sum_{s\in\mathcal{N}_j} \gamma_{js}(t)(x_j(t) - x_s(t)) = \sum_{p=1}^{\mathcal{F}} \zeta_p(t)[\bar{E}]_{jp}z_p(t),$$

$$\sum_{h\in\mathcal{N}_i} \gamma_{ih}(t)(x_i(t) - x_h(t)) = \sum_{p=1}^{\mathcal{F}} \zeta_p(t)[\bar{E}]_{ip}z_p(t)$$

and (b) follows from the fact that

$$[E'\bar{E}]_{lp} = \sum_{s=1}^{N} [E]_{sl}[\bar{E}]_{sp} = [E]_{jl}[\bar{E}]_{jp} + [E]_{il}[\bar{E}]_{ip} = [\bar{E}]_{jp} - [\bar{E}]_{ip}.$$

Let $z = [z_1', z_2', \ldots, z_{\mathcal{F}}']'$, then we have

$$\begin{aligned} z(t+1) &= (I \otimes A + E'\bar{E}\zeta(t) \otimes BK)z(t) \\ &= (I \otimes A + \mathcal{L}_e\zeta(t) \otimes BK)z(t), \end{aligned} \tag{8.10}$$

where $\zeta(t) = \mathrm{diag}(\zeta_1(t), \ldots, \zeta_{\mathcal{F}}(t))$.

Suppose there is a directed cycle in $\mathcal{G}$, the sum of edge states on the directed cycle always equals zero, which imposes a constraint on the edge state z. We can further verify that as long as there is a cycle in the underlying graph of $\mathcal{G}$, such constraints always exist. Thus, not all edge states are free variables. This is illustrated in the following proposition.

Proposition 8.6 If $\mathcal{G}$ contains a directed spanning tree, then $z_c = (S' \otimes I)z_\tau$, where z_τ is the edge state on the directed spanning tree and z_c is the remaining edge state.

Proof: Suppose the edges in $\mathcal{G}$ are indexed such that $E = [E_\tau, E_c]$ and $\bar{E} = [\bar{E}_\tau, \bar{E}_c]$. The edge states can be partitioned correspondingly as $z = [z_\tau', z_c']'$. From the definition of the CIM E, we know that the edge states z and the node states x are related by $z = (E' \otimes I)x$. Thus, we have $[z_\tau', z_c']' = ([E_\tau, E_c]' \otimes I)x$, $z_\tau = (E_\tau' \otimes I)x$ and $z_c = (E_c' \otimes I)x$. In view of Proposition 8.4, we have $E_c = E_\tau S$. Then $z_c = ((S'E_\tau') \otimes I)x = (S' \otimes I)(E_\tau' \otimes I)x = (S' \otimes I)z_\tau$. The proof is completed. □

For brevity, we call z_c the cycle edge states since the edges associated with z_c necessarily complete cycles in the underlying graph of $\mathcal{G}$. Proposition 8.6 implies that cycle edge states can be reconstructed from the tree edge states. Thus, we can make a decomposition and further simplify the edge dynamics (8.10). Since $z = [z_\tau', z_c']'$, we have from (8.10) that

$$\begin{aligned} z_\tau(t+1) &= (I \otimes A)z_\tau(t) + (E_\tau'\bar{E}_\tau\zeta_\tau(t) \otimes BK)z_\tau(t) + (E_\tau'\bar{E}_c\zeta_c(t) \otimes BK)z_c(t) \\ &\overset{(a)}{=} (I \otimes A + (E_\tau'\bar{E}_\tau\zeta_\tau(t) + E_\tau'\bar{E}_c\zeta_c(t)S') \otimes BK)z_\tau(t) \\ &= (I \otimes A + M\zeta(t)R' \otimes BK)z_\tau(t), \end{aligned} \tag{8.11}$$

where ζ_τ, ζ_c represent the fading noise on directed spanning tree edges and cycle edges, respectively, and (a) follows from Proposition 8.6.

Since the graph contains a directed spanning tree, in view of the definition of the edge state z, if (8.10) is mean-square stable, mean-square consensus can be achieved. Based on Proposition 8.6, the stability property of (8.10) is determined by (8.11). Thus in the following, we shall focus on studying the mean-square stability of (8.11). In the subsequent analysis, we make the following assumption about the fading noise ζ_i, $i = 1, \dots, F$.

Assumption 8.3 The channel fading sequence $\{\zeta_i(t)\}_{t\geq 0}$ is i.i.d. with mean μ_i and variance σ_i^2 for all $i = 1, 2, \dots, \mathcal{F}$.

Analogous to the proof of Lemma 7.5 in Chapter 7, we can show that a necessary and sufficient condition to ensure the mean-square stabilizability of (8.11) is given as below.

Lemma 8.2 *Under Assumptions 8.1 and 8.3, (8.11) is mean-square stable if and only if there exist $P > 0$ and K such that*

$$P > (I \otimes A + M\Lambda R' \otimes BK)'P(I \otimes A + M\Lambda R' \otimes BK) + (R' \otimes K)'G(R' \otimes K) \tag{8.12}$$

with $G = (\Sigma \otimes \mathbf{11}') \odot ((M \otimes B)'P(M \otimes B))$, $\Sigma = [\sigma_{ij}]_{\mathcal{F}\times\mathcal{F}}$, $\sigma_{ij} = \mathbb{E}\{(\zeta_i - \mu_i)(\zeta_j - \mu_j)\}$ for $i \neq j$, $\sigma_{ii} = \sigma_i^2$ and $\Lambda = \mathrm{diag}(\mu_1, \mu_2, \dots, \mu_{\mathcal{F}})$.

The condition (8.12) is not easy to verify. In the following, we provide a simplified sufficient condition, which can be solved via a feasibility problem over real numbers.

Theorem 8.4 *Under Assumptions 8.1 and 8.3, the MAS (8.1) is mean-square consensusable by the protocol (8.2) under a directed communication topology if there exists $k \in \mathbb{R}$, such that*

$$k\left(M\Lambda R' + R\Lambda M'\right) + k^2 R(W \odot \Lambda M' M \Lambda)R' < -\tau_c I, \tag{8.13}$$

where $W = \mathbf{1}\mathbf{1}' + \Lambda^{-1}\Sigma\Lambda^{-1}$ and τ_c is defined in Lemma 7.6. Moreover, if such k exists, there exists a solution $P_0 > 0$ to (7.26), with τ being the smallest eigenvalue of $-k\left(M\Lambda R' + R\Lambda M'\right) - k^2 R(W \odot \Lambda M' M \Lambda)R'$, and a control gain that ensures the mean-square consensus can be given by $K = k(B'P_0B)^{-1}B'P_0A$.

Proof: If there exists $k \in \mathbb{R}$, such that (8.13) holds, in view of the solvability of (7.26), one can show that there exists $P_0 > 0$ to the matrix inequality

$$\begin{aligned} I \otimes P_0 > {} & I \otimes A'P_0A + (k\left(M\Lambda R' + R\Lambda M'\right) \\ & + k^2 R(W \odot \Lambda M' M \Lambda)R') \otimes A'P_0B(B'P_0B)^{-1}B'P_0A. \end{aligned} \tag{8.14}$$

Since $W \odot \Lambda M' M \Lambda = \Lambda M' M \Lambda + \Sigma \odot M'M$, we have from (8.14) that

$$I \otimes P_0 > I \otimes A'P_0A + H \otimes A'P_0B(B'P_0B)^{-1}B'P_0A \tag{8.15}$$

with $H = k^2(R\Lambda M' M \Lambda R' + R(\Sigma \odot M'M)R') + k\left(M\Lambda R' + R\Lambda M'\right)$. The inequality (8.15) is (8.12) with $K = k(B'P_0B)^{-1}B'P_0A$ and $P = I \otimes P_0 > 0$. In view of Lemma 8.2, the proof is completed. □

Remark 8.3 Since $W \geq 0$ and $\Lambda M' M \Lambda \geq 0$, in view of Theorem 5.2.1 in Horn and Johnson [1991], we have $W \odot \Lambda M' M \Lambda \geq 0$, thus, $R(W \odot \Lambda M' M \Lambda)R' \geq 0$. Let V be the Cholesky decomposition of $R(W \odot \Lambda M' M \Lambda)R'$, i.e. $R(W \odot \Lambda M' M \Lambda)R' = VV'$, then the sufficient condition in Theorem 8.4 can be numerically verified by the following LMI feasibility problem:

$$\exists\, k \quad s.t. \quad \begin{bmatrix} -I & kV' \\ kV & k(M\Lambda R' + R\Lambda M') + \tau_c I \end{bmatrix} < 0.$$

Remark 8.4 If the fading networks are identical, i.e. $\zeta_i(t) = \zeta_0(t), \forall i = 1, 2, \ldots, \mathcal{F}$, $\mathbb{E}\{\zeta_0(t)\} = \mu$ and $\mathbb{E}\{(\zeta_0(t) - \mu)^2\} = \sigma^2$, and $\mathcal{G}$ is an undirected tree, i.e. $R = I$ and $M = M' = \mathcal{L}_e = \mathcal{L}_e'$, then (8.13) is equivalent to $\min_k \max_{i \in \{2,\ldots,N\}} k^2(\mu^2 + \sigma^2)\lambda_i^2 + 2k\mu\lambda_i < -\tau_c$ with $\lambda_2, \ldots, \lambda_N$ being the nonzero real eigenvalues of $\mathcal{L}$ arranged in an ascending order, which can result in the sufficient mean-square consensus condition given by $\frac{\mu^2}{\mu^2+\sigma^2}\left[1 - \left(\frac{\lambda_N - \lambda_2}{\lambda_N + \lambda_2}\right)^2\right] > \tau_c$. This is consistent with Theorem 7.1, where it is also shown to be necessary for mean-square consensus when the agents are with scalar dynamics.

In the following, we try to derive closed-form consensus conditions for some specific fading networks.

8.5.1 $\Lambda = \mu I$

Since $\tau_c I + k^2 R(W \odot \Lambda M' M \Lambda) R' > 0$, when $\Lambda = \mu I$, a necessary condition to ensure the feasibility of (8.13) is that there exists k, such that $k(MR' + RM') < 0$. Since $\mathrm{tr}(MR' + RM') = 2\mathrm{tr}(MR') = 2\sum_i \lambda_i(MR') \overset{(a)}{=} 2\sum_i \lambda_i(\mathcal{L}_e) \overset{(b)}{=} 2\sum_i \lambda_i(\mathcal{L}) > 0$, where (a) follows from Proposition 8.5 and (b) follows from Proposition 8.3, we know that at least one eigenvalue of $MR' + RM'$ should be positive. Thus, if $k(MR' + RM')$ is required to be negative definite, k should be selected to be negative and $MR' + RM'$ should be positive definite. Thus, we make the assumption that $MR' + RM' > 0$ during the following analysis, which is an implicitly required graph property for (8.13) to hold.

Corollary 8.1 *Under Assumptions 8.1 and 8.3, if $\Lambda = \mu I$ and $MR' + RM' > 0$, the MAS (8.1) is mean-square consensusable by the protocol (8.2) under a directed communication topology, if the following condition is satisfied*

$$\tau_3 := \frac{\mu^2}{\mu^2 + \max_i \sigma_i^2} \times \frac{\lambda_{\min}^2\left(\frac{MR'+RM'}{2}\right)}{\rho(RR')\rho(M'M)} > \tau_c, \tag{8.16}$$

where τ_c is defined in Lemma 7.6. Moreover, if (8.16) holds, there exists a solution $P_0 > 0$ to (7.26) with $\tau = \tau_2$, and a control gain that ensures mean-square consensus can be given by $K = k_2(B'P_0B)^{-1}B'P_0A$ with

$$k_2 = -\frac{\mu\lambda_{\min}\left(\frac{MR'+RM'}{2}\right)}{[\mu^2 + \max_i \sigma_i^2]\rho(RR')\rho(M'M)}.$$

Proof: Since $W \geq 0$, $M'M \geq 0$ and $W \odot M'M \geq 0$, in view of Theorem 5.3.4 in Horn and Johnson [1991], we know that $0 \leq \lambda(W \odot M'M) \leq \max_i [W]_{ii} \times \rho(M'M) = \max_i \left(1 + \frac{\sigma_i^2}{\mu^2}\right)\rho(M'M)$ with $\lambda(W \odot M'M)$ being any eigenvalue of $W \odot M'M$. Thus, we have that $R(W \odot M'M)R' \leq \rho(W \odot M'M)RR' \leq \max_i \left(1 + \frac{\sigma_i^2}{\mu^2}\right)\rho(M'M)RR'$. Further from Weyl's inequality Bernstein [2009], we have that $\rho(R(W \odot M'M)R') \leq \max_i \left(1 + \frac{\sigma_i^2}{\mu^2}\right)\rho(M'M)\rho(RR')$. Since $RR' = I + SS' > 0$, we have $\rho(RR') > 0$. Besides, when $\mathcal{G}$ contains a directed spanning tree, in view of Lemma 1.1 and Proposition 8.2, $E_\tau E_\tau' = \mathcal{L}_\tau > 0$ with $\mathcal{L}_\tau$ being the graph Laplacian for the underlying graph of a directed spanning tree in $\mathcal{G}$. Since $M'M = \bar{E}'E_\tau E_\tau'\bar{E}$, we know that $M'M > 0$ and thus $\rho(M'M) > 0$. Since $MR' + RM' > 0$, if there exists k such that

$$k^2\left[\mu^2 + \max_i \sigma_i^2\right]\rho(RR')\rho(M'M) + 2k\mu\lambda_{\min}\left(\frac{MR' + RM'}{2}\right) < -\tau_c, \tag{8.17}$$

the sufficient condition (8.13) can be satisfied. Since the minimum of the left-hand side of (8.17) is achieved at $k = k_2$, with the minimal value $-\tau_3$, we can then obtain the sufficient consensus condition (8.16). The proof is completed. □

The sufficient condition (8.16) implies that the mean-square consensusability is determined by the channel fading, the network topology, and the agent dynamics. Besides, the mean-square consensusability is affected by the channel with the largest fading variance. Moreover, the effect of the network topology on the mean-square consensusability is reflected on the term α with

$$\alpha := \frac{\lambda_{\min}^2\left(\frac{MR'+RM'}{2}\right)}{\rho(RR')\rho(M'M)}.$$

In view of (8.16), a large α is always preferred to compensate the fading variance and tolerate unstable agent dynamics. In the following, we will use α as a measure to study how certain network topology affects the mean-square consensusability. First of all, we have the following proposition about the range of α.

Proposition 8.7 If $\mathcal{G}$ contains a directed spanning tree and $MR' + RM' > 0$, then $0 < \alpha \leq 1$.

Proof: It is trivial to have $\alpha > 0$. In the sequel, we will show that $\lambda_{\min}^2\left(\frac{MR'+RM'}{2}\right) \leq \rho(RR')\rho(M'M)$. Since when $MR' + RM' > 0$, we have $\lambda_{\min}^2\left(\frac{MR'+RM'}{2}\right) \leq \mathrm{Re}^2(\lambda(MR'))$ with $\lambda(MR')$ being any eigenvalue of MR' from Bendixson's theorem Bernstein [2009]. In view of the Browne's theorem Bernstein [2009], we have that $|\lambda(MR')|^2 \leq \rho(RM'MR')$, thus $\lambda_{\min}^2\left(\frac{MR'+RM'}{2}\right) \leq \rho(RM'MR') \leq \rho(RR')\rho(M'M)$. The proof is completed. □

We give some examples of different communication graphs as follows:

8.5.1.1 Star Graphs

If the graph is a star as shown in Figure 8.2a, we have that $R = I$ and $M = \mathcal{L}_e = I_{N-1}$. Evidently, $\frac{MR'+RM'}{2} = I > 0$ and $\lambda_{\min}^2\left(\frac{MR'+RM'}{2}\right) = \rho(M'M) = \rho(RR') = 1$. Thus, $\alpha = 1$, which means that scaling on the number of agents in the MAS does not affect the mean-square consensus for star graphs. Moreover, from Proposition 8.7, if we use α as an indicator to select the network topology, star graph is the most favorable in the sense that it has the largest possible value of α.

By adding an edge to the star graph, we obtain the graph in Figure 8.2b, which contains a cycle in its underlying graph. Then we have

$$M = \left[I_{(N-1)\times(N-1)}, Q\right], \quad R = \left[I_{(N-1)\times(N-1)}, S\right],$$

with $Q = [0,1,0,\ldots,0]'$ and $S = [-1,1,0,\ldots,0]'$. We can show that $MR' + RM' > 0$, $\lambda_{\min}(MR' + RM') = 3 - \sqrt{2}$, $\rho(M'M) = 2$ and $\rho(RR') = 3$. Thus,

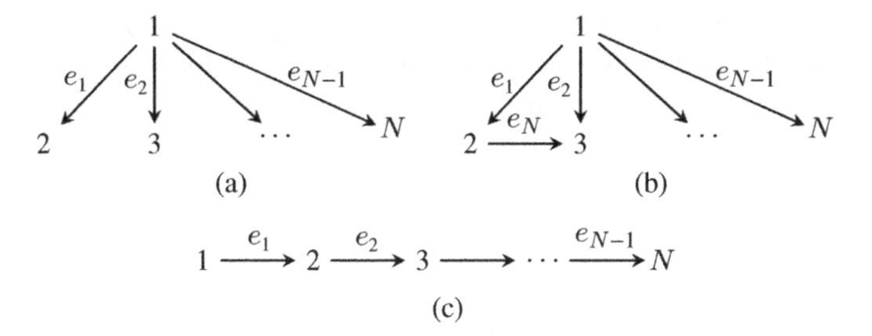

Figure 8.2 (a) A star graph, (b) a directed graph with a cycle in its underlying graph, (c) a directed path graph. Source: Xu et al. (2018)/with permission of Elsevier.

$\alpha = \frac{(3-\sqrt{2})^2}{24}$. Since $\frac{(3-\sqrt{2})^2}{24} < 1$, more edges are not always beneficial to the mean-square consensus. This can be interpreted from (8.11). Even though mean-square consensus is determined by edge states on a directed spanning tree, the fading noise on cycle edges still affects mean-square consensus as seen from (8.11). Thus, the insertion of an edge also introduces the associated fading noise into the tree edge state dynamics, which may pose negative effects on the mean-square consensus.

8.5.1.2 Directed Path Graphs

If the directed graph is a path as denoted in Figure 8.2c, then $R = I$ and

$$M = \begin{bmatrix} 1 & 0 & \dots & \dots & 0 \\ -1 & 1 & 0 & \dots & 0 \\ \vdots & & \ddots & & \vdots \\ 0 & \dots & 0 & -1 & 1 \end{bmatrix}.$$

Since $MR' + RM'$ is a tri-diagonal matrix, in view of Kulkarni et al. [1999], we know that the eigenvalues of $MR' + RM'$ are $2 - 2\cos\frac{l\pi}{N}$, $l = 1,2,\dots,N-1$. Thus, $MR' + RM' > 0$ and $\lambda_{\min}(MR' + RM') = 2 - 2\cos\frac{\pi}{N}$. Since $RM'MR' = MR' + RM' + D$ with

$$D = \begin{bmatrix} 0 & \dots & 0 \\ \vdots & \ddots & \vdots \\ 0 & \dots & -1 \end{bmatrix},$$

the eigenvalue perturbation theorem [Horn and Johnson, 1985] implies that $\lambda_1(D) \le \rho(RM'MR') - \rho(MR' + RM') \le \lambda_{N-1}(D)$ with $\lambda_i(D)$ being the i-th smallest eigenvalues of D. Since $\lambda_1(D) = -1$ and $\lambda_2(D) = \dots = \lambda_{N-1}(D) = 0$, and $\rho(MR' + RM') = 2 - 2\cos\frac{(N-1)\pi}{N}$, we have that $1 - 2\cos\frac{(N-1)\pi}{N} \le \rho(RM'MR') = \rho(RR')\rho(M'M) \le 2 - 2\cos\frac{(N-1)\pi}{N}$. When N is sufficiently large, the ratio α is lower and upper bounded, respectively, by

$$\frac{(1-\cos\frac{\pi}{N})^2}{2 - 2\cos\frac{(N-1)\pi}{N}} \le \alpha \le \frac{(1-\cos\frac{\pi}{N})^2}{1 - 2\cos\frac{(N-1)\pi}{N}}.$$

With the increasing number of agents, α will eventually converge to zero. Thus, consensus is hard to achieve. This is consistent with our intuition: for consensus over a path graph, more agents means that the consensus is harder to achieve. This is different from the star graph, where scaling does not affect the consensus condition.

8.5.2 $\Lambda \neq \mu I$

When $\Lambda \neq \mu I$, we have the following sufficient consensus condition. The proof is similar to that of Corollary 8.1 and is omitted here.

Corollary 8.2 *Under Assumptions 8.1 and 8.3, if $M\Lambda R' + R\Lambda M' > 0$, the MAS (8.1) is mean-square consensusable by the protocol (8.2) under a directed communication topology, if the following condition is satisfied*

$$\tau_4 := \frac{\lambda_{\min}^2\left(\frac{M\Lambda R'+R\Lambda M'}{2}\right)}{\max_i\left(1+\frac{\sigma_i^2}{\mu_i^2}\right)\rho(RR')\rho(\Lambda M'M\Lambda)} > \tau_c, \tag{8.18}$$

where τ_c is defined in Lemma 7.6. Moreover, if (8.18) holds, there exists a solution $P_0 > 0$ to (7.26) with $\tau = \tau_4$, and a control gain that ensures mean-square consensus can be given by

$$K = -\frac{\lambda_{\min}\left(\frac{M\Lambda R'+R\Lambda M'}{2}\right)}{\max_i\left(1+\frac{\sigma_i^2}{\mu_i^2}\right)\rho(RR')\rho(\Lambda M'M\Lambda)}(B'P_0B)^{-1}B'P_0A.$$

Remark 8.5 When $\Lambda = \mu I$, (8.18) recovers (8.16). Next, consider the case that $\Lambda = \mu I$ and the graph is an undirected tree, then $R = I$ and $M = M' = \mathcal{L}_e = \mathcal{L}_e'$. Thus, we have $\lambda_{\min}\left(\frac{MR'+RM'}{2}\right) = \lambda_2$ and $\rho(RR')\rho(M'M) = \lambda_N^2$, with λ_2 and λ_N being the smallest and the largest nonzero eigenvalues of the graph Laplacian for the undirected graph. Then a sufficient condition to ensure mean-square consensus for nonidentical fading networks with undirected tree graph from (8.16) is $\frac{\mu^2}{\mu^2+\max_i\sigma_i^2}\frac{\lambda_2^2}{\lambda_N^2} > \tau_c$. Since $\max_i\sigma_i^2 = \max_i\sigma_{ii} \leq \rho(\Sigma)$, Corollary 8.1 recovers Corollary 7.1. Similarly, we can also show that Corollary 8.2 recovers Corollary 7.2 for the case of $\Lambda \neq \mu I$.

8.6 Simulations

In this section, simulations are conducted to validate the derived results. We consider two different scenarios, i.e. identical fading networks and nonidentical fading networks with nonequal fading means. In simulations, the agents are

Figure 8.3 Communication graphs used in simulations: (a) a directed graph, (b) applying an orientation to the bidirectional edge in (a). Source: Xu et al. (2018)/with permission of Elsevier.

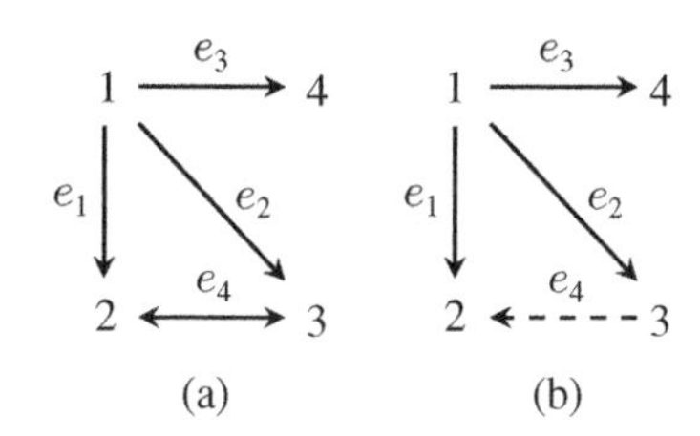

assumed to have the system parameters as in Chapter 7. The initial state of each agent is uniformly and randomly generated from the interval $(0, 0.5)$. We assume that there are four agents and the directed communication topology among agents is given in Figure 8.3a, where the CIIM and CIM are constructed from Figure 8.3b. The channel fadings are assumed to follow Rayleigh distribution with probability density function $f(x; \sigma_p) = \frac{x}{\sigma_p^2} e^{-x^2/(2\sigma_p^2)}$, $x \geq 0$. The additive noises are set to have standard normal distributions. The simulation results are presented by averaging over 1000 runs.

First, suppose that all the channel fadings are identical and follow Rayleigh distributions with $\sigma_p = 5$. Then the sufficient consensus condition in Theorem 8.1 is satisfied and one admissible control gain is $K = [6.7757, -8.1021, 1.2307]$ mean-square consensus errors for agent 1 are plotted in Figure 8.4. It is clear

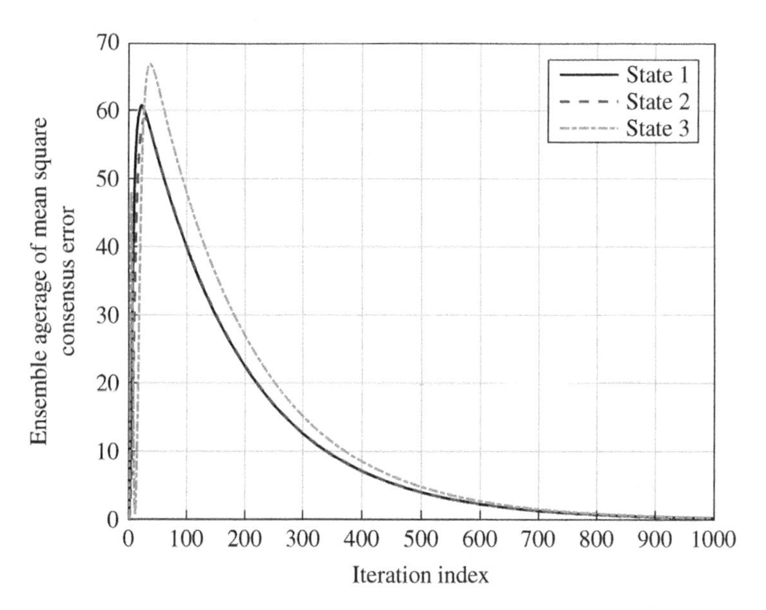

Figure 8.4 Mean-square consensus error for agent 1 under a directed topology with identical fading networks.

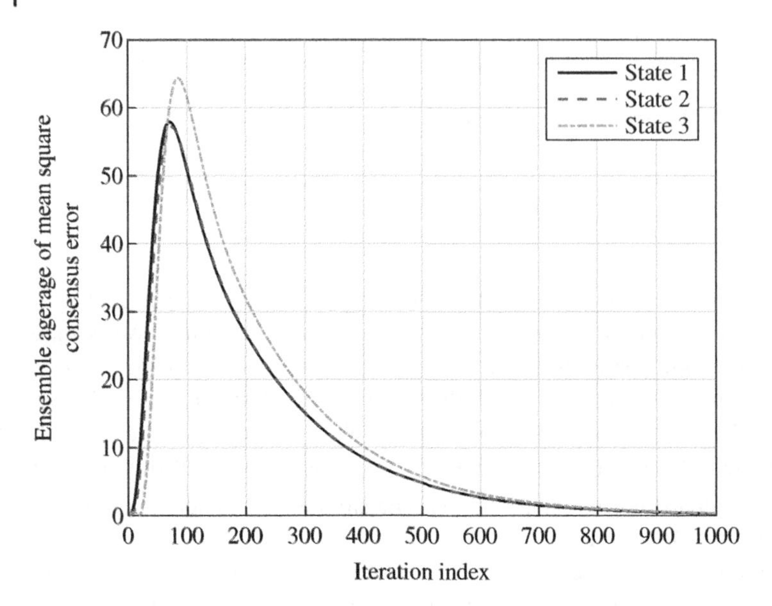

Figure 8.5 Mean-square consensus error for agent 1 under a directed topology with nonidentical fading networks and nonequal mean value. Source: Xu et al. (2018)/with permission of Elsevier.

that mean-square consensus of the MAS is achieved. Now, suppose the fading parameters for the four edges in Figure 8.3a are $\sigma_{p12} = 5$, $\sigma_{p13} = 4.9$, $\sigma_{p14} = 4.8$, $\sigma_{p23} = 4.7$. Then the fading on different edges have different mean value. With such fading parameters, the sufficient condition in Corollary 8.2 is satisfied and an admissible control gain is given by $K = [0.3750, -0.4686, 0.0868]$. Mean-square consensus errors for agent 1 are plotted in Figure 8.5, which also shows that the mean-square consensus is achieved. Since the consensus parameter K is designed for mean-square stabilization and not for performance, there are overshots in both simulations.

8.7 Conclusions

This chapter studies the mean-square consensus problem of discrete-time linear MASs over analog fading networks with directed graphs. Sufficient conditions are first provided for mean-square consensus over identical fading networks with directed graphs. It is shown that the sufficient condition is necessary when agents are with scalar dynamics. For consensus over nonidentical fading networks with directed graphs, CIIM, CIM, and CEL are proposed to facilitate the modeling and consensus analysis. It is shown that the mean-square consensusability is solely determined by the edge state dynamics on a directed spanning tree. As a result, sufficient conditions are provided for mean-square consensus over nonidentical fading networks with directed graphs in terms of fading parameters, the network topology, and the agent dynamics. Moreover, the role of network topology on the mean-square consensusability is discussed. In the end, simulations are conducted to verify the derived results.

Bibliography

D. S. Bernstein. *Matrix mathematics: theory, facts, and formulas*. Princeton University Press, Princeton, NJ, 2009.

S. Dey, A. S. Leong, and J. S. Evans. Kalman filtering with faded measurements. *Automatica*, 45(10):2223–2233, 2009.

N. Elia. Remote stabilization over fading channels. *Systems & Control Letters*, 54(3):237–249, 2005.

C. Godsil and G. Royle. *Algebraic graph theory*. Springer, New York, 2001.

R. A. Horn and C. R. Johnson. *Matrix analysis*. Cambridge University Press, Cambridge, 1985.

R. A. Horn and C. R. Johnson. *Topics in matrix analysis*. Cambridge University Press, New York, 1991.

D. Kulkarni, D. Schmidt, and S. K. Tsui. Eigenvalues of tridiagonal pseudo-Toeplitz matrices. *Linear Algebra and its Applications*, 297(1):63–80, 1999.

W. Ren and R. W. Beard. *Distributed consensus in multi-vehicle cooperative control: theory and applications*. Springer-Verlag London Limited, London, 2008.

L. Schenato, B. Sinopoli, M. Franceschetti, K. Poolla, and S. S. Sastry. Foundations of control and estimation over lossy networks. *Proceedings of the IEEE*, 95(1):163–187, 2007.

N. Xiao, L. Xie, and L. Qiu. Feedback stabilization of discrete-time networked systems over fading channels. *IEEE Transactions on Automatic Control*, 57(9):2176–2189, 2012.

K. You and L. Xie. Network topology and communication data rate for consensusability of discrete-time multi-agent systems. *IEEE Transactions on Automatic Control*, 56(10):2262–2275, 2011.

K. You, Z. Li, and L. Xie. Consensus condition for linear multi-agent systems over randomly switching topologies. *Automatica*, 49(10):3125–3132, 2013.

D. Zelazo and M. Mesbahi. Edge agreement: Graph-theoretic performance bounds and passivity analysis. *IEEE Transactions on Automatic Control*, 56(3):544–555, 2011.

Z. Zeng, X. Wang, and Z. Zheng. Convergence analysis using the edge Laplacian: Robust consensus of nonlinear multi-agent systems via ISS method. *International Journal of Robust and Nonlinear Control*, 26(5):1051–1072, 2016a.

Z. Zeng, X. Wang, and Z. Zheng. Edge agreement of multi-agent system with quantised measurements via the directed edge Laplacian. *IET Control Theory & Applications*, 10(13):1583–1589, 2016b.

9

Distributed Consensus over Networks with Communication Delay and Packet Dropouts

9.1 Introduction

This chapter continues to investigate the mean-square consensus problem of discrete-time multi-agent systems (MASs) over directed networks with constant communication delay and packet dropouts. Note that the approaches given in the aforementioned chapters dealing with fadings only cannot be directly applied to the case when both communication delays and packet dropouts exist in the communication channels. New ideas and techniques are required. Two cases are discussed. When the packet dropouts in different communication channels are identical, the consensusability problem is transformed into a simultaneous mean-square stabilization problem of systems with constant time delay and packet dropout. Sufficient consensusability conditions are obtained in this case which demonstrate how the delay, the packet dropout rate, the communication network topology, and the agent dynamics interplay with each other to allow the existence of a linear distributed consensus protocol. When the packet dropouts are nonidentical, we employ the notion of compressed edge Laplacian (CEL), which is initially proposed in Chapter 8 to facilitate the analysis of consensusability of MASs over fading channels with directed graphs. Then with the aid of such a notion, we can build the dynamics over edges to separate the packet dropouts from the network topology. A sufficient consensusability condition in terms of the communication delay, the packet dropout rates, the communication network topology, and the agent dynamics is provided. Under specific configurations, closed-form consensusability conditions are further obtained.

The remainder of this chapter is organized as follows: in Section 9.2, we present the problem formulation. The consensusability conditions for MASs with delay and identical packet dropouts are provided in Section 9.3, while

Control over Communication Networks: Modeling, Analysis, and Design of Networked Control Systems and Multi-Agent Systems over Imperfect Communication Channels, First Edition.
Jianying Zheng, Liang Xu, Qinglei Hu, and Lihua Xie.

the consensusability conditions for MASs with delay and nonidentical packet dropouts are stated in Section 9.4. Simulations are given in Section 9.5. A conclusion follows in Section 9.6.

9.2 Problem Formulation

In this chapter, we consider a multi-agent system with N agents. A directed graph $\mathcal{G} = (\mathcal{V}, \mathcal{E})$ is used to characterize the interaction among agents as in Section 1.4.1. Each agent is described by the following discrete-time dynamics:

$$x_i(k+1) = Ax_i(k) + Bu_i(k), \tag{9.1}$$

where $x_i \in \mathbb{R}^n$ and $u_i \in \mathbb{R}^m$ are the state and input of agent i, respectively. The agents exchange information over the directed communication graph $\mathcal{G}$. As what often happens in practical applications, the communication channels are not perfect but corrupted by communication delay and packet dropouts. Specifically, the consensus protocol is proposed as

$$u_i(k) = K \sum_{j \in \mathcal{N}_i} \epsilon_{ij}(k)(x_j(k-\tau) - x_i(k-\tau)), \tag{9.2}$$

where τ is the constant communication delay, and the packet dropout is characterized by an i.i.d. Bernoulli process $\epsilon_{ij}(k)$ with probability distribution $P(\epsilon_{ij}(k) = 0) = p_{ij}$ and $P(\epsilon_{ij}(k) = 1) = 1 - p_{ij}$, where $0 \leq p_{ij} \leq 1$ is called the packet dropout rate.

The discrete-time MAS (9.1) is said to be mean-square consensusable under protocol (9.2) if there exists a control gain K such that

$$\lim_{k \to \infty} \mathbb{E}\left\{\|x_i(k) - x_j(k)\|^2\right\} = 0$$

for any $i, j \in \{1, \dots, N\}$.

We aim to find conditions on the delay, the packet dropout rates, the communication topology, and the agent dynamics under which there exists a control gain K such that the discrete-time MAS (9.1) is mean-square consensusable. We make the following standard assumptions on the system matrices A, B and the network topology, as clarified in Yu et al. [2010].

Assumption 9.1

(1) All the eigenvalues of A are either on or outside the unit circle, and $[A|B]$ is controllable.

(2) $\mathcal{G}$ has a directed spanning tree.

In the rest of this chapter, we will study the mean-square consensusability problem under two scenarios: (i) communication channels with delay and identical packet dropouts (i.e. $\epsilon_{ij}(k) = \epsilon(k)$); (ii) communication channels with delay and nonidentical packet dropouts.

9.3 Consensusability with Delay and Identical Packet Dropouts

The assumption of identical packet dropouts means that at every moment, the information exchange among agents all fails or succeeds. This may happen when the agents are located in a small local area.

9.3.1 Stability Criterion of NCSs with Delay and Multiplicative Noise

To analyze the consensusability, we first establish the mean-square stability of an networked-control system (NCS) with delay and multiplicative noise, which can be seen as an extension of Lemma 7.1

Lemma 9.1 *The stochastic system*

$$x(k+1) = Ax(k) + \epsilon(k)BKx(k-\tau),$$

where $\tau \geq 0$ and the i.i.d. random process $\epsilon(k)$ has mean μ and variance σ^2, is mean-square stable if there exist $P > 0$ and K such that

$$\begin{aligned}&(A+\mu BK)'P(A+\mu BK) + (\tau\mu^2+\sigma^2)K'B'PBK \\ &\quad + \tau(A+\mu BK-I)'P(A+\mu BK-I) < P.\end{aligned} \tag{9.3}$$

Proof: Inequality (9.3) can be rewritten as

$$\begin{aligned}\Pi(P) \triangleq &(\tau+1)(A+\mu BK-I)'P(A+\mu BK-I) + (A+\mu BK-I)'P \\ &+ P(A+\mu BK-I) + (\tau\mu^2+\sigma^2)K'B'PBK < 0.\end{aligned}$$

Let $z(k) = x(k) + \sum_{j=k-\tau}^{k-1} \mu BKx(j)$. Then it holds that

$$\begin{aligned}z(k+1) &= x(k+1) + \sum_{j=k+1-\tau}^{k} \mu BKx(j) \\ &= Ax(k) + \epsilon(k)BKx(k-\tau) + \sum_{j=k+1-\tau}^{k} \mu BKx(j) \\ &= z(k) + (A-I+\mu BK)x(k) + (\epsilon(k)-\mu)BKx(k-\tau).\end{aligned}$$

Let $V(k) = V_1(k) + V_2(k)$, where

$$\begin{aligned}V_1(k) &= z'(k)Pz(k), \\ V_2(k) &= \sum_{s=-\tau}^{-1}\sum_{j=k+s}^{k-1} \mu^2 x'(j)K'B'PBKx(j).\end{aligned}$$

Then we have

$$\begin{aligned}
&\mathbb{E}\left\{V_1(k+1)\right\}\\
&=\mathbb{E}\left\{z'(k+1)Pz(k+1)\right\}\\
&=\mathbb{E}\left\{[z(k)+(A-I+\mu BK)x(k)+(\epsilon-\mu)BKx(k-\tau)]'P\right.\\
&\quad\left.\times[z(k)+(A-I+\mu BK)x(k)+(\epsilon-\mu)BKx(k-\tau)]\right\}\\
&=\mathbb{E}\left\{z'(k)Pz(k)\right\}+\mathbb{E}\left\{z'(k)P(A-I+\mu BK)x(k)\right\}\\
&\quad+\mathbb{E}\left\{x'(k)(A-I+\mu BK)'Pz(k)\right\}\\
&\quad+\mathbb{E}\left\{x'(k)(A-I+\mu BK)'P(A-I+\mu BK)x(k)\right\}\\
&\quad+\mathbb{E}\left\{\sigma^2x'(k-\tau)K'B'PBKx(k-\tau)\right\}\\
&=\mathbb{E}\left\{V_1(k)\right\}+\mathbb{E}\left\{x'(k)\left[(A-I+\mu BK)'P(A-I+\mu BK)\right.\right.\\
&\quad\left.\left.+P(A-I+\mu BK)+\ (A-I+\mu BK)'P\right]x(k)\right\}\\
&\quad+2\mathbb{E}\left\{\sum_{j=k-\tau}^{k-1}\mu x'(j)K'B'P(A-I+\mu BK)x(k)\right\}\\
&\quad+\mathbb{E}\left\{\sigma^2x'(k-\tau)K'B'PBKx(k-\tau)\right\}.
\end{aligned}$$

Two facts are recalled, i.e.

$$2x'y\le\beta x'x+\frac{1}{\beta}y'y,\forall\beta>0,$$

$$\left(\sum_{j=i}^{i+s}x'(j)\right)P\left(\sum_{j=i}^{i+s}x(j)\right)\le(s+1)\sum_{j=i}^{i+s}x'(j)Px(j).$$

It follows that

$$\begin{aligned}
&2\sum_{j=k-\tau}^{k-1}\mu x'(j)K'B'P(A-I+\mu BK)x(k)\\
&\le\frac{1}{\tau}\left(\sum_{j=k-\tau}^{k-1}\mu x'(j)K'B'\right)P\left(\sum_{j=k-\tau}^{k-1}\mu BKx(j)\right)\\
&\quad+\tau x'(k)(A-I+\mu BK)'P(A-I+\mu BK)x(k)\\
&\le\sum_{j=k-\tau}^{k-1}\mu^2x'(j)K'B'PBKx(j)+\tau x'(k)(A-I+\mu BK)'P(A-I+\mu BK)x(k).
\end{aligned}$$

Thus, we have

$$\begin{aligned}
\mathbb{E}\left\{V_1(k+1)\right\}-\mathbb{E}\left\{V_1(k)\right\}\le{}&\mathbb{E}\left\{x'(k)\left[(\tau+1)(A-I+\mu BK)'P(A-I+\mu BK)\right.\right.\\
&\left.\left.+P(A-I+\mu BK)+(A-I+\mu BK)'P\right]x(k)\right\}\\
&+\mathbb{E}\left\{\sum_{j=k-\tau}^{k-1}\mu^2x'(j)K'B'PBKx(j)\right\}\\
&+\mathbb{E}\left\{\sigma^2x'(k-\tau)K'B'PBKx(k-\tau)\right\}.
\end{aligned}$$

On the other hand,

$$\begin{aligned}
&\mathbb{E}\left\{V_2(k+1)\right\} - \mathbb{E}\left\{V_2(k)\right\} \\
&= \mathbb{E}\left\{\sum_{s=-\tau}^{-1}\left(\sum_{j=k+1+s}^{k} \mu^2 x'(j)K'B'PBKx(j) - \sum_{j=k+s}^{k-1} \mu^2 x'(j)K'B'PBKx(j)\right)\right\} \\
&= \mathbb{E}\left\{\sum_{s=-\tau}^{-1}\left[\mu^2 x'(k)K'B'PBKx(k) - \mu^2 x'(k+s)K'B'PBKx(k+s)\right]\right\} \\
&= \mathbb{E}\left\{\tau\mu^2 x'(k)K'B'PBKx(k)\right\} - \mathbb{E}\left\{\sum_{j=k-\tau}^{k-1} \mu^2 x'(j)K'B'PBKx(j)\right\}.
\end{aligned}$$

Therefore, we get that

$$\begin{aligned}
\mathbb{E}\left\{V(k+1)\right\} - \mathbb{E}\left\{V(k)\right\}) \le \mathbb{E}\Big\{x'(k)\big[&(\tau+1)(A-I+\mu BK)'P(A-I+\mu BK) \\
&+ P(A-I+\mu BK) + (A-I+\mu BK)'P + \tau\mu^2 K'B'PBK\big]x(k)\Big\} \\
&+ \mathbb{E}\left\{\sigma^2 x'(k-\tau)K'B'PBKx(k-\tau)\right\}.
\end{aligned}$$

Let

$$\begin{aligned}
P_1 = &(\tau+1)(A-I+\mu BK)'P(A-I+\mu BK) + P(A-I+\mu BK) \\
&+ (A-I+\mu BK)'P + \tau\mu^2 K'B'PBK
\end{aligned}$$

and

$$\Delta_P = \sigma^2 K'B'PBK.$$

Then

$$\mathbb{E}\left\{V(k+1)\right\} \le \mathbb{E}\left\{V(k)\right\} + \mathbb{E}\left\{x'(k)P_1 x(k)\right\} + \mathbb{E}\left\{x'(k-\tau)\Delta_P x(k-\tau)\right\}.$$

Given any constant $S \ge 1$, it holds that

$$\begin{aligned}
S^{k+1}\mathbb{E}\left\{V(k+1)\right\} - S^k\mathbb{E}\left\{V(k)\right\} \le (S^{k+1} - S^k)\mathbb{E}\left\{V(k)\right\} + S^{k+1}\mathbb{E}\left\{x'(k)P_1 x(k)\right\} \\
+ S^{k+1}\mathbb{E}\left\{x'(k-\tau)\Delta_P x(k-\tau)\right\}.
\end{aligned}$$

Then it follows that

$$\begin{aligned}
S^{k+1}\mathbb{E}\left\{V(k+1)\right\} \le \mathbb{E}\left\{V(0)\right\} &+ \sum_{j=0}^{k}(S^{j+1} - S^j)\mathbb{E}\left\{V(j)\right\} + \sum_{j=0}^{k} S^{j+1}\mathbb{E}\left\{x'(j)P_1 x(j)\right\} \\
&+ \sum_{j=0}^{k} S^{j+1}\mathbb{E}\left\{x'(j-\tau)\Delta_P x(j-\tau)\right\} \\
\le \mathbb{E}\left\{V(0)\right\} &+ \sum_{j=0}^{k}(S^{j+1} - S^j)\mathbb{E}\left\{V(j)\right\} + \sum_{j=0}^{k} S^{j+1}\mathbb{E}\left\{x'(j)P_1 x(j)\right\} \\
&+ \sum_{j=0}^{k+\tau} S^{j+1}\mathbb{E}\left\{x'(j-\tau)\Delta_P x(j-\tau)\right\}
\end{aligned}$$

$$
\begin{aligned}
&\le \mathbb{E}\{V(0)\} + \sum_{j=0}^{k}(S^{j+1} - S^{j})\mathbb{E}\{V(j)\} + \sum_{j=0}^{k} S^{j+1}\mathbb{E}\left\{x'(j)P_1x(j)\right\} \\
&\quad + S^{\tau}\sum_{j=-\tau}^{-1} S^{j+1}\mathbb{E}\left\{x'(j)\Delta_P x(j)\right\} + S^{\tau}\sum_{j=0}^{k} S^{j+1}\mathbb{E}\left\{x'(j)\Delta_P x(j)\right\} \\
&= C_0(S) + \sum_{j=0}^{k}(S^{j+1} - S^{j})\mathbb{E}\{V(j)\} + \sum_{j=0}^{k} S^{j+1}\mathbb{E}\left\{x'(j)P_2(S)x(j)\right\}.
\end{aligned}
$$

where

$$
\begin{aligned}
C_0(S) &= \mathbb{E}\{V(0)\} + S^{\tau}\sum_{j=-\tau}^{-1} S^{j+1}\mathbb{E}\left\{x'(j)\Delta_P x(j)\right\}, \\
P_2(S) &= P_1 + S^{\tau}\Delta_P.
\end{aligned}
$$

By the definition of $V(k)$, we have that

$$
\begin{aligned}
V(k) &= \left[x(k) + \sum_{j=k-\tau}^{k-1}\mu BKx(j)\right]' P \left[x(k) + \sum_{j=k-\tau}^{k-1}\mu BKx(j)\right] \\
&\quad + \sum_{s=-\tau}^{-1}\sum_{j=k+s}^{k-1}\mu^2 x'(j)K'B'PBKx(j) \\
&\le 2x'(k)Px(k) + 2\sum_{j=k-\tau}^{k-1}\mu x'(j)K'B'P\sum_{j=k-\tau}^{k-1}\mu BKx(j) \\
&\quad + \sum_{s=-\tau}^{-1}\sum_{j=k+s}^{k-1}\mu^2 x'(j)K'B'PBKx(j) \\
&\le 2x'(k)Px(k) + 3\tau\sum_{j=k-\tau}^{k-1}\mu^2 x'(j)K'B'PBKx(j) \\
&\le 2\|P\|\|x(k)\|^2 + 3\tau\mu^2\|P\|\|BK\|^2\sum_{j=k-\tau}^{k-1}\|x(j)\|^2.
\end{aligned}
$$

Hence,

$$
\begin{aligned}
S^{k+1}\mathbb{E}\{V(k+1)\} &\le C_0(S) + 2(1 - S^{-1})\|P\|\sum_{j=0}^{k} S^{j+1}\mathbb{E}\left\{\|x(j)\|^2\right\} \\
&\quad + 3\tau\mu^2\|P\|\|BK\|^2(1 - S^{-1})\sum_{j=0}^{k} S^{j+1}\sum_{i=j-\tau}^{j-1}\mathbb{E}\left\{\|x(i)\|^2\right\} \\
&\quad + \sum_{j=0}^{k} S^{j+1}\mathbb{E}\left\{x'(j)P_2(S)x(j)\right\}.
\end{aligned}
$$

Since

$$\sum_{j=0}^{k} S^{j+1} \sum_{i=j-\tau}^{j-1} \mathbb{E}\left\{\|x(i)\|^2\right\} \leq \sum_{j=-\tau}^{-1} \mathbb{E}\left\{\|x(j)\|^2\right\} \sum_{i=0}^{j+\tau} S^{i+1} + \sum_{j=0}^{k-1} \mathbb{E}\left\{\|x(j)\|^2\right\} \sum_{i=j+1}^{j+\tau} S^{i+1}$$

$$\leq \tau S^{\tau} \sum_{j=-\tau}^{-1} \mathbb{E}\left\{\|x(j)\|^2\right\} + \tau S^{\tau} \sum_{j=0}^{k-1} S^{j+1} \mathbb{E}\left\{\|x(j)\|^2\right\},$$

it holds that

$$S^{k+1}\mathbb{E}\{V(k+1)\} \leq C_0(S) + 2(1-S^{-1})\|P\| \sum_{j=0}^{k} S^{j+1}\mathbb{E}\left\{\|x(j)\|^2\right\}$$

$$+ 3\tau\mu^2\|P\|\|BK\|^2(1-S^{-1})\left[\tau S^{\tau} \sum_{j=-\tau}^{-1} \mathbb{E}\left\{\|x(j)\|^2\right\} + \tau S^{\tau} \sum_{j=0}^{k-1} S^{j+1}\mathbb{E}\left\{\|x(j)\|^2\right\}\right]$$

$$+ \sum_{j=0}^{k} S^{j+1}\mathbb{E}\left\{x'(j)P_2(S)x(j)\right\}$$

$$\leq C_1(S) + C_2(S)(1-S^{-1})\sum_{j=0}^{k} S^{j+1}\mathbb{E}\left\{\|x(j)\|^2\right\} + \sum_{j=0}^{k} S^{j+1}\mathbb{E}\left\{x'(j)P_2(S)x(j)\right\},$$

where

$$C_1(S) = C_0(S) + 3\tau^2\mu^2 S^{\tau}\|P\|\|BK\|^2 \sum_{j=-\tau}^{-1} \mathbb{E}\left\{\|x(j)\|^2\right\},$$

$$C_2(S) = 2\|P\| + 3\tau^2\mu^2 S^{\tau}\|P\|\|BK\|^2.$$

It is easy to see that

$$P_2(S) = P_1 + \Delta_P + (S^{\tau}-1)\Delta_P \leq -\lambda_{\min}(-\Pi_p)I_m + (S^{\tau}-1)\|\Delta_P\|I_m,$$

which implies that

$$S^{k+1}\mathbb{E}\{V(k+1)\} \leq C_1(S) + H(S)\sum_{j=0}^{k} S^{j+1}\mathbb{E}\left\{\|x(j)\|^2\right\},$$

where

$$H(S) = C_2(S)(1-S^{-1}) - \lambda_{\min}(-\Pi_p) + (S^{\tau}-1)\|\Delta_P\|.$$

Evidently, $H(S)$ is monotonically increasing. Moreover, $H(1) = -\lambda_{\min}(-\Pi_p) < 0$ by inequality (9.3) and $H(S) > 0$ when

$$S > S_0 \triangleq \left(\frac{\lambda_{\min}(-\Pi_P) + \|\Delta_P\|}{\|\Delta_P\|}\right)^{1/\tau}$$

due to that $C_2(S)(1-S^{-1}) > 0$ for any $S > 1$. Then there must exist $\overline{S} \in (1, S_0)$ such that $H(\overline{S}) = 0$. Select $S^* \in (1, \overline{S})$, then $H(S^*) < 0$. Hence,

$$0 \leq S^{*k+1}\mathbb{E}\{V(k+1)\} \leq C_1(S^*) + H(S^*)\sum_{j=0}^{k} S^{*j+1}\mathbb{E}\left\{\|x(j)\|^2\right\},$$

which implies

$$-H(S^*)\sum_{j=0}^{k} S^{*j+1}\mathbb{E}\left\{\|x(j)\|^2\right\} \le C_1(S^*).$$

Select $\gamma > 0$ such that $S^* = e^{\gamma}$. Then we can get that $e^{\gamma k}\mathbb{E}\left\{\|x(k)\|^2\right\} \le C_3$ for some constant C_3. This implies the mean-square stability. □

Remark 9.1 When there is no delay, i.e. $\tau = 0$, inequality (9.3) is reduced to

$$(A + \mu BK)'P(A + \mu BK) - P + \sigma^2 K'B'PBK < 0,$$

which is the stabilizability condition for NCSs over fading channel [Xiao et al., 2012].

9.3.2 Consensusability Conditions

We now formally state the assumption of identical packet dropouts.

Assumption 9.2 $\epsilon_{ij}(k) = \epsilon(k)$, where $\epsilon(k)$ is an i.i.d. Bernoulli process with p being the packet dropout rate.

Denote by $\mu \triangleq 1 - p$ and $\sigma^2 \triangleq p(1 - p)$. Let

$$X(k) = \begin{bmatrix} x_1'(k) & \cdots & x_N'(k) \end{bmatrix}'.$$

Substitute (9.2) into (9.1), and we can get that

$$X(k + 1) = (I_N \otimes A)X(k) - \epsilon(k)(\mathcal{L}_{\mathcal{G}} \otimes BK)X(k - \tau).$$

Let $r = [r_i]$ be the left eigenvector of $\mathcal{L}_{\mathcal{G}}$ associated with the zero eigenvalue satisfying $r'\mathbf{1}_N = 1$ and define the consensus error as $\delta(k) = \begin{bmatrix} \delta_1'(k) & \cdots & \delta_N'(k) \end{bmatrix}'$ where $\delta_i(k) = x_i(k) - \sum_{j=1}^{N} r_j x_j(k)$. Then

$$\delta(k) = X(k) - [(\mathbf{1}_N r') \otimes I_n]X(k),$$

which evolves as

$$\begin{aligned}
\delta(k + 1) &= [I_N \otimes I_n - (\mathbf{1}_N r') \otimes I_n]X(k + 1) \\
&= [I_N \otimes I_n - (\mathbf{1}_N r') \otimes I_n](I_N \otimes A)X(k) \\
&\quad - \epsilon(k)[I_N \otimes I_n - (\mathbf{1}_N r') \otimes I_n](\mathcal{L}_{\mathcal{G}} \otimes BK)X(k - \tau) \\
&= (I_N \otimes A)[I_N \otimes I_n - (\mathbf{1}_N r') \otimes I_n]X(k) \\
&\quad - \epsilon(k)(\mathcal{L}_{\mathcal{G}} \otimes BK)[I_N \otimes I_n - (\mathbf{1}_N r') \otimes I_n]X(k - \tau) \\
&= (I_N \otimes A)\delta(k) - \epsilon(k)(\mathcal{L}_{\mathcal{G}} \otimes BK)\delta(k - \tau).
\end{aligned}$$

Since $\mathcal{G}$ has a directed spanning tree, according to Ren and Beard [2008], there exist matrices S, Y such that $\Phi = \begin{bmatrix} \mathbf{1}_N & Y \end{bmatrix}$ and $\Phi^{-1} = \begin{bmatrix} r' \\ S \end{bmatrix}$ satisfying $\Phi^{-1}\mathcal{L}_{\mathcal{G}}\Phi = \text{diag}\{0, \Delta_1, \dots, \Delta_{N'}\}$, where $\Delta_i, i = 1, \dots, N'$, are Jordan blocks of $\mathcal{L}$. Define the coordinate transformation

$$\hat{\delta}(k) = [\hat{\delta}_1'(k), \hat{\delta}_2'(k), \dots, \hat{\delta}_N'(k)]' = (\Phi^{-1} \otimes I)\delta(k),$$

then it follows that

$$\begin{aligned}
\hat{\delta}(k+1) &= (\Phi^{-1} \otimes I_n)\delta(k+1) \\
&= (\Phi^{-1} \otimes I_n)(I_N \otimes A)(\Phi \otimes I_n)\hat{\delta}(k) \\
&\quad - \epsilon(k)(\Phi^{-1} \otimes I_n)(\mathcal{L}_{\mathcal{G}} \otimes BK)(\Phi \otimes I_n)\hat{\delta}(k-\tau) \\
&= (I_N \otimes A)\hat{\delta}(k) - \epsilon(k)(\text{diag}\{0, \Delta_1, \dots, \Delta_{N'}\} \otimes BK)\hat{\delta}(k-\tau).
\end{aligned}$$

It is easy to see that $\lim_{k\to\infty}\mathbb{E}\left\{\|x_i(k) - x_j(k)\|^2\right\} = 0$ for all $i, j \in \{1, \dots, N\}$ amounts to

$$\lim_{k\to\infty}\mathbb{E}\left\{\|\delta_i(k)\|^2\right\} = \lim_{k\to\infty}\mathbb{E}\left\{\|\hat{\delta}_i(k)\|^2\right\} = 0$$

for all $i \in \{1, \dots, N\}$. Since

$$\hat{\delta}_1(k) = (r' \otimes I_n)\delta(k) = (r' \otimes I_n)[I_N \otimes I_n - (\mathbf{1}_N r') \otimes I_n]X(k) \equiv 0,$$

where the last equality holds due to $r'\mathbf{1}_N = 1$, following a similar analysis of Lemma 8.1, the simultaneous mean-square stability of the following $N-1$ systems:

$$\bar{\delta}_i(k+1) = A\bar{\delta}_i(k) - \epsilon(k)\lambda_i BK\bar{\delta}_i(k-\tau), \quad i = 2, \dots, N \tag{9.4}$$

implies the mean-square stability of $\hat{\delta}(k)$. Hence, the mean-square consensusability problem is boiled down to finding conditions under which there exists some K such that systems (9.4) are simultaneously mean-square stable. In view of Lemma 9.1, we have the following consensusability conditions.

Theorem 9.1 *Under Assumptions 9.1 and 9.2, the MAS (9.1) is mean-square consensusable over $\mathcal{G}$ under the protocol (9.2) provided that*

$$\eta \triangleq \frac{(\tau+1)(1-p)}{2\tau(1-p)+1}\left[1 - \min_{\omega\in\mathbb{R}}\max_{i=2,\dots,N}|1-\omega\lambda_i|^2\right] > \tilde{\gamma}_c, \tag{9.5}$$

where $\tilde{\gamma}_c$ is the critical value such that for any $\gamma > \tilde{\gamma}_c$, there exists a positive definite solution P to the modified Riccati inequality

$$P > \tilde{A}'P\tilde{A} - \gamma\tilde{A}'PB(B'PB)^{-1}B'P\tilde{A} \tag{9.6}$$

with $\tilde{A} = [(\tau+1)A - \tau I]$. In this situation, a control gain K that ensures the mean-square consensus is given as

$$K = \frac{\omega^*}{2\tau(1-p)+1}(B'PB)^{-1}B'P\tilde{A},$$

where $\omega^* = \arg\min_{\omega \in \mathbb{R}} \max_{i=2,\ldots,N} |1 - \omega\lambda_i|^2$ *and* $P > 0$ *is a solution to the modified Riccati inequality (9.6) with* $\gamma = \eta$.

In addition, when $\mathcal{G}$ *is undirected, the MAS (9.1) is mean-square consensusable under the protocol (9.2) if*

$$\frac{(\tau+1)(1-p)}{2\tau(1-p)+1}\left[1-\left(\frac{\lambda_N-\lambda_2}{\lambda_N+\lambda_2}\right)^2\right] > \tilde{\gamma}_c \tag{9.7}$$

with

$$K = \frac{2}{(\lambda_2+\lambda_N)(2\tau(1-p)+1)}(B'PB)^{-1}B'P\tilde{A}.$$

Proof: Let

$$\gamma = \frac{\omega^*}{2\tau(1-p)+1} = \frac{\omega^*\mu}{\mu^2(2\tau+1)+\sigma^2}.$$

Then

$$\begin{aligned}
-\eta &= -\frac{(\tau+1)(1-p)}{2\tau(1-p)+1}\left[1-\min_{\omega\in\mathbb{R}}\max_{i=2,\ldots,N}|1-\omega\lambda_i|^2\right] \\
&= -\frac{\mu^2(\tau+1)}{\mu^2(2\tau+1)+\sigma^2}\left[1-\max_{i=2,\ldots,N}|1-\omega^*\lambda_i|^2\right] \\
&= \frac{\mu^2(\tau+1)}{\mu^2(2\tau+1)+\sigma^2}\left[\max_{i=2,\ldots,N}\left|1-\frac{\gamma\left[\mu^2(2\tau+1)+\sigma^2\right]}{\mu}\lambda_i\right|^2-1\right] \\
&= \max_{i=2,\ldots,N}\gamma^2(\tau+1)\left[\mu^2(2\tau+1)+\sigma^2\right]\lambda_i^*\lambda_i - 2\gamma\mu(\tau+1)\mathrm{Re}(\lambda_i).
\end{aligned}$$

Since $[\tilde{A}|B]$ is controllable, which is implied by the controllability of $[A|B]$, in view of Lemma 7.1, inequality (9.5) implies that there exists some $P > 0$ such that

$$\begin{aligned}
P > \tilde{A}'P\tilde{A} + &\left\{\max_{i=2,\ldots,N}\gamma^2(\tau+1)[\mu^2(2\tau+1)+\sigma^2]\lambda_i^*\lambda_i - 2\gamma\mu(\tau+1)\mathrm{Re}(\lambda_i)\right\} \\
&\times \tilde{A}'PB(B'PB)^{-1}B'P\tilde{A}.
\end{aligned}$$

With $K = \gamma(B'PB)^{-1}B'P\tilde{A}$, we have that

$$\begin{aligned}
&(\tau+1)^2\left(A-\mu\lambda_iBK-\frac{\tau}{\tau+1}I\right)'P\left(A-\mu\lambda_iBK-\frac{\tau}{\tau+1}I\right) \\
&\quad + (\tau+1)(\tau\mu^2+\sigma^2)\lambda_i^*\lambda_iK'B'PBK \\
&= [\tilde{A}-\mu(\tau+1)\lambda_iBK]'P[\tilde{A}-\mu(\tau+1)\lambda_iBK] \\
&\quad + (\tau+1)(\tau\mu^2+\sigma^2)\lambda_i^*\lambda_iK'B'PBK \\
&= \tilde{A}'P\tilde{A} + \left\{(\tau+1)\left[\mu^2(2\tau+1)+\sigma^2\right]\lambda_i^*\lambda_i\gamma^2 - \mu(\tau+1)\lambda_i\gamma - \mu(\tau+1)\lambda_i^*\gamma\right\} \\
&\quad \times \tilde{A}'PB(B'PB)^{-1}B'P\tilde{A}
\end{aligned}$$

$$
\begin{aligned}
&= \tilde{A}'P\tilde{A} + \left\{\gamma^2(\tau+1)\left[\mu^2(2\tau+1)+\sigma^2\right]\lambda_i^*\lambda_i - 2\gamma\mu(\tau+1)\mathrm{Re}(\lambda_i)\right\} \\
&\quad \times \tilde{A}'PB(B'PB)^{-1}B'P\tilde{A} \\
&\leq \tilde{A}'P\tilde{A} + \left\{\max_{i=2,\ldots,N}\gamma^2(\tau+1)\left[\mu^2(2\tau+1)+\sigma^2\right]\lambda_i^*\lambda_i - 2\gamma\mu(\tau+1)\mathrm{Re}(\lambda_i)\right\} \\
&\quad \times \tilde{A}'PB(B'PB)^{-1}B'P\tilde{A} \\
&< P,
\end{aligned}
$$

which is further equivalent to that

$$
\begin{aligned}
0 &> (\tau+1)\left(A - \mu\lambda_i BK - \frac{\tau}{\tau+1}I\right)' P \left(A - \mu\lambda_i BK - \frac{\tau}{\tau+1}I\right) \\
&\quad + (\tau\mu^2+\sigma^2)\lambda_i^*\lambda_i K'B'PBK - \frac{1}{\tau+1}P \\
&= (A-\mu\lambda_i BK)'P(A-\mu\lambda_i BK) + (\tau\mu^2+\sigma^2)\lambda_i^*\lambda_i K'B'PBK \\
&\quad + \tau(A-\mu\lambda_i BK - I)'P(A-\mu\lambda_i BK - I) - P. \qquad (9.8)
\end{aligned}
$$

Implied by Lemma 9.1, systems (9.4) are simultaneously mean-square stable with K, i.e. the MAS (9.1) can achieve mean-square consensus with the designed K.

When $\mathcal{G}$ is undirected, we have $\lambda_i > 0, i = 2, \ldots, N$. Then

$$
\min_{\omega\in\mathbb{R}}\max_{i=2,\ldots,N}|1-\omega\lambda_i|^2 = \left(\frac{\lambda_N-\lambda_2}{\lambda_N+\lambda_2}\right)^2
$$

with $\omega^* = \frac{2}{\lambda_2+\lambda_N}$. We can obtain (9.7) from (9.5), which completes the proof. □

Remark 9.2 Suppose that the inequality (9.5) holds for some τ and p. Since η is monotonically decreasing with respect to p, a smaller packet dropout rate is beneficial to the MAS's mean-square consensusability. On the other hand, inequality (9.5) implies inequality (9.8) whose RHS is monotonically increasing with respect to τ. Therefore, the MAS (9.1) will achieve mean-square consensus under any delay smaller than τ.

It is easy to show that $0 \leq \eta \leq 1$, where the latter equality holds when $\tau = p = 0$ and $\mathcal{G}$ is complete. Note that when $\eta = 1$, the MAS (9.1) always achieves consensus. Since $\tilde{\gamma}_c$ in (9.5) is a function of the time delay τ and the system matrix, generally, it is difficult to obtain some explicit consensusability condition where the system dynamics and the time delay are expressed in a decoupled way. However, when $\mathrm{rank}(B) = n$, we can decouple the system dynamics and the time delay by decomposing $\tilde{\gamma}_c$. Thus, the delay, the packet dropout rate and the network topology together give a tolerable range of the instability degree of the MAS, as stated in the following corollary.

Corollary 9.1 *Under Assumptions 9.1 and 9.2, if rank(B) = n and $0 < \eta < 1$, the MAS (9.1) is mean-square consensusable over $\mathcal{G}$ under the protocol (9.2) if*

$$1 \leq \rho(A) < \frac{\tau\sqrt{1-\eta}+1}{(\tau+1)\sqrt{1-\eta}}. \tag{9.9}$$

Proof: When rank$(B) = n$, by Lemma 7.1, we have

$$\tilde{\gamma}_c = 1 - \frac{1}{\rho(\tilde{A})^2} = 1 - \frac{1}{[(\tau+1)\rho(A)-\tau]^2}.$$

It suffices to show the equivalence of inequalities (9.5) and (9.9). Note that inequality (9.5) can be written as

$$\eta > 1 - \frac{1}{[(\tau+1)\rho(A)-\tau)]^2},$$

which is equal to

$$[(\tau+1)\rho(A)-\tau)]^2 - \frac{1}{1-\eta} < 0,$$

i.e.

$$\left[\rho(A) - \frac{\tau}{\tau+1}\right]^2 < \frac{1}{(\tau+1)^2(1-\eta)}.$$

Then it must hold that

$$\frac{\tau}{\tau+1} - \frac{1}{(\tau+1)\sqrt{1-\eta}} < \rho(A) < \frac{\tau}{\tau+1} + \frac{1}{(\tau+1)\sqrt{1-\eta}} = \frac{\tau\sqrt{1-\eta}+1}{(\tau+1)\sqrt{1-\eta}}.$$

Since

$$\tau - \frac{1}{\sqrt{1-\eta}} < \tau + 1$$

and $\rho(A) \geq 1$, inequality (9.5) amounts to

$$1 \leq \rho(A) < \frac{\tau\sqrt{1-\eta}+1}{(\tau+1)\sqrt{1-\eta}},$$

which completes the proof. □

Remark 9.3 Our results can be easily extended to the case when communication channels are with delay and multiplicative noises. Consider the protocol (9.2) where $\epsilon(k)$ is not a Bernoulli process but a random process with mean μ and variance σ^2. Then under Assumptions 9.1 and 9.2, the MAS (9.1) is mean-square consensusable over $\mathcal{G}$ if

$$\frac{\mu^2(\tau+1)}{\mu^2(2\tau+1)+\sigma^2}\left[1 - \min_{\omega\in\mathbb{R}} \max_{i=2,\ldots,N} |1-\omega\lambda_i|^2\right] > \tilde{\gamma}_c.$$

When $\tau = 0$, the above result covers Theorem 8.1 in Chapter 8. Further, when there is no stochastic uncertainty, the result is consistent with those stated in You and Xie [2010]. In some papers [Ni and Li, 2013, Zong et al., 2016, 2018], the intensity functions of the multiplicative noises are related to the absolute values of relative states, and there are different time delays in deterministic and stochastic terms of the designed protocol, i.e.

$$u_i(k) = K \sum_{j \in \mathcal{N}_i} \{ [x_j(k-\tau) - x_i(k-\tau)] + \sigma\omega(k)|x_j(k-\tau_1) - x_i(k-\tau_1)| \},$$

where $\tau \geq 0$, $\tau_1 \geq 0$, $\sigma \geq 0$, $\mathbb{E}\{\omega(k)\} = 0$ and $\mathbb{E}\{\omega(k)^2\} = 1$. As shown in Zong et al. [2016], the mean-square stability of the associated NCS is independent of τ_1. Hence, τ_1 does not affect the MAS's mean-square consensusability. Indeed, under Assumptions 9.1 and 9.2, the MAS (9.1) is mean-square consensusable over $\mathcal{G}$ if

$$\frac{\tau+1}{2\tau+1+\sigma^2}\left[1 - \min_{\omega\in\mathbb{R}} \max_{i=2,\dots,N} |1-\omega\lambda_i|^2\right] > \tilde{\gamma}_c.$$

9.4 Consensusability with Delay and Nonidentical Packet Dropouts

We further consider the case when the packet dropouts of communication channels are nonidentical. Then the error dynamics is given by

$$\delta(k+1) = (I_N \otimes A)\delta(k) + (\mathcal{L}(k) \otimes BK)\delta(k-\tau),$$

where $\mathcal{L}_{ij}(k) = \mathcal{L}_{ij}\epsilon_{ij}(k)$ when $i \neq j$ and $\mathcal{L}_{ii}(k) = -\sum_{j=1}^{N} \mathcal{L}_{ij}\epsilon_{ij}(k)$. In this case, $\epsilon_{ij}(k)$ cannot be separated from the Laplacian matrix, which makes it difficult to analyze this error dynamics. To tackle this problem, we assume that if two agents can communicate with each other, then they communicate through the same channel. This is a reasonable assumption, especially when $\mathcal{G}$ is undirected. Based on this assumption, we adopt the notions of compressed in-incidence matrix (CIIM), compressed incidence matrix (CIM), and CEL for $\mathcal{G}$ given in Chapter 8 and model the dynamics on the edges to separate the packet dropouts from the network topology.

Denote the set of compressed edges of $\mathcal{G}$, CIIM, CIM, and CEL by $\overline{\mathcal{E}}$, $\bar{E}$, E and $\mathcal{L}_e$, respectively. Since two bidirectional edges are treated as one, $\epsilon_{ij}(k) = \epsilon_{ji}(k)$ if (i,j) is bidirectional. Define the state on the edge $e_s = (i,j) \in \overline{\mathcal{E}}$ as $z_s = x_i - x_j$. Let $\zeta_s = \epsilon_{ji}$ represent the packet dropout on e_s with packet dropout rate p_s. It is not hard to see that

$$\begin{aligned} u_i(k) &= K \sum_{j \in \mathcal{N}_i} \epsilon_{ij}(k)(x_j(k-\tau) - x_i(k-\tau)) \\ &= -K \sum_{l=1}^{|\overline{\mathcal{E}}|} \zeta_s(k)[\bar{E}]_{il} z_l(k-\tau). \end{aligned}$$

Then it holds that for any edge $e_s = (i,j) \in \overline{\mathcal{E}}$,

$$\begin{aligned} z_s(k+1) &= Az_s(k) + B[u_i(k) - u_j(k)] \\ &= Az_s(t) - BK\sum_{l=1}^{|\overline{\mathcal{E}}|}\zeta_l(k)([\bar{E}]_{il} - [\bar{E}]_{jl})z_l(k-\tau) \\ &= Az_s(k) - BK\sum_{l=1}^{|\overline{\mathcal{E}}|}\zeta_l(t)[\mathcal{L}_e]_{sl}z_l(k-\tau), \end{aligned}$$

where the last equality holds due to that

$$[\mathcal{L}_e]_{sl} = \sum_{q=1}^{N}[E]_{qs}[\bar{E}]_{ql} = [E]_{is}[\bar{E}]_{il} + [E]_{js}[\bar{E}]_{jl} = [\bar{E}]_{il} - [\bar{E}]_{jl}.$$

Let

$$Z(k) = [z_1'(k), \dots, z_{|\overline{\mathcal{E}}|}'(k)]'$$

and

$$\zeta(k) = \mathrm{diag}\{\zeta_1(k), \dots, \zeta_{|\overline{\mathcal{E}}|}(k)\}.$$

For brevity, denote the mean of $\zeta(k)$ by $\Lambda = \mathrm{diag}\{1-p_1, \dots, 1-p_{|\overline{\mathcal{E}}|}\}$ and the covariance matrix by $\Sigma = [\sigma_{ij}] \in \mathbb{R}^{|\overline{\mathcal{E}}|\times|\overline{\mathcal{E}}|}$, where $\sigma_{ij} = \mathbb{E}\{(\zeta_i - p_i + 1)(\zeta_j - p_j + 1)\}$ for $i \neq j$ and $\sigma_{ii} = p_i(1-p_i)$. We have that

$$Z(k+1) = (I_{|\overline{\mathcal{E}}|} \otimes A)Z(k) - (\mathcal{L}_e\zeta(k) \otimes BK)Z(k-\tau). \tag{9.10}$$

Suppose that $\mathcal{G}$ contains a directed spanning tree. With appropriate indexing of edges, $\bar{E} = [\bar{E}_t, \bar{E}_r]$ and $E = [E_t, E_r]$, where $\bar{E}_t, E_t$ correspond to the edges on the spanning tree and $\bar{E}_r, E_r$ correspond to the remaining edges. Then, as shown in Chapter 8, we have that

(1) there exists a matrix T such that $E_r = E_tT$;
(2) $\mathcal{L}_e$ is similar to

$$\begin{bmatrix} MR' & M\theta \\ \mathbf{0} & \mathbf{0}_{(|\overline{\mathcal{E}}|-N+1)\times(|\overline{\mathcal{E}}|-N+1)} \end{bmatrix},$$

where $R = [I, T]$, $M = E_t'\bar{E}$, and θ is the orthonormal basis of the null space of E. The nonzero eigenvalues of $\mathcal{L}_e$ equal that of MR'.

Correspondingly, $Z(k) = [Z_t'(k), Z_r'(k)]'$, where Z_t are the edge states on the directed spanning tree and Z_r are the remaining edge states. Following a similar approach in Chapter 8, we have the following result.

Lemma 9.2 *If $\mathcal{G}$ contains a directed spanning tree, then $Z_r = (T' \otimes I)Z_t$.*

Lemma 9.2 further implies that

$$\begin{aligned}
Z_t(k+1) &= (I \otimes A)Z_t(k) - ([E]_t'[\bar{E}]_t\zeta_t \otimes BK)Z_t(k-\tau) \\
&\quad - ([E]_t'[\bar{E}]_r\zeta_r(k) \otimes BK)Z_r(k-\tau) \\
&= (I \otimes A)Z_t(k) - \left[\left([E]_t'[\bar{E}]_t\zeta_t(k) + [E]_t'[\bar{E}]_r\zeta_r(k)T'\right) \otimes BK\right] Z_t(k-\tau) \\
&= (I \otimes A)Z_t(k) - (M\zeta(k)R' \otimes BK)Z_t(k-\tau). \qquad (9.11)
\end{aligned}$$

Hence, when system (9.11) is mean-square stable with some K, in view of Lemma 9.2, system (9.10) must be mean-square stable with K, which amounts to that the MAS (9.1) achieves mean-square consensus with K. Therefore, to derive consensusability conditions, we first extend Lemma 9.1 to the case with delay and multiple multiplicative noises.

Lemma 9.3 *The stochastic system*

$$x(k+1) = Ax(k) + B\epsilon(k)Kx(k-\tau),$$

where the i.i.d. random process $\epsilon(k) = diag\{\epsilon_1(k), \dots, \epsilon_n(k)\}$ has mean Λ and covariance Σ, is mean-square stable if there exist $P > 0$ and K such that

$$\begin{aligned}
&(A + B\Lambda K)'P(A + B\Lambda K) + K'(\Sigma \odot B'PB)K \\
&\quad + \tau(A + B\Lambda K - I)'P(A + B\Lambda K - I) < P.
\end{aligned}$$

Now we state a sufficient consensusability condition, which can be solved via a feasibility problem over real numbers.

Theorem 9.2 *Under Assumption 9.1, the MAS (9.1) is mean-square consensusable over $\mathcal{G}$ under the protocol (9.2) provided that there exists some $\kappa \in \mathbb{R}$ such that*

$$\kappa(M\Lambda R' + R\Lambda M') - \kappa^2\left[(2\tau+1)R\Lambda M'M\Lambda R' + R(\Sigma \odot M'M)R'\right] > \frac{\tilde{\gamma}_c}{\tau+1}I, \qquad (9.12)$$

where $\tilde{\gamma}_c$ is the critical value for the modified Riccati inequality (9.6).

Proof: According to Lemma 7.1, the sufficient condition (9.12) implies that there exists some $P > 0$ such that

$$\begin{aligned}
I \otimes P > I \otimes \tilde{A}'P\tilde{A} - \{&\kappa(\tau+1)(M\Lambda R' + R\Lambda M') \\
- \kappa^2(\tau+1)&\left[(2\tau+1)R\Lambda M'M\Lambda R' + R(\Sigma \odot M'M)R'\right]\} \otimes \tilde{A}'PB(B'PB)^{-1}B'P\tilde{A}.
\end{aligned}$$

Let $K = \kappa(B'PB)^{-1}B'P\tilde{A}$ and $\mathcal{P} = I \otimes P$. Then the above inequality is equivalent to

$$\begin{aligned}&[I \otimes \tilde{A} - (\tau+1)M\Lambda R' \otimes BK]'\mathcal{P}[I \otimes \tilde{A} - (\tau+1)M\Lambda R' \otimes BK]\\&\quad + \tau(\tau+1)(M\Lambda R' \otimes BK)'\mathcal{P}M\Lambda R' \otimes BK\\&\quad + (\tau+1)(R \otimes K)'\left[(\Sigma \otimes \mathbf{11}') \odot (M \otimes B)'\mathcal{P}(M \otimes B)\right](R \otimes K) < \mathcal{P}.\end{aligned}$$

Substitute $\tilde{A} = (\tau+1)A - \tau I$ into the preceding inequality and do similar computations as shown in the proof of Theorem 9.1 to get that

$$\begin{aligned}&(I \otimes A - M\Lambda R' \otimes BK)'\mathcal{P}(I \otimes A - M\Lambda R' \otimes BK)\\&\quad + \tau(M\Lambda R' \otimes BK)'\mathcal{P}(M\Lambda R' \otimes BK)\\&\quad + (R \otimes K)'\left[(\Sigma \otimes \mathbf{11}') \odot (M \otimes B)'\mathcal{P}(M \otimes B)\right](R \otimes K)\\&\quad + \tau(I \otimes A - M\Lambda R' \otimes BK - I)'\mathcal{P}(I \otimes A - M\Lambda R' \otimes BK - I) < \mathcal{P}.\end{aligned}$$

In view of Lemma 9.3, the system (9.11) is mean-square stable with K, which implies that the MAS (9.1) is mean-square consensusable with this designed control gain K. □

Closed-form consensusability conditions can be obtained under some specific configurations.

Corollary 9.2 *Under Assumption 9.1, when $M\Lambda R' + R\Lambda M' > 0$, the MAS (9.1) is mean-square consensusable over $\mathcal{G}$ under the protocol (9.2) if*

$$\frac{(\tau+1)\lambda^2_{\min}(M\Lambda R' + R\Lambda M')}{4[(2\tau+1)\rho(RR')\rho(\Lambda M'M\Lambda) + \max_i \sigma_{ii}\rho(M'M)\rho(RR')]} \geq \tilde{\gamma}_c. \tag{9.13}$$

Moreover, when $\Lambda = (1-p)I$, the MAS (9.1) is mean-square consensusable if

$$\frac{(\tau+1)(1-p)\lambda^2_{\min}(MR' + RM')}{4[(2\tau+1)(1-p)\rho(RR')\rho(M'M) + p\rho(M'M)\rho(RR')]} \geq \tilde{\gamma}_c.$$

Proof: When $M\Lambda R' + R\Lambda M' > 0$, we have $\lambda_{\min}(M\Lambda R' + R\Lambda M') > 0$. Let

$$\begin{aligned}f(\kappa) &= \kappa\lambda_{\min}(M\Lambda R' + R\Lambda M')\\&\quad - \kappa^2\left[(2\tau+1)\rho(RR')\rho(\Lambda M'M\Lambda) + \max_i \sigma_{ii}\rho(M'M)\rho(RR')\right].\end{aligned}$$

Then

$$\max_{\kappa\in\mathbb{R}} f(\kappa) = \frac{\lambda^2_{\min}(M\Lambda R' + R\Lambda M')}{4[(2\tau+1)\rho(RR')\rho(\Lambda M'M\Lambda) + \max_i\sigma_{ii}\rho(M'M)\rho(RR')]}$$

which is achieved at

$$\kappa^* = \frac{\lambda_{\min}(M\Lambda R' + R\Lambda M')}{2[(2\tau+1)\rho(RR')\rho(\Lambda M'M\Lambda) + \max_i\sigma_{ii}\rho(M'M)\rho(RR')]}.$$

Since $\kappa^* > 0$, it holds that

$$\begin{aligned}
&\kappa^*(M\Lambda R' + R\Lambda M') - \kappa^{*2}\left[(2\tau+1)R\Lambda M'M\Lambda R' + R(\Sigma \odot M'M)R'\right] \\
&\geq \kappa^* \lambda_{\min}(M\Lambda R' + R\Lambda M') - \kappa^{*2}\Big[(2\tau+1)\rho(RR')\rho(\Lambda M'M\Lambda) \\
&\quad + \max_i \sigma_{ii}\rho(M'M)\rho(RR')\Big] \\
&= \frac{\lambda_{\min}^2(M\Lambda R' + R\Lambda M')}{4[(2\tau+1)\rho(RR')\rho(\Lambda M'M\Lambda) + \max_i \sigma_{ii}\rho(M'M)\rho(RR')]} \\
&\geq \frac{\tilde{\gamma}_c}{\tau+1} I,
\end{aligned}$$

where the last inequality holds since inequality (9.13) is true. Hence, the MAS (9.1) is mean-square consensusable by Theorem 9.2. When $\Lambda = (1-p)I$, inequality (9.13) can be simplified. This completes the proof. □

We call $\overline{\mathcal{G}} = (\mathcal{V}, \overline{\mathcal{E}})$ a compressed graph of $\mathcal{G}$. When $\overline{\mathcal{G}}$ is a directed spanning tree, then $M = \mathcal{L}_e$ and $R = I$. Further result is derived.

Corollary 9.3 *Under Assumption 9.1, when $\mathcal{L}_e\Lambda + \Lambda\mathcal{L}_e' > 0$ and $\overline{\mathcal{G}}$ is a directed spanning tree, the MAS (9.1) is mean-square consensusable over $\mathcal{G}$ under the protocol (9.2) if*

$$\frac{(\tau+1)\lambda_{\min}^2(\mathcal{L}_e\Lambda + \Lambda\mathcal{L}_e')}{4\left[\max\limits_{i=1,\dots,|\overline{\mathcal{E}}|}(2\tau+1)(1-p_i)^2 + \max\limits_{i=1,\dots,|\overline{\mathcal{E}}|} p_i(1-p_i)\right]\rho(\mathcal{L}_e'\mathcal{L}_e)} \geq \tilde{\gamma}_c.$$

When $\Lambda = (1-p)I$ and $\mathcal{L}_e + \mathcal{L}_e' > 0$, the MAS (9.1) is mean-square consensusable if

$$\frac{(\tau+1)(1-p)}{4[2\tau(1-p)+1]}\frac{\lambda_{\min}^2(\mathcal{L}_e + \mathcal{L}_e')}{\rho(\mathcal{L}_e'\mathcal{L}_e)} \geq \tilde{\gamma}_c.$$

Moreover, when $\Lambda = (1-p)I$ and G is an undirected tree, the MAS (9.1) is mean-square consensusable if

$$\frac{(\tau+1)(1-p)}{[2\tau(1-p)+1]}\frac{\lambda_2^2}{\lambda_N^2} \geq \tilde{\gamma}_c.$$

Remark 9.4 When rank$(B) = n$, by Lemma 7.1, $\tilde{\gamma}_c = 1 - \frac{1}{[(\tau+1)\rho(A)-\tau)]^2}$. Then following a similar approach to Corollary 9.1, we can also give a tolerable range of the spectra radius of the MAS in terms of the delay, the packet dropout rate and the network topology to ensure its mean-square consensusability.

9.5 Illustrative Examples

Concrete examples are provided to illustrate the derived consensusability conditions. We consider an MAS with five agents. Each agent is described by a one-dimensional state-space model

$$x_i(k+1) = ax_i(k) + u_i(k),$$

where $a \geq 1$. The communication graph $\mathcal{G}$ is shown in Figure 9.1. Clearly, Assumption 9.1 is satisfied.

We first compute the tolerable range of a ensuring the mean-square consensusability according to Corollary 9.1 when the communication channels are with delay and identical packet dropouts. Four different combinations of τ and p are taken into account, i.e. $\{\tau = 0, p = 0\}$, $\{\tau = 0, p = 0.1\}$, $\{\tau = 1, p = 0\}$, and $\{\tau = 1, p = 0.1\}$. Then we consider the case when the packet dropouts are nonidentical, but with identical packet dropout rate, i.e. $\Lambda = (1-p)I$ for some p. Note that $\overline{\mathcal{G}}$ can be chosen as a directed spanning tree. Then we can perform the computations of the tolerable a by Corollary 9.3 and Remark 9.4 for the previous four sets of τ and p. The results are shown in Table 9.1, which clearly indicate that the existence of delays and packet dropouts greatly restricts the stability degree of the MAS, especially when the packet dropouts are nonidentical. It is also worth noting that by comparing the tolerable ranges of a in both cases with

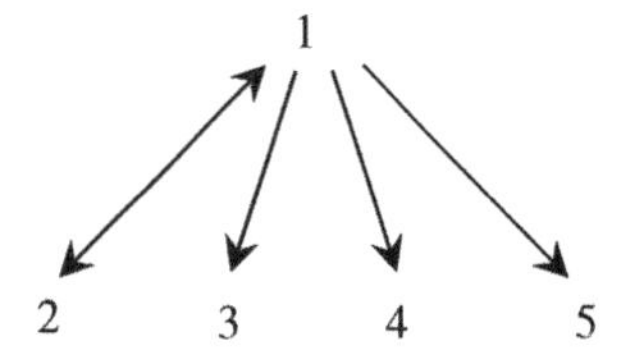

Figure 9.1 The communication graph $\mathcal{G}$. Source: Zheng et al. (2019)/with permission of IEEE.

Table 9.1 The tolerable a ensuring mean-square consensus.

	Identical packet dropouts	**Non-identical packet dropouts with identical packet dropout rate**
$\tau = 0, p = 0$	$1 \leq a < 3$	$1 \leq a < 1.0172$
$\tau = 0, p = 0.1$	$1 \leq a < 2.2361$	$1 \leq a < 1.0154$
$\tau = 1, p = 0$	$1 \leq a < 1.2833$	$1 \leq a < 1.0057$
$\tau = 1, p = 0.1$	$1 \leq a < 1.2638$	$1 \leq a < 1.0055$

Source: Zheng et al. (2019)/with permission of IEEE.

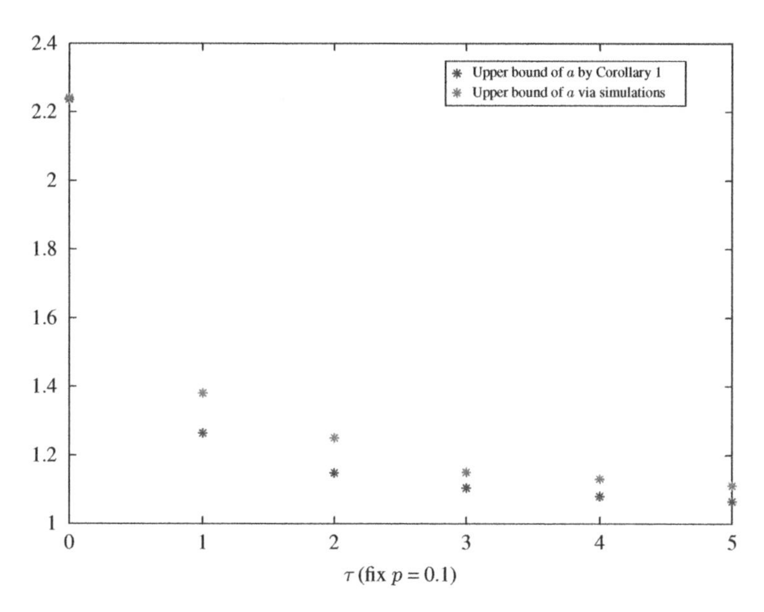

Figure 9.2 Upper bounds of tolerable a from Corollary 1 and via simulations when $p = 0.1$. Source: Zheng et al. (2019)/with permission of IEEE.

$p = 0$, we can see that Corollary 1 is less conservative than Corollary 3 since the former one is derived from a more special case which results in a better result.

Through a large number of simulations, we can approximate the real tolerable range of a for a predefined set of time delay and packet dropout rate. Consider the case when the communication channels are with delay and identical packet dropouts. Fixing $p = 0.1$ and taking $\tau = 0,1, \ldots, 5$, we depict the upper bounds of tolerable a from Corollary 1 and via simulations in Figure 9.2. It can be observed that when $\tau = 0$, the upper bounds of tolerable a from Corollary 9.1 and via simulations are consistent. This is due to that when there is no delay, the consensusability condition provided by Theorem 9.1 for a scalar MAS is not only sufficient but also necessary. Fixing $\tau = 1$ and taking $p = 0.05(k-1), k = 1, \ldots, 21$, we depict the upper bounds of tolerable a from Corollary 1 and via simulations in Figure 9.3. In principle, the real upper bound of tolerable a should also be monotonically decreasing with respect to p. However, as indicated in Figure 9.3, there are some fluctuations of the upper bounds via simulations since they are computed based on a finite number of simulations. It can be observed that when $p = 1$, the upper bounds of tolerable a from Corollary 1 and via simulations are equal to 1, i.e. there does not exist some scalar $a \geq 1$ such that the MAS can achieve consensus. This is clear since when there is no information exchange among the agents, the isolated and unstable agents will never reach a consensus.

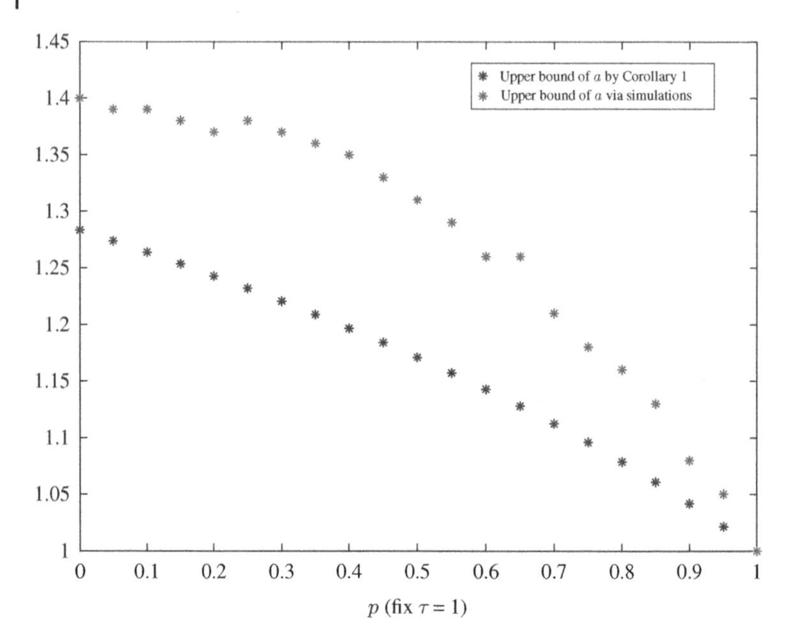

Figure 9.3 Upper bounds of tolerable a from Corollary 1 and via simulations when $\tau = 1$. Source: Zheng et al. (2019)/with permission of IEEE.

9.6 Conclusions

In this chapter, the mean-square consensusability problem of discrete-time linear MASs over directed networks with constant communication delay and packet dropouts is studied. When the packet dropouts in different communication channels are identical, the consensusability problem is transformed into a simultaneous mean-square stabilization problem of systems with constant time delay and packet dropout. When the packet dropouts are nonidentical, the dynamics over edges is built based on the notion of CEL for directed graphs to separate the packet dropouts from the network topology. Sufficient consensusability conditions are obtained for both cases in terms of the communication delay, the packet dropout rates, the communication network topology, and the agent dynamics.

Bibliography

Y. H. Ni and X. Li. Consensus seeking in multi-agent systems with multiplicative measurement noises. *Systems & Control Letters*, 62(5):430–437, 2013.

W. Ren and R. W. Beard. *Distributed consensus in multi-vehicle cooperative control: theory and applications*. Springer-Verlag London Limited, London, 2008.

N. Xiao, L. Xie, and L. Qiu. Feedback stabilization of discrete-time networked systems over fading channels. *IEEE Transactions on Automatic Control*, 57(9):2176–2189, 2012.

K. You and L. Xie. Minimum data rate for mean square stabilization of discrete LTI systems over lossy channels. *IEEE Transactions on Automatic Control*, 55(10):2373–2378, 2010.

W. Yu, G. Chen, and M. Cao. Some necessary and sufficient conditions for second-order consensus in multi-agent dynamical systems. *Automatica*, 46(6):1089–1095, 2010.

X. Zong, T. Li, and J. Zhang. Consensus control of discrete-time multi-agent systems with time-delays and multiplicative measurement noises. *Scientia Sinica Mathematica*, 46(10):1617–1636, 2016.

X. Zong, T. Li, G. Yin, L. Wang, and J. Zhang. Stochastic consentability of linear systems with time delays and multiplicative noises. *IEEE Transactions on Automatic Control*, 63(4):1059–1074, 2018.

10

Distributed Consensus over Markovian Packet Loss Channels

10.1 Introduction

The distributed consensus over independent and identically distributed (i.i.d.) fading channels has been studied in Chapters 7 and 8. However, the i.i.d. assumption fails to capture the temporal correlation of channel fadings. The finite-state Markov channel (FSMC) model is a simple model that captures the main features of fading channel [Goldsmith, 2005], where the channel fading is approximated as a discrete-time Markov process. In this chapter, we consider the Markovian packet loss channel, which is a special type of FSMCs with only two states representing the reception of the packet or not. We try to provide consensusability conditions for general multi-agent systems by directly analyzing the solvability of certain matrix inequalities.

This chapter is organized as follows: the problem formulation is stated in Section 10.2. The consensusability results for the cases with identical Markovian and nonidentical Markovian packet losses are discussed in Section 10.3 and Section 10.4, respectively. Numerical simulations are provided in Section 10.5. This chapter ends with some concluding remarks in Section 10.6.

10.2 Problem Formulation

In this chapter, we assume that each agent has the homogeneous dynamics

$$x_i(t+1) = Ax_i(t) + Bu_i(t), \quad i = 1, \dots, N, \tag{10.1}$$

where $x_i \in \mathbb{R}^n$ is the system state; $u_i \in \mathbb{R}^m$ is the control input; and (A, B) is controllable and B has full-column rank. The interaction among agents is characterized by an undirected connected graph $\mathcal{G} = \{\mathcal{V}, \mathcal{E}\}$. The consensus protocol is given by

$$u_i(t) = \sum_{j\in\mathcal{N}_i} \gamma_{ij}(t)K(x_i(t) - x_j(t)), \tag{10.2}$$

Control over Communication Networks: Modeling, Analysis, and Design of Networked Control Systems and Multi-Agent Systems over Imperfect Communication Channels, First Edition.
Jianying Zheng, Liang Xu, Qinglei Hu, and Lihua Xie.

where $\gamma_{ij}(t) \in \{0, 1\}$ models the lossy effect of the communication channel from agent j to agent i, which satisfies that $\gamma_{ij}(t) = 1$ when the transmission is successful at time t, and 0 otherwise.

Throughout the chapter, we say that the MAS (10.1) is mean-square consensusable by the protocol (10.2) if there exists K such that the MAS (10.1) can achieve mean-square consensus under the protocol (10.2), i.e. for all $i, j \in \mathcal{V}$, $\lim_{t\to\infty} \mathbb{E}\{\|x_i(t) - x_j(t)\|^2\} = 0$. The following assumption is made as in You and Xie [2011]:

Assumption 10.1

1. All the eigenvalues of A are either on or outside the unit disk and (A, B) is controllable.
2. The undirected graph $\mathcal{G}$ is connected.

10.3 Identical Markovian Packet Loss

In the section, we consider the case that $\gamma(t)$ follows a Markov process and make the following assumption.

Assumption 10.2 $\gamma_{ij}(t) = \gamma(t)$ for all $(i, j) \in \mathcal{E}$ and $t \geq 0$. Moreover, $\{\gamma(t)\}_{t\geq 0}$ is a time-homogeneous Markov process with two states $\{0, 1\}$ and the transition probability matrix Q is

$$Q = \begin{bmatrix} 1-q & q \\ p & 1-p \end{bmatrix}, \tag{10.3}$$

where $0 < p < 1$ represents the failure rate and $0 < q < 1$ denotes the recovery rate.

Remark 10.1 Markov models are widely used to capturing temporal correlations of channel conditions [Goldsmith, 2005, Huang and Dey, 2007]. However, due to the correlations of packet losses over time, the methods used to deal with the i.i.d. channel fading in Chapter 7 cannot be applied to the Markovian packet loss case.

Remark 10.2 Note that in general, the packet losses in a multi-agent system may not be identical since different channels are involved in communications among agents. However, there are some situations where the packet losses among channels can be considered identical. For example, the packet losses are caused by a malicious jammer which randomly jams the communication channels or global positioning system (GPS) position signals. On the other hand, the assumption allows us to characterize how the communication channel, the network topology, and the agent dynamics interplay with each other in the consensus problem, which is challenging for general nonidentical packet losses cases.

Let $x(t) = [x_1'(t), \dots, x_N'(t)]'$. Define the consensus error as

$$\delta(t) = \left(I - \frac{1}{N}\mathbf{1}\mathbf{1}'\right)x(t).$$

Following similar derivations as in Chapter 7, the consensus error dynamics is given by

$$\delta(t+1) = (I \otimes A + \gamma(t)\mathcal{L} \otimes BK)\delta(t), \tag{10.4}$$

where $\mathcal{L}$ is the graph Laplacian of $\mathcal{G}$. If there exists K such that system (10.4) is mean-square stable, i.e. $\lim_{t\to\infty}\mathbb{E}\{\delta(t)\delta'(t)\} = 0$, the MAS can achieve mean-square consensus. It has been proved in Chapter 7 that the mean-square stability of (10.4) is equivalent to the simultaneous mean-square stability of

$$\delta_i(t+1) = (A + \lambda_i\gamma(t)BK)\delta_i(t), \quad i = 2, \dots, N, \tag{10.5}$$

where λ_i is the nonzero positive eigenvalue of $\mathcal{L}$ with $\lambda_2 \le \cdots \le \lambda_N$.

Since $\{\gamma(t)\}_{t\ge0}$ is a Markov process, the consensusability is equivalent to the simultaneous mean-square stabilizability of the $N-1$ Markov jump linear systems (10.5). In view of Theorem 3.9 in Costa et al. [2005] describing the stability of Markov jump linear systems, we can obtain the following consensusability condition.

Theorem 10.1 *Under Assumptions 10.1 and 10.2, the MAS* (10.1) *is mean-square consensusable by the protocol* (10.2) *if and only if either of the following conditions holds*

1. *There exist K, $P_{i,1} > 0$, $P_{i,2} > 0$ with $i = 2, \dots, N$, such that*

$$P_{i,1} - (1-q)A'P_{i,1}A - q(A + \lambda_i BK)'P_{i,2}(A + \lambda_i BK) > 0,$$
$$P_{i,2} - pA'P_{i,1}A - (1-p)(A + \lambda_i BK)'P_{i,2}(A + \lambda_i BK) > 0.$$

2. *There exists K such that*

$$\rho\left(\mathcal{H}_i\right) < 1,$$

for all $i = 2, \dots, N$ with

$$\mathcal{H}_i = \begin{bmatrix} (1-q)A \otimes A & p(A + \lambda_i BK) \otimes (A + \lambda_i BK) \\ qA \otimes A & (1-p)(A + \lambda_i BK) \otimes (A + \lambda_i BK) \end{bmatrix}.$$

With similar transformations as in the proof of Theorem 10.2, the consensus criterion in Theorem 10.1 (1) can be shown to be equivalent to a feasibility problem with bilinear matrix inequality (BMI) constraints. It is well known that checking the solvability of a BMI is generally NP-hard [Toker and Ozbay, 1995]. Therefore, in the sequel, we propose a sufficient consensus condition in terms of the feasibility of linear matrix inequalities (LMIs) by a fixed $P_{i,1}$ and $P_{i,2}$.

Theorem 10.2 *Under Assumptions 10.1 and 10.2, if there exist $Q_1 > 0$, $Q_2 > 0$, Z_1, Z_2 such that the following LMIs hold,*

$$\begin{bmatrix} Q_1 & * & * & * \\ \sqrt{qc}(AQ_1 + BZ_1) & Q_2 & * & * \\ \sqrt{q(1-c)}AQ_1 & 0 & Q_2 & * \\ \sqrt{1-q}AQ_1 & 0 & 0 & Q_1 \end{bmatrix} > 0, \tag{10.6}$$

$$\begin{bmatrix} Q_2 & * & * & * \\ \sqrt{(1-p)c}(AQ_2 + BZ_2) & Q_2 & * & * \\ \sqrt{(1-p)(1-c)}AQ_2 & 0 & Q_2 & * \\ \sqrt{p}AQ_2 & 0 & 0 & Q_1 \end{bmatrix} > 0, \tag{10.7}$$

where $c = 1 - \left(\frac{\lambda_N - \lambda_2}{\lambda_N + \lambda_2}\right)^2 > 0$, then the MAS (10.1) is mean-square consensusable by the protocol (10.2) and an admissible control gain is given by

$$K = -\frac{2}{\lambda_2 + \lambda_N}(B'Q_2^{-1}B)^{-1}B'Q_2^{-1}A.$$

Proof: If there exist $Q_1 > 0$, $Q_2 > 0$, Z_1, Z_2 such that (10.6) and (10.7) hold, then there exist $P_1 = Q_1^{-1} > 0$, $P_2 = Q_2^{-1} > 0$, $K_1 = Z_1P_1$ and $K_2 = Z_2P_2$ such that

$$\begin{bmatrix} P_1^{-1} & * & * & * \\ \sqrt{qc}(A + BK_1)P_1^{-1} & P_2^{-1} & * & * \\ \sqrt{q(1-c)}AP_1^{-1} & 0 & P_2^{-1} & * \\ \sqrt{1-q}AP_1^{-1} & 0 & 0 & P_1^{-1} \end{bmatrix} > 0, \tag{10.8}$$

$$\begin{bmatrix} P_2^{-1} & * & * & * \\ \sqrt{(1-p)c}(A + BK_2)P_2^{-1} & P_2^{-1} & * & * \\ \sqrt{(1-p)(1-c)}AP_2^{-1} & 0 & P_2^{-1} & * \\ \sqrt{p}AP_2^{-1} & 0 & 0 & P_1^{-1} \end{bmatrix} > 0. \tag{10.9}$$

Left and right multiply (10.8) with diag(P_1, I, I, I), and left and right multiply (10.9) with diag(P_2, I, I, I), we obtain

$$\begin{bmatrix} P_1 & * & * & * \\ \sqrt{qc}(A+BK_1) & P_2^{-1} & * & * \\ \sqrt{q(1-c)}A & 0 & P_2^{-1} & * \\ \sqrt{1-q}A & 0 & 0 & P_1^{-1} \end{bmatrix} > 0,$$
$$\begin{bmatrix} P_2 & * & * & * \\ \sqrt{(1-p)c}(A+BK_2) & P_2^{-1} & * & * \\ \sqrt{(1-p)(1-c)}A & 0 & P_2^{-1} & * \\ \sqrt{p}A & 0 & 0 & P_1^{-1} \end{bmatrix} > 0.$$

In view of Schur complement lemma, we know that

$$\begin{aligned} &P_1 - (1-q)A'P_1A - q(1-c)A'P_2A \\ &\quad - qc(A+BK_1)'P_2(A+BK_1) > 0, \end{aligned} \tag{10.10}$$
$$\begin{aligned} &P_2 - pA'P_1A - (1-p)(1-c)A'P_2A \\ &\quad - (1-p)c(A+BK_2)'P_2(A+BK_2) > 0. \end{aligned} \tag{10.11}$$

For any $P_2 > 0$ and K, we have

$$\begin{aligned} (A+BK)'P_2(A+BK) &= A'P_2A - A'P_2B(B'P_2B)^{-1}B'P_2A \\ &\quad + (K + (B'P_2B)^{-1}B'P_2A)'(B'P_2B)(K + (B'P_2B)^{-1}B'P_2A), \end{aligned}$$

which implies

$$A'P_2B(B'P_2B)^{-1}B'P_2A \geq A'P_2A - (A+BK)'P_2(A+BK).$$

Therefore,

$$-cA'P_2A + c(A+BK)'P_2(A+BK) \geq -cA'P_2B(B'P_2B)^{-1}B'P_2A,$$

for any K and $P_2 > 0$.

In view of the above result and (10.10), (10.11), we have,

$$\frac{P_1 - (1-q)A'P_1A}{q} > A'P_2A - cA'P_2B(B'P_2B)^{-1}B'P_2A, \tag{10.12}$$
$$\frac{P_2 - pA'P_1A}{1-p} > A'P_2A - cA'P_2B(B'P_2B)^{-1}B'P_2A. \tag{10.13}$$

Since $-c = \min_k \max_i (\lambda_i^2 k^2 + 2\lambda_i k)$ and the optimal k to the min-max problem is $\check{k} = -\frac{2}{\lambda_2+\lambda_N}$, we know that

$$\frac{P_1 - (1-q)A'P_1A}{q} > A'P_2A + (\lambda_i^2 \check{k}^2 + 2\lambda_i \check{k})A'P_2B(B'P_2B)^{-1}B'P_2A,$$

$$\frac{P_2 - pA'P_1A}{1-p} > A'P_2A + (\lambda_i^2 \check{k}^2 + 2\lambda_i \check{k})A'P_2B(B'P_2B)^{-1}B'P_2A$$

hold for all $i = 2, \ldots, N$. Therefore, Theorem 10.1.1 is satisfied with

$$P_{i,1} = P_1, P_{i,2} = P_2, K = \check{k}(B'P_2B)^{-1}B'P_2A.$$

The proof is completed. □

10.3.1 Analytic Consensus Conditions

The criterion stated in Theorem 10.2 is easy to verify. However, it fails to provide insights into the consensusability problem. In the following, we provide analytical consensusability conditions, which show directly how the channel properties, the network topology, and the agent dynamics interplay with each other to allow the existence of a distributed consensus controller. The following lemma is needed in proving the main result and is stated first.

Lemma 10.1 **(Schenato et al. [2007])** *Under Assumption 10.1, if (A, B) is controllable, then*

$$P > A'PA - \gamma A'PB(B'PB)^{-1}B'PA \tag{10.14}$$

admits a solution $P > 0$, if and only if γ is greater than a critical value $\gamma_c > 0$. The calculation of γ_c is discussed in Remark 7.2.

Theorem 10.3 *Under Assumptions 10.1 and 10.2, the MAS (10.1) is mean-square consensusable by the protocol (10.2) if*

$$\gamma_1 = \min\{q, 1-p\}\left[1 - \left(\frac{\lambda_N - \lambda_2}{\lambda_N + \lambda_2}\right)^2\right] > \gamma_c, \tag{10.15}$$

where γ_c is given in Lemma 10.1. Moreover, if (10.15) holds, an admissible control gain is given by

$$K = -\frac{2}{\lambda_2 + \lambda_N}(B'PB)^{-1}B'PA,$$

where P is the solution to (10.14) with $\gamma = \gamma_1$.

Proof: If (10.15) holds, in view of Lemma 10.1, there exists a $P > 0$ to (10.14) with $\gamma = \gamma_1$, such that

$$P > A'PA - qcA'PB(B'PB)^{-1}B'PA,$$
$$P > A'PA - (1-p)cA'PB(B'PB)^{-1}B'PA.$$

Since $-c = \max_i(\lambda_i^2\check{k}^2 + 2\lambda_i\check{k})$ with $\check{k} = -\frac{2}{\lambda_2+\lambda_N}$, we have

$$P > A'PA + q(2\lambda_i\check{k} + \lambda_i^2\check{k}^2)A'PB(B'PB)^{-1}B'PA,$$
$$P > A'PA + (1-p)(2\lambda_i\check{k} + \lambda_i^2\check{k}^2)A'PB(B'PB)^{-1}B'PA$$

for all $i = 2, \dots, N$, which is the condition in Theorem 10.1.1 with

$$P_{i,1} = P_{i,2} = P, \quad K = \check{k}(B'PB)^{-1}B'PA.$$

The proof is completed. □

Remark 10.3 Theorem 10.2 is obtained by letting $P_{i,1} = P_1$, $P_{i,2} = P_2$. Theorem 10.3 is obtained by letting $P_{i,1} = P_{i,2} = P$. Since the latter is more restrictive than the former, we can expect that Theorem 10.3 is more restrictive than Theorem 10.2, which will be illustrated by a simulation example in Subsection 10.3.2.

In conjunction with the analytic sufficient consensusability condition in Theorem 10.3, we also provide an explicit necessary consensusability condition as stated below.

Theorem 10.4 *Under Assumptions 10.1 and 10.2, the MAS* (10.1) *is mean-square consensusable by the protocol* (10.2) *only if there exists K such that*

$$(1-q)^{\frac{1}{2}}\rho(A) < 1, \tag{10.16}$$

$$(1-p)^{\frac{1}{2}}\rho(A + \lambda_i BK) < 1 \tag{10.17}$$

for all $i = 2, \dots, N$. Moreover, when the agent is with single input, i.e. $m = 1$, the MAS (10.1) *is mean-square consensusable by the protocol* (10.2) *only if*

$$(1-q)^{\frac{1}{2}}\rho(A) < 1, \tag{10.18}$$

$$(1-p)^{\frac{n}{2}}\det(A)\frac{\lambda_N - \lambda_2}{\lambda_N + \lambda_2} < 1. \tag{10.19}$$

Proof: If the MAS can achieve mean-square consensus, in view of Theorem 10.1.1, we have that there exist $P_{i,1} > 0, P_{i,2} > 0$, and K such that

$$P_{i,1} > (1-q)A'P_{i,1}A,$$
$$P_{i,2} > (1-p)(A+\lambda_i BK)'P_{i,2}(A+\lambda_i BK),$$

for all $i = 2, \dots, N$. Further from Lyapunov stability theory, we can obtain the necessary conditions (10.16), (10.17).

When the agent is with single input, following similar line of argument as in the necessity proof of Lemma 3.1 in You and Xie [2011], we can obtain the necessary condition (10.19) from (10.17). The proof is completed. □

10.3.2 Critical Consensus Condition for Scalar Agent Dynamics

When all the agents are with scalar dynamics, we can obtain a closed-form consensusability condition. The following lemma is needed in the proof of the main result and is stated first.

Lemma 10.2 **(Xu et al. [2017])** *Let Q be defined in* (10.3)*; $D = \begin{bmatrix} 1 & 0 \\ 0 & \delta \end{bmatrix}$ with $0 < q, p, \delta < 1$; $\lambda \in \mathbb{R}$, $|\lambda| \geq 1$. The following conditions are equivalent:*

1.

$$\lambda^2 \rho(Q'D) < 1,$$

2.

$$1 - \lambda^2(1-q) > 0, \tag{10.20}$$

$$\lambda^2 \delta \left[1 + \frac{p(\lambda^2 - 1)}{1 - \lambda^2(1-q)}\right] < 1. \tag{10.21}$$

Without loss of generality, for scalar agent dynamics, i.e. $n = m = 1$, we let $A = a \in \mathbb{R}, B = 1, K = k \in \mathbb{R}$. The main result is stated as follows.

Theorem 10.5 *Under Assumptions 10.1 and 10.2, the MAS* (10.1) *with scalar agent dynamics is mean-square consensusable by the protocol* (10.2) *if and only if*

$$(1-q)a^2 < 1, \tag{10.22}$$

$$a^2 \left(\frac{\lambda_N - \lambda_2}{\lambda_N + \lambda_2}\right)^2 \left[1 + \frac{p(a^2-1)}{1 - a^2(1-q)}\right] < 1. \tag{10.23}$$

Proof: In view of Theorem 10.1.2, for scalar agent dynamics, the MAS (10.1) is mean-square consensusable by the protocol (10.2) if and only if there exists k such that

$$a^2\rho\left(Q' \times \begin{bmatrix} 1 & 0 \\ 0 & \frac{(a+\lambda_i k)^2}{a^2} \end{bmatrix}\right) < 1,$$

for all $i = 2, \dots, N$. Further from Lemma 10.2, a necessary and sufficient consensus condition is that if there exists k such that for all $i = 2, \dots, N$.

$$(1-q)a^2 < 1, \tag{10.24}$$

$$\left(a+\lambda_i k\right)^2 \left[1+\frac{p(a^2-1)}{1-a^2(1-q)}\right] < 1. \tag{10.25}$$

Since (10.25) holds for all i, we have that

$$\min_k \max_i (a+\lambda_i k)^2 \left[1+\frac{p(a^2-1)}{1-a^2(1-q)}\right] < 1.$$

Moreover, since

$$\min_k \max_i (a+\lambda_i k)^2 = a^2\left(\frac{\lambda_N-\lambda_2}{\lambda_N+\lambda_2}\right)^2,$$

we can obtain the necessary and sufficient consensusability condition (10.22), (10.23) from (10.24), (10.25). The proof is completed. □

Interestingly, we can show that when the agent dynamics is scalar, the sufficient condition indicated in Theorem 10.2 is also necessary. Theorem 10.2 is equivalent to check the solvability of (10.12) and (10.13). For scalar systems with $A = a \in \mathbb{R}$, $B = 1$, (10.12) and (10.13) change to

$$[1-(1-q)a^2]P_1 > qa^2(1-c)P_2, \tag{10.26}$$

$$[1-(1-p)(1-c)a^2]P_2 > pa^2P_1. \tag{10.27}$$

We can show that the necessary and sufficient condition to guarantee the solvability of the above inequality is given by (10.22) and (10.23). Since $P_1 > 0$ and $qa^2(1-c)P_2 > 0$, we have from (10.26) that $1-(1-q)a^2 > 0$, which gives (10.22). Let $\theta = 1-c = \left(\frac{\lambda_N-\lambda_2}{\lambda_N+\lambda_2}\right)^2$. We can obtain a lower bound of P_1 from (10.26) and substitute this bound into (10.27) to obtain

$$[1-(1-p)\theta a^2]P_2 > pa^2\frac{qa^2\theta P_2}{[1-(1-q)a^2]}.$$

Since $P_2 > 0$, we further have that

$$[1-(1-p)\theta a^2][1-(1-q)a^2] - pqa^4\theta > 0,$$

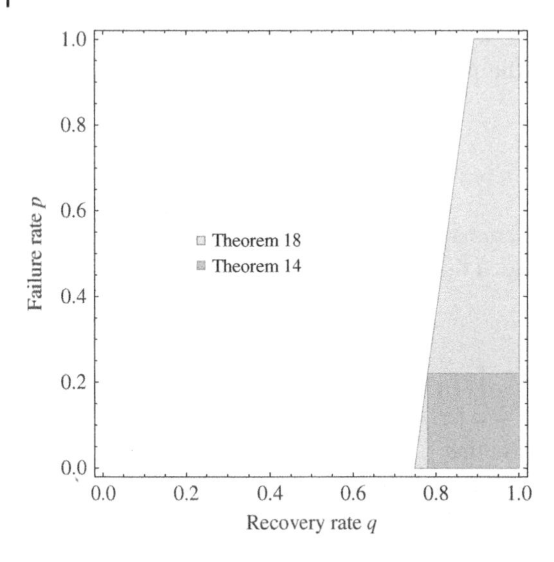

Figure 10.1 Tolerable failure rate and recovery rate. Source: Xu et al. (2020)/with permission of IEEE.

which implies

$$a^2\theta[(1-p) - a^2(1-p-q)] < 1 - a^2(1-q).$$

Dividing both sides by $1 - a^2(1-q)$, we can obtain (10.23).

In contrast to the tightness of Theorem 10.2 for scalar systems, Theorem 10.3 is generally not necessary. Consider the case that $A = 2$, $B = 1$, $\lambda_2 = 2$ and $\lambda_N = 3$, then the tolerable (p, q) from Theorem 10.5 are given by

$$q > \frac{3}{4}, \quad p < 7 \times \left(q - \frac{3}{4}\right).$$

While the sufficiency indicated by Theorem 10.3 is given by

$$q > \frac{25}{32}, \quad p < \frac{7}{32}.$$

The tolerable failure rate and recovery rate are plotted in Figure 10.1. It is clear that the result in Theorem 10.3 is conservative in the case of scalar agent dynamics.

The assumption of identical channel loss distributions is somewhat restrictive and less practical. However, it is the simplest case in studying the consensus problem over Markovian packet loss channels and is expected to shed light on solutions to more general nonidentical cases, which is studied in the subsequent section.

10.4 Nonidentical Markovian Packet Loss

In the presence of nonidentical packet losses, the consensus error dynamics of δ is given by $\delta(t+1) = (I \otimes A + \mathcal{L}(t) \otimes BK)\,\delta(t)$ with $\mathcal{L}(t)$ modeling both the communication topology and the packet losses. Since the packet loss is coupled with the communication topology in $\mathcal{L}(t)$, the analysis of the mean-square consensus is difficult. Therefore, the edge Laplaican [Zelazo and Mesbahi, 2011] is used to model the consensus error dynamics as in Chapter 7, which allows to separate the lossy effect from the network topology to facilitate the consensusability analysis by building dynamics on edges rather than on vertexes.

The following graph definitions are needed in introducing the edge Laplacian. A virtual orientation of the edge in an undirected graph is an assignment of directions to the edge (i,j) such that one vertex is chosen to be the initial node and the other to be the terminal node. The incidence matrix E for an oriented graph $\mathcal{G}$ is a $\{0, 1, -1\}$-matrix with rows and columns indexed by vertices and edges of $\mathcal{G}$, respectively, such that

$$[E]_{ik} = \begin{cases} +1, & \text{if } i \text{ is the initial node of edge } k, \\ -1, & \text{if } i \text{ is the terminal node of edge } k, \\ 0, & \text{otherwise.} \end{cases}$$

The graph Laplacian $\mathcal{L}$ and edge Laplacian $\mathcal{L}_e$ can be constructed from the incidence matrix, respectively, as $\mathcal{L} = EE'$, $\mathcal{L}_e = E'E$ [Zelazo and Mesbahi, 2011].

Since fading is mostly caused by path loss and shadowing from obstacles, for simplicity, we assume that the fadings (packet losses) on the same edge are equal, i.e. $\gamma_{ij}(t) = \gamma_{ji}(t)$ if j and i are connected, which makes sense in some practical applications [Dey et al., 2009]. For general channel fading models, where $\gamma_{ij} \neq \gamma_{ji}$, the directed edge Laplacian [Zeng et al., 2016a,b] or the compressed edge Laplacian [Xu et al., 2018] can be used to formulate the consensus dynamics and similar analysis methods proposed in this section can be applicable to the study of the consensusability problem. Define the state on the ith edge as $z_i = x_j - x_k$, with j, k representing the initial node and the terminal node of the ith edge, respectively. Following the definition of incidence matrix, the controller (10.2) can be alternatively represented as

$$u_j(t) = K \sum_{k=1}^{l} e_{jk} \zeta_k(t) z_k(t),$$

where l is the total number of edges in $\mathcal{G}$, e_{jk} is the jkth element of E, and ζ_k denotes the packet loss effect on the kth edge, i.e. $\zeta_k = \gamma_{ij}$, where i, j are the initial node and terminal node of the kth edge. If we define $z = [z_1', z_2', \dots, z_l']'$, then the following similar steps as in Chapter 7, the closed-loop dynamics on edges can be calculated as follows:

$$z(t+1) = \left(I \otimes A + \mathcal{L}_e \zeta(t) \otimes BK\right) z(t) \tag{10.28}$$

with $\zeta(t) = \text{diag}(\zeta_1(t), \zeta_2(t), \dots, \zeta_l(t))$.

With appropriate indexing of edges, we can write the incidence matrix E as $E = [E_\tau, E_c]$, where edges in E_τ are on a spanning tree and edges in E_c complete cycles in $\mathcal{G}$. We further have that when $\mathcal{G}$ is connected, there exists a matrix T, such that $E_c = E_\tau T$ [Zelazo and Mesbahi, 2011]. Moreover, with such indexing of edges, we can decompose the edge state z as $z = [z_\tau', z_c']'$, where z_τ is the edge state on the spanning tree and z_c is the remaining edge state. Besides, it is straightforward to verify that $z_c = (T' \otimes I)z_\tau$, since $z = [z_\tau', z_c']' = (E' \otimes I)x = ([E_\tau, E_c]' \otimes I)x$ and $E_c = E_\tau T$. Let $M = E_\tau' E$ and $R = [I, T]$, we have that

$$\begin{aligned} z_\tau(t+1) &= (I \otimes A)z_\tau(t) + ((E_\tau' E_\tau \zeta_\tau(t)) \otimes (BK))z_\tau(t) \\ &\quad + ((E_\tau' E_c \zeta_c(t)) \otimes (BK))z_c(t) \\ &= (I \otimes A + (E_\tau' E_\tau \zeta_\tau(t) + E_\tau' E_c \zeta_c(t) T') \otimes (BK))z_\tau(t) \\ &= (I \otimes A + (M\zeta(t)R') \otimes (BK))z_\tau(t), \end{aligned} \tag{10.29}$$

where ζ_τ, ζ_c represent the packet losses on tree edges and cycle edges, respectively. The MAS can achieve mean-square consensus if and only if (10.29) is mean-square stable.

The possible sample space of $\zeta(t)$ is $\Phi = \{\Lambda_0, \dots, \Lambda_{2^l-1}\}$, where the ith element Λ_i is $\Lambda_i = \text{diag}(\eta_1, \dots, \eta_l)$ with $\eta_j \in \{0, 1\}$, $j = 1, \dots, l$, being the jth component of the binary expansion of i, i.e. $i = \eta_l 2^{l-1} + \cdots + \eta_1 2^0$. We make the following assumption for the packet loss matrix $\zeta(t)$.

Assumption 10.3 The packet loss process $\{\zeta(t)\}_{t\geq 0}$ is a time-homogeneous Markov stochastic process, which has o states $\{\Gamma_1, \dots, \Gamma_o\}$, where $\Gamma_i \in \Phi$. The probability transition matrix Q is an $o \times o$ matrix with the (i, j)th element being p_{ij}.

Remark 10.4 It is possible that certain outcomes in Φ are unlikely to happen. For example, if two agents are close to each other, the communication between them can be reliable. It is unlikely that the communication link would undergo packet losses. In such cases, the sample space of $\zeta(t)$ would be a subset of Φ. Therefore, in Assumption 10.3, we use o to denote the carnality of the actual sample space of $\zeta(t)$, which might be smaller than 2^l.

Therefore, (10.29) is a Markov jump linear system. In view of Theorem 3.9 in Costa et al. [2005], we have the following consensus result.

Theorem 10.6 *Under Assumptions 10.1 and 10.3, the MAS* (10.1) *is mean-square consensusable by the protocol* (10.2) *if and only if either of the following conditions holds, where* $S_i(K) = (I \otimes A + M\Gamma_i R' \otimes BK)$

1. *there exist* $P_i > 0$, $i = 1, \dots, o$ *and* K *such that*

$$P_i > \sum_{j=1}^{o} p_{ij} S_j(K)' P_j S_j(K)$$

   for all $i = 1, \dots, o$.
2. *there exists* K *such that*

$$\rho\left((Q' \otimes I)\text{diag}\left(S_i(K) \otimes S_i(K)\right)\right) < 1.$$

We can show that the consensus criterion in Theorem 10.6.1 is equivalent to a feasibility problem with BMI constraints. Therefore, checking the conditions in Theorem 10.6 is generally not easy. In the following, a numerically easy testable condition in terms of the feasibility of LMIs is proposed.

Theorem 10.7 *Under Assumptions 10.1 and 10.3, the MAS* (10.1) *is mean-square consensusable by the protocol* (10.2) *if there exists* $\kappa \in \mathbb{R}$ *such that the following LMIs are feasible,*

$$\begin{bmatrix} -I & \kappa V_i' \\ \kappa V_i & \kappa N_i + \gamma_c I \end{bmatrix} < 0 \tag{10.30}$$

for all $i = 1, \dots, o$, *where* γ_c *is given in Lemma 10.1,* $N_i = \sum_{j=1}^{o} p_{ij}(R\Gamma_j M' + M\Gamma_j R')$, $M_i = \sum_{j=1}^{o} p_{ij} R\Gamma_j M' M\Gamma_j R'$ *and* V_i *is the Cholesky decomposition of* M_i, *i.e.* $M_i = V_i V_i'$. *Moreover, if* (10.30) *is satisfied, a control gain is given by* $K = \kappa(B'PB)^{-1}B'PA$, *where* P *is the solution of* (10.14) *with* $\gamma = \min_i \lambda_{\min}(-\kappa N_i - \kappa^2 M_i)$.

Proof: If (10.30) holds, there exists κ such that $\kappa N_i + \kappa^2 M_i < -\gamma_c I$ for all $i = 1, \dots, o$. Since $\kappa N_i + \kappa^2 M_i$ is real and symmetric, it is diagonalizable by an orthogonal matrix Ψ, i.e. $\Psi'(\kappa N_i + \kappa^2 M_i)\Psi = \Upsilon$ and Υ is diagonal. Then we have that $\Upsilon < -\gamma_c I$. In view of Lemma 10.1, we can find $P > 0$ such that

$$I \otimes P > I \otimes A'PA + \Upsilon \otimes A'PB(B'PB)^{-1}B'PA.$$

Left and right multiply the above inequality with $\Psi \otimes I$ and $\Psi' \otimes I$, we have that

$$I \otimes P > I \otimes A'PA + (\kappa N_i + \kappa^2 M_i) \otimes A'PB(B'PB)^{-1}B'PA.$$

From the definitions of N_i and M_i and the relation that $\sum_{j=1}^{o} p_{ij} = 1$, we further have that

$$I \otimes P > \sum_{j=1}^{o} p_{ij}(I \otimes A'PA + (\kappa R\Gamma_j M' + \kappa M\Gamma_j R' + \kappa^2 R\Gamma_j M'M\Gamma_j R') \otimes A'PB(B'PB)^{-1}B'PA),$$

which is the sufficient condition given in Theorem 10.6.1 with $P_1 = \ldots = P_o = I \otimes P$ and $K = \kappa(B'PB)^{-1}B'PA$. The proof is completed. □

Remark 10.5 This chapter only discusses the consensusability problem over undirected graphs. For the consensusability problem with directed graphs, the compressed edge Laplacian defined in Chapter 8 or the directed edge Laplacian Zeng et al. [2016a,b] can be used to model the consensus error dynamics. Then following similar derivations as in this section, consensus conditions over directed graphs in the presence of Markovian packet losses can be obtained.

10.5 Numerical Simulations

In this section, simulations are conducted to verify the derived results. In simulations, agents are assumed to have system parameters

$$A = \begin{bmatrix} 1.1830 & -0.1421 & -0.0399 \\ 0.1764 & 0.8641 & -0.0394 \\ 0.1419 & -0.1098 & 0.9689 \end{bmatrix}, \quad B = \begin{bmatrix} 0.1697 & 0.3572 \\ 0.5929 & 0.5165 \\ 0.1355 & 0.9659 \end{bmatrix}.$$

The initial state of each agent is uniformly and randomly generated from the interval $(0, 0.5)$. We assume that there are four agents, and the undirected communication topology among agents is given in Figure 10.2a. We first consider the consensus with identical Markovian packet losses. The Markov packet losses in transmission channels are assumed to have parameters $p = 0.2, q = 0.7$. With such configurations, the LMIs in Theorem 10.2 are feasible and an admissible control parameter is given by

$$K = \begin{bmatrix} 2.0423 & -1.3094 & -0.0885 \\ -0.5723 & 0.2934 & -0.3335 \end{bmatrix}.$$

The simulation results are presented by averaging over 1000 runs. Mean-square consensus errors for agent 1 are plotted in Figure 10.3, which shows that the mean-square consensus is achieved.

Second, we consider the consensus over nonidentical Markovian packet loss networks. We index the edges and apply a virtual orientation to each edge as in Figure 10.2b. Denote the packet loss processes in these edges as

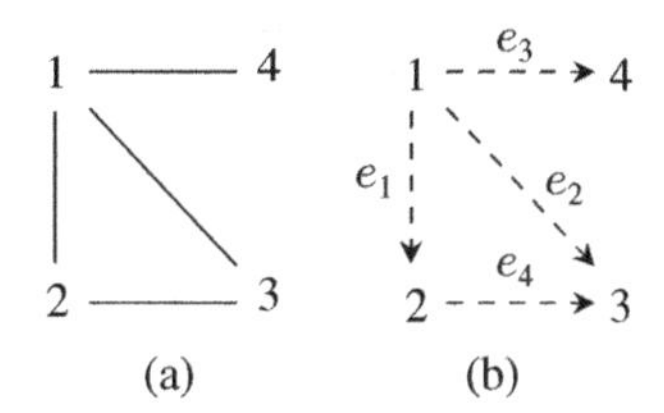

Figure 10.2 Communication graphs used in simulations: (a) an undirected graph (b) applying an orientation to edges in (a). Source: Xu et al. (2020)/with permission of IEEE.

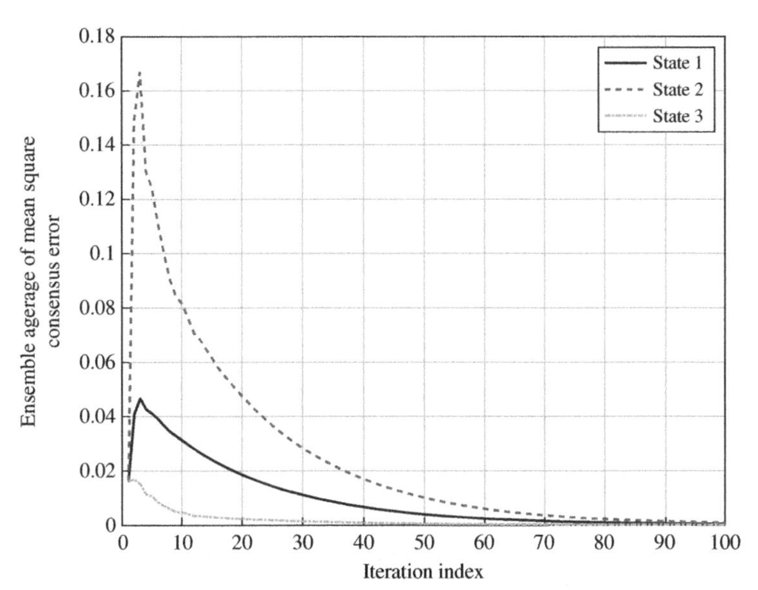

Figure 10.3 Mean-square consensus error for agent 1 under identical packet losses. Source: Xu et al. (2020)/with permission of IEEE.

$\zeta_1(t), \zeta_2(t), \zeta_3(t), \zeta_4(t)$. Suppose the time-homogeneous Markov packet loss process $\{\zeta(t)\}_{t\geq 0}$ with $\zeta(t) = \text{diag}(\zeta_1(t), \zeta_2(t), \zeta_3(t), \zeta_4(t))$ has three states $\Gamma_1 = \text{diag}(1, 0, 1, 0)$, $\Gamma_2 = \text{diag}(0, 1, 0, 1)$, $\Gamma_3 = \text{diag}(1, 1, 1, 1)$ and is with the probability transition matrix

$$Q = \begin{bmatrix} 0.3811 & 0.1446 & 0.4743 \\ 0.2445 & 0.5121 & 0.2434 \\ 0.5390 & 0.0215 & 0.4395 \end{bmatrix}.$$

With such settings, we can show that (10.30) is feasible, and an admissible control gain is given by

$$K = \begin{bmatrix} 1.7394 & -1.3873 & 0.0771 \\ -0.2133 & 0.2212 & -0.5269 \end{bmatrix}.$$

The simulation results are presented by averaging over 1000 runs. The consensus error for agent 1 is given in Figure 10.4, which shows that the mean-square consensus is achieved.

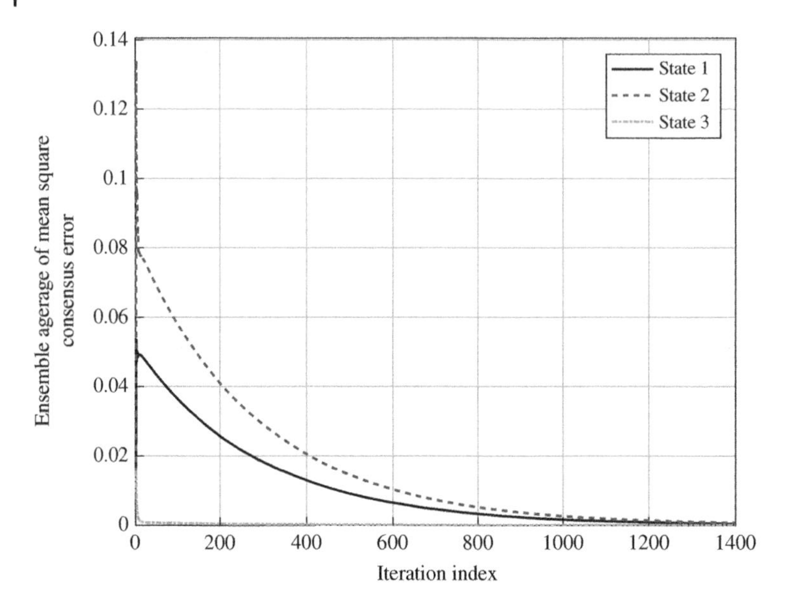

Figure 10.4 Mean-square consensus error for agent 1 under nonidentical packet losses. Source: Xu et al. (2020)/with permission of IEEE.

10.6 Conclusions

This chapter studies the consensusability problem of MASs, where agents communicate with each other through Markovian packet loss channels. We try to determine conditions under which there exists a linear distributed consensus controller such that the MAS can achieve mean-square consensus. We first study the case with identical Markovian packet losses. A necessary and sufficient consensus condition is first derived based on the stability of Markov jump linear systems. Then a numerically verifiable consensus criterion in terms of the feasibility of LMIs is proposed. Furthermore, analytic sufficient conditions and necessary conditions for mean-square consensusability are provided for general MASs. The case with nonidentical packet loss is studied subsequently. The necessary and sufficient consensus condition and a sufficient consensus condition in terms of LMIs are proposed. In the end, numerical simulations are conducted to verify the derived results. The derived results show how the agent dynamics, the network topology, and the channel loss interplay with each other to allow the existence of a linear distributed consensus controller.

Bibliography

O. L. V. Costa, M. D. Fragoso, and R. P. Marques. *Discrete-time Markov jump linear systems*. Probability and its applications. Springer, London, 2005.

S. Dey, A. S. Leong, and J. S. Evans. Kalman filtering with faded measurements. *Automatica*, 45(10):2223–2233, 2009.

A. Goldsmith. *Wireless communications*. Cambridge University Press, Cambridge, 2005.

M. Huang and S. Dey. Stability of Kalman filtering with Markovian packet losses. *Automatica*, 43(4):598–607, 2007.

L. Schenato, B. Sinopoli, M. Franceschetti, K. Poolla, and S. S. Sastry. Foundations of control and estimation over lossy networks. *Proceedings of the IEEE*, 95(1):163–187, 2007.

O. Toker and H. Ozbay. On the NP-hardness of solving bilinear matrix inequalities and simultaneous stabilization with static output feedback. In *Proceedings of the 1995 American Control Conference*, volume 4, pages 2525–2526. IEEE, 1995.

L. Xu, L. Xie, and N. Xiao. Mean square stabilization over Gaussian finite-state Markov channels. *IEEE Transactions on Control of Network Systems*, 5(4):1830–1840, 2017.

L. Xu, J. Zheng, N. Xiao, and L. Xie. Mean square consensus of multi-agent systems over fading networks with directed graphs. *Automatica*, 95:503–510, 2018.

K. You and L. Xie. Consensusability of discrete-time multi-agent systems over directed graphs. In *Proceedings of the 30th Chinese Control Conference*, pages 6413–6418, 2011.

D. Zelazo and M. Mesbahi. Edge agreement: Graph-theoretic performance bounds and passivity analysis. *IEEE Transactions on Automatic Control*, 56(3):544–555, 2011.

Z. Zeng, X. Wang, and Z. Zheng. Convergence analysis using the edge Laplacian: Robust consensus of nonlinear multi-agent systems via ISS method. *International Journal of Robust and Nonlinear Control*, 26(5):1051–1072, 2016a.

Z. Zeng, X. Wang, and Z. Zheng. Edge agreement of multi-agent system with quantised measurements via the directed edge Laplacian. *IET Control Theory & Applications*, 10(13):1583–1589, 2016b.

11

Synchronization of the Delayed Vicsek Model

11.1 Introduction

This chapter studies the synchronization problem of a special class of multi-agent systems, called the Vicsek model [Vicsek et al., 1995], which can be used to characterize the collective behaviors of multi-agent systems, such as flocking birds, shoaling fish, and marching ants. The Vicsek model is a discrete-time model consisting of a population of agents moving in the plane with the same speed but with different headings. At each time step, the heading of each agent is updated based on the average of those of its neighbors, which are the agents located in a circle of a prespecified radius centered at the given agent's current position. Simulation results show that all agents, using only the local rule, might eventually move in the same direction, exhibiting the synchronization behavior. The first result on mathematical analysis of the Vicsek model was given in Jadbabaie et al. [2003], where the consensus is proved under some connectivity assumptions on the whole evolution of dynamical neighbor graphs. Following the work of Jadbabaie et al. [2003], a substantial body of literature has appeared, see, Ren and Beard [2005], Moreau [2005], Bertsekas and Tsitsiklis [2007], and Cao et al. [2012], to name only a few. Most of the results mentioned above are based on some connectivity conditions on the dynamically changing neighbor graphs. It is often either nontrivial or impossible to check a priori the conditions, and thus such results are difficult to apply in practice. Having observed this fact [Liu and Guo, 2008] derive a sufficient condition depending only on the initial state to guarantee the synchronization of the Vicsek model, and the smallest possible radius for synchronization is given in Chen et al. [2012]. Along the line of the work in Liu and Guo [2008], several extensions have been made, including extensions to a continuous-time Vicsek model [Zhu et al., 2013], to a high-dimensional Vicsek model [Liu, 2014], and to a simple time-delayed Vicsek model [Chen et al., 2016].

Control over Communication Networks: Modeling, Analysis, and Design of Networked Control Systems and Multi-Agent Systems over Imperfect Communication Channels, First Edition.
Jianying Zheng, Liang Xu, Qinglei Hu, and Lihua Xie.

Inspired by the aforementioned work, this chapter will provide a rigorous theoretical analysis of the synchronization problem of a Vicsek model. Compared to the classical Vicsek model, we study the more practical delayed Vicsek model where the time delays appear in both position and heading transmissions. Moreover, the derived synchronization conditions depend only on the initial conditions and time delay, which can be easily verified.

The rest of this chapter is organized as follows: some basic definitions in graph theory are introduced in Section 11.2. The problem formulation is given in Section 11.3. The delay-dependent synchronization conditions on the initial state for the delayed linear and nonlinear Vicsek models are given in Sections 11.4 and 11.5, respectively. Some simulation results are provided in Section 11.6. Finally, a conclusion follows in Section 11.7.

11.2 Directed Graphs

It is convenient to model the neighborhood relationship between agents by directed graphs. However, some definitions of graph adopted in this chapter is slightly different from that in Section 1.4.1. In addition, some new concepts from graph theory are employed. Thus, the preliminaries of graph theory used in this chapter are stated below.

A digraph $\mathcal{G} = (\mathcal{V}, \mathcal{E})$ consists of a finite set $\mathcal{V} = \{1, \dots, k\}$ of vertices and a set $\mathcal{E} \subset \mathcal{V} \times \mathcal{V}$ of arcs. If $(j, i) \in \mathcal{E}$, one usually says that j is a neighbor of i. The neighbor set of vertex i is $\mathcal{N}_i := \{j : (j, i) \in \mathcal{E}\}$. Note that in this chapter, self-loop is allowed. That is, if $i \in \mathcal{N}_i$, i.e. $(i, i) \in \mathcal{E}$, we understand that $\mathcal{G}$ has a self-loop at vertex i. A path in $\mathcal{G}$ from i_0 to i_p is a sequence $i_0, i_1, \dots, i_p$ of distinct vertices such that each successive pair of vertices is an arc of the digraph. The integer p (the number of its arcs) is the length of the path. If there exists a path from i to j, then vertex j is said to be reachable from vertex i, and we define the distance from i to j, $\text{dist}(i,j)$, as the length of a shortest path from i to j. A digraph $\mathcal{G}$ is said to be strongly connected if each vertex can be reached from any other vertex. A weaker connected concept is rooted digraph, i.e. $\mathcal{G}$ is said to be a rooted digraph if we can find a vertex (called a root) such that any other vertex of $\mathcal{G}$ is reachable from it.

Consider a rooted digraph $\mathcal{G}$. For each root ℓ of $\mathcal{G}$, the quantity $\max_{j\in\mathcal{V}} \text{dist}(\ell, j)$ can be treated as the depth of the spanning tree of $\mathcal{G}$ rooted at ℓ. The smallest depth of $\mathcal{G}$ is

$$d := \min_{\ell \text{ is a root}} \max_{j\in\mathcal{V}} \text{dist}(\ell, j).$$

It is trivial that $d \leq k - 1$. Moreover, $d \leq k/2$ when $\mathcal{G}$ is undirected [Aouchiche et al., 2009]. We understand that $d = \infty$ when $\mathcal{G}$ is not rooted.

11.3 Problem Formulation

Consider a population of k autonomous agents, labeled $1, \ldots, k$, moving in the plane with the same absolute velocity v ($v > 0$). For each agent i, its position $(x_i(t), y_i(t)) \in \mathbb{R}^2$ at time $t \in \mathbb{N}$ is updated according to the following equation:

$$\begin{aligned} x_i(t+1) &= x_i(t) + v\cos\theta_i(t), \\ y_i(t+1) &= y_i(t) + v\sin\theta_i(t), \end{aligned} \tag{11.1}$$

where $\theta_i(t)$ is the heading angle of agent i at time t. The neighbor set of agent i at time t is denoted by

$$\mathcal{N}_i(t) := \{j : \ d_{ij}(t) < r_i\},$$

where r_i is the neighborhood radius of agent i and

$$d_{ij}(t) := \sqrt{[x_i(t) - x_j(t - \tau_{ij}(t))]^2 + [y_i(t) - y_j(t - \tau_{ij}(t))]^2}$$

with $\tau_{ij}(\cdot)$ being the communication delay from agent j to agent i at that time. The neighbor sets generate the neighbor graph $\mathcal{G}(t) = (\mathcal{V}, \mathcal{E}(t))$ in a natural way: $\mathcal{V} = \{1, \ldots, k\}$ and $(j, i) \in \mathcal{E}(t)$ if and only if $j \in \mathcal{N}_i(t)$. Note that $\mathcal{G}(t)$ is a directed graph generally since different agents may have different radii r_i.

In the delayed linear Vicsek model, agent i adjusts its heading according to the average of the headings of its neighbors:

$$\theta_i(t+1) = \frac{1}{n_i(t)} \sum_{j\in\mathcal{N}_i(t)} \theta_j(t - \tau_{ij}(t)), \tag{11.2}$$

where $n_i(t)$ is the cardinality of $\mathcal{N}_i(t)$. In the delayed nonlinear Vicsek model, $\theta_i(t)$ is updated according to the following equation:

$$\theta_i(t+1) = \arctan\frac{\sum_{j\in\mathcal{N}_i(t)} \sin\theta_j(t - \tau_{ij}(t))}{\sum_{j\in\mathcal{N}_i(t)} \cos\theta_j(t - \tau_{ij}(t))}. \tag{11.3}$$

Note that in these two models, the communication delays appear not only in heading transmission but also in position transmission which defines the neighborhood relationship. Throughout this chapter, we assume that the time-varying delays $\tau_{ij}(\cdot)$ are bounded, i.e. there exists a constant $\tau \geq 0$ such that $0 \leq \tau_{ij}(t) \leq \tau$ for all $t \in \mathbb{N}$ and $i, j \leq k$. Self-delay is not allowed, i.e. $\tau_{ii}(\cdot) \equiv 0$ for all $i \leq k$. The models studied here generalize the classical Vicsek model [Vicsek et al., 1995], in two aspects, including considering the existence of communication delays and allowing the agents to have different neighborhood radii. The objective is to derive some conditions, imposed only on the initial configuration, to guarantee that the agents can realize synchronization in the following sense:

$$\lim_{t\to\infty} \max_{1\leq i,j\leq k} |\theta_i(t) - \theta_j(t)| = 0.$$

To state the main results, the concept of robust subgraphs of a digraph is defined. For any $\rho > 0$, the robust subgraph $\mathcal{H}_\rho$ of the initial neighbor digraph $\mathcal{G}(0) = (\mathcal{V}, \mathcal{E}(0))$ is defined as $\mathcal{H}_\rho := (\mathcal{V}, \mathcal{E}_\rho)$ so that $(j, i) \in \mathcal{E}_\rho$ if and only if

$$\Gamma_{ij}(0) + \rho \le r_i,$$

where

$$\Gamma_{ij}(0) := \max_{-\tau \le s \le 0} \sqrt{[x_i(0) - x_j(s)]^2 + [y_i(0) - y_j(s)]^2}.$$

In addition, denote by d_ρ the smallest depth of $\mathcal{H}_\rho$. Note that $d_\rho = \infty$ if $\mathcal{H}_\rho$ is not rooted.

Remark 11.1 By the definition of the robust subgraph $\mathcal{H}_\rho$, we see that it is a subgraph of the initial neighbor graph $G(0)$. Roughly speaking, the neighbors of agent i in $\mathcal{H}_\rho$ are restricted within the circle of the radius $r_i - \rho$ at any time $s \in [-\tau, 0]$, equivalently to say, they are still neighbors of agent i even when they have moved away a distance ρ from agent i.

11.4 Synchronization of Delayed Linear Vicsek Model

We now present the synchronization condition for the delayed linear Vicsek model (11.1) and (11.2).

Theorem 11.1 *Consider the delayed linear Vicsek model (11.1) and (11.2). Let $\theta_i(s) \in [0,2\pi)$ for all $-\tau \le s \le 0$ and $i \le k$. Then the agents reach synchronization exponentially fast provided that*

$$v \cdot \mathcal{D}(0) \le \sup_{\rho > 0} \frac{\delta\rho}{\alpha}, \tag{11.4}$$

where

$$\begin{aligned} \alpha &:= d_\rho \cdot (\tau + 1) + \tau, \\ \delta &:= \frac{1}{2k^\alpha}, \\ \mathcal{D}(0) &:= \max_{\substack{1 \le i,j \le k \\ -\tau \le s \le 0}} (\theta_i(s) - \theta_j(s)). \end{aligned}$$

To prove Theorem 11.1, the following definitions and lemmas are given. Define

$$\overline{\theta}(t) := \max_{\substack{1 \le i \le k \\ t-\tau \le s \le t}} \theta_i(s), \quad \underline{\theta}(t) := \min_{\substack{1 \le i \le k \\ t-\tau \le s \le t}} \theta_i(s) \tag{11.5}$$

and let $\mathcal{D}(t) := \overline{\theta}(t) - \underline{\theta}(t)$. We next give some stepping stones toward the proof.

Lemma 11.1 *For the delayed linear Vicsek model (11.1) and (11.2), $\overline{\theta}(t)$ is nonincreasing and $\underline{\theta}(t)$ is nondecreasing.*

Proof: We first show that $\overline{\theta}(t+1) \leq \overline{\theta}(t)$ for all $t \in \mathbb{N}$. It follows from (11.2) that for all $i \leq k$,

$$\theta_i(t+1) = \frac{1}{n_i(t)} \sum_{j \in \mathcal{N}_i(t)} \theta_j(t - \tau_{ij}(t)) \leq \frac{1}{n_i(t)} \sum_{j \in \mathcal{N}_i(t)} \overline{\theta}(t) = \overline{\theta}(t),$$

which implies that

$$\overline{\theta}(t+1) = \max_{\substack{1 \leq i \leq k \\ t+1-\tau \leq s \leq t+1}} \theta_i(s) \leq \max\{\overline{\theta}(t), \max_{1 \leq i \leq k} \theta_i(t+1)\} = \overline{\theta}(t).$$

This shows that $\overline{\theta}(t)$ is nonincreasing. Likewise, we can show that $\underline{\theta}(t)$ is nondecreasing. □

Lemma 11.2 *Consider the delayed linear Vicsek model (11.1) and (11.2). Assume that for $t_2 \geq t_1 \geq 0$, there exists a constant $c > 0$ such that*

$$\theta_i(t_2) \leq \overline{\theta}(t_1) - c.$$

Then, for all $t \geq t_2$,

$$\theta_i(t) \leq \overline{\theta}(t_1) - \frac{c}{k^{t-t_2}}.$$

Proof: By (11.2), we have

$$\begin{aligned}
\theta_i(t_2+1) &= \frac{1}{n_i(t_2)} \sum_{j \in \mathcal{N}_i(t_2)} \theta_j(t_2 - \tau_{ij}(t_2)) \\
&\leq \frac{\theta_i(t_2)}{n_i(t_2)} + \frac{n_i(t_2)-1}{n_i(t_2)} \overline{\theta}(t_1) \\
&\leq \frac{1}{n_i(t_2)} \left(\overline{\theta}(t_1) - c\right) + \frac{n_i(t_2)-1}{n_i(t_2)} \overline{\theta}(t_1) \\
&\leq \overline{\theta}(t_1) - \frac{c}{k},
\end{aligned}$$

where the first inequality is from Lemma 11.1. We now finish the proof by induction. □

Lemma 11.3 *Let the hypothesis of Lemma 11.2 be satisfied with $i \in \mathcal{N}_j(t_2 + \tau)$. Then,*

$$\theta_j(t_2 + \tau + 1) \leq \overline{\theta}(t_1) - \frac{c}{k^{\tau+1}}.$$

Proof: Note that

$$
\begin{aligned}
\theta_j(t_2+\tau+1) &= \frac{1}{n_j(t_2+\tau)} \sum_{l\in\mathcal{N}_j(t_2+\tau)} \theta_l(t_2+\tau-\tau_{jl}(t_2)) \\
&\leq \frac{\theta_i(t_2+\tau-\tau_{ji}(t_2))}{n_j(t_2+\tau)} + \frac{n_j(t_2+\tau)-1}{n_j(t_2+\tau)}\overline{\theta}(t_1) \\
&\leq \frac{1}{n_j(t_2+\tau)}\left(\overline{\theta}(t_1) - \frac{c}{k^\tau}\right) + \frac{n_j(t_2+\tau)-1}{n_j(t_2+\tau)}\overline{\theta}(t_1) \\
&\leq \overline{\theta}(t_1) - \frac{c}{k^{\tau+1}},
\end{aligned}
$$

where the second inequality is from Lemma 11.2. □

Now, we are ready to prove Theorem 11.1.

Proof of Theorem 11.1: If (11.4) holds, we can find some $\rho > 0$ such that

$$v \cdot \mathcal{D}(0) \leq \frac{\delta\rho}{\alpha} \tag{11.6}$$

which also implies that $\mathcal{H}_\rho$ is rooted because, otherwise, $\mathcal{D}(0) = 0$. It follows from Lemma 11.1 that $\mathcal{D}(t)$ is nonincreasing. We next show that, for all $t \in \mathbb{N}$,

$$\mathcal{D}(t\alpha) \leq (1-\delta)^t \mathcal{D}(0)$$

by induction. It is trivially true for $t = 0$. Now, assume that it holds true for all $t \leq t^*$. This, together with the monotonicity of $\mathcal{D}(t)$, implies that, for all $t\alpha \leq s < (t+1)\alpha$ with $0 \leq t \leq t^*$,

$$\mathcal{D}(s) \leq \mathcal{D}(t\alpha) \leq (1-\delta)^t \mathcal{D}(0). \tag{11.7}$$

We need to show it is also true for t^*+1. Let $(j,i) \in \mathcal{E}_\rho$. For $t^*\alpha < t \leq (t^*+1)\alpha$, we have

$$
\begin{aligned}
d_{ij}(t) &= \sqrt{(x_i(t)-x_j(t-\tau_{ij}(t)))^2 + (y_i(t)-y_j(t-\tau_{ij}(t)))^2} \\
&= \Big[\big(x_i(t-1)-x_j(t-1-\tau_{ij}(t))\big) + \big(y_i(t-1)-y_j(t-1-\tau_{ij}(t))\big) \\
&\quad + v\big(\cos\theta_i(t-1)-\cos\theta_j(t-1-\tau_{ij}(t))\big)^2 \\
&\quad + v\big(\sin\theta_i(t-1)-\sin\theta_j(t-1-\tau_{ij}(t))\big)^2\Big]^{\frac{1}{2}} \\
&\leq \left[(x_i(t-1)-x_j(t-1-\tau_{ij}(t)))^2 + (y_i(t-1)-y_j(t-1-\tau_{ij}(t)))^2\right]^{\frac{1}{2}} \\
&\quad + v\left[(\cos\theta_i(t-1)-\cos\theta_j(t-1-\tau_{ij}(t)))^2\right. \\
&\quad \left.+ (\sin\theta_i(t-1)-\sin\theta_j(t-1-\tau_{ij}(t)))^2\right]^{\frac{1}{2}}.
\end{aligned}
$$

It can be verified that

$$\begin{aligned}&\left[(\cos\theta_i(t-1)-\cos\theta_j(t-1-\tau_{ij}(t)))^2+(\sin\theta_i(t-1)-\sin\theta_j(t-1-\tau_{ij}(t)))^2\right]^{\frac{1}{2}}\\&=2\left|\sin\frac{\theta_i(t-1)-\theta_j(t-1-\tau_{ij}(t))}{2}\right|\le|\theta_i(t-1)-\theta_j(t-1-\tau_{ij}(t))|.\end{aligned}$$

Therefore, we arrive at

$$\begin{aligned}d_{ij}(t)&\le\left[(x_i(t-1)-x_j(t-1-\tau_{ij}(t)))^2+(y_i(t-1)-y_j(t-1-\tau_{ij}(t)))^2\right]^{\frac{1}{2}}\\&\quad+v|\theta_i(t-1)-\theta_j(t-1-\tau_{ij}(t))|\\&\quad\vdots\\&\le\left[(x_i(0)-x_j(-\tau_{ij}(t)))^2+(y_i(0)-y_j(-\tau_{ij}(t)))^2\right]^{\frac{1}{2}}\\&\quad+v\sum_{s=0}^{t-1}|\theta_i(s)-\theta_j(s-\tau_{ij}(t))|\\&\le\Gamma_{ij}(0)+v\sum_{m=0}^{t^*}\sum_{s=m\alpha}^{(m+1)\alpha}\mathcal{D}(s)\\&\le\Gamma_{ij}(0)+v\alpha\sum_{m=0}^{t^*}(1-\delta)^m\mathcal{D}(0)\\&<\Gamma_{ij}(0)+v\frac{\alpha\mathcal{D}(0)}{\delta}\\&\le\Gamma_{ij}(0)+\rho\\&\le r_i,\end{aligned}$$

where the fourth inequality is from (11.7) and the sixth one is from (11.6). This shows that $(j,i)\in\mathcal{E}(t)$, i.e. $\mathcal{H}_\rho\subseteq\mathcal{G}(t)$ for all $t^*\alpha<t\le(t^*+1)\alpha$.

Take a root ℓ of $\mathcal{H}_\rho$. Without loss of generality, we assume that

$$\theta_\ell(t^*\alpha)\le\frac{\overline{\theta}(t^*\alpha)+\underline{\theta}(t^*\alpha)}{2}=\overline{\theta}(t^*\alpha)-\frac{\mathcal{D}(t^*\alpha)}{2}.$$

For the inverse one, the analysis is very similar to the following and thus omitted. We know that ℓ is also a root of $\mathcal{G}(t)$ for all $t^*\alpha<t\le(t^*+1)\alpha$. We apply Lemma 11.2 to obtain that, for $t\ge t^*\alpha$

$$\theta_\ell(t)\le\overline{\theta}(t^*\alpha)-\frac{\mathcal{D}(t^*\alpha)}{2k^{t-t^*\alpha}}. \tag{11.8}$$

Consider any agent $j\ne\ell$. Then there exists a path $\ell=i_0,i_1,\dots,i_p=j$ from ℓ to j in $\mathcal{H}_\rho$. Moreover, we have $p\le d_\rho$ and this path is preserved in $\mathcal{G}(t)$ for all $t^*\alpha<t\le(t^*+1)\alpha$ since $\mathcal{H}_\rho\subseteq\mathcal{G}(t)$. For i_1, noting the assumption on ℓ, Proposition 11.3 implies that

$$\theta_{i_1}(t^*\alpha+\tau+1)\le\overline{\theta}(t^*\alpha)-\frac{\mathcal{D}(t^*\alpha)}{2k^{\tau+1}}.$$

Iteratively, applying Proposition 11.3 p times to arrive at

$$\theta_j(t^*\alpha + p(\tau+1)) \leq \overline{\theta}(t^*\alpha) - \frac{\mathcal{D}(t^*\alpha)}{2k^{p(\tau+1)}}.$$

This, together with Lemma 11.2, finally gives that, for all $t \geq t^*\alpha + p(\tau+1)$

$$\theta_j(t) \leq \overline{\theta}(t^*\alpha) - \frac{\mathcal{D}(t^*\alpha)}{2k^{t-t^*\alpha}}. \tag{11.9}$$

Recalling that $\alpha = d_\rho(\tau+1)+\tau$, we have

$$t^*\alpha + p(\tau+1) \leq (t^*+1)\alpha - \tau.$$

Using (11.8), we conclude that (11.9) holds for all $j \leq k$ and $t \geq (t^*+1)\alpha - \tau$. Therefore, one has

$$\overline{\theta}((t^*+1)\alpha) \leq \overline{\theta}(t^*\alpha) - \frac{\mathcal{D}(t^*\alpha)}{2k^\alpha} = \overline{\theta}(t^*\alpha) - \delta\mathcal{D}(t^*\alpha)$$

which immediately implies that

$$\begin{aligned}
\mathcal{D}((t^*+1)\alpha) &= \overline{\theta}((t^*+1)\alpha) - \underline{\theta}((t^*+1)\alpha) \\
&\leq \overline{\theta}(t^*\alpha) - \delta\mathcal{D}(t^*\alpha) - \underline{\theta}(t^*\alpha) \\
&= (1-\delta)\mathcal{D}(t^*\alpha) \\
&\leq (1-\delta)^{t^*+1}\mathcal{D}(0),
\end{aligned}$$

where the last inequality follows from our reduction assumption. This completes the proof. □

Remark 11.2 It immediately follows from our sufficient condition (11.4) that the initial digraph $\mathcal{G}(0)$ is rooted generally unless in the trivial case of $\mathcal{D}(0) = 0$, i.e. all the agents already move in the same direction at the initial time.

We now provide a way to compute the supremum in the right-hand side of (11.4). Given the initial neighbor digraph $\mathcal{G}(0) = (\mathcal{V}, \mathcal{E}(0))$, we define the set

$$S := \left\{ r_i - \Gamma_{ij}(0) > 0 : \ (j,i) \in \mathcal{E}(0), 1 \leq i,j \leq k \right\}.$$

If S is nonempty, we list the elements of the finite set S in the ascending order

$$0 = \rho_0 < \rho_1 < \rho_2 < \cdots < \rho_n.$$

Note that

$$\rho_1 = \min_{(j,i)\in\mathcal{E}(0)} (r_i - \Gamma_{ij}(0)). \tag{11.10}$$

For the graph-valued mapping $f : [0,\infty) \mapsto \mathcal{H}_\rho$, we can see that $f(\rho) = \mathcal{H}_{\rho_i}$ for all $\rho_{i-1} < \rho \leq \rho_i$ and $1 \leq i \leq n$. Moreover, $f(\rho_i)$ is a subgraph of $f(\rho_{i-1})$ for all $2 \leq i \leq n$ and $f(\rho) = \emptyset$ for all $\rho > \rho_n$. Let

$$\overline{n} := \max\{ i \leq n : \ \mathcal{H}_{\rho_i} \text{ is rooted} \}.$$

Then, the computation of the supremum in the right-hand side of (11.4) reduces to computing $\overline{n}$ and the quantities $\frac{\delta_i\rho_i}{\alpha_i}$, where

$$\alpha_i = d_{\rho_i} \cdot (\tau + 1) + \tau, \ \delta_i = \frac{1}{2k^{\alpha_i}},$$

and then taking the maximum among them.

Remark 11.3 Theorems 11.1 indicates that it is beneficial to have a small delay and small velocity in order to achieve synchronization. The key is that they guarantee the preservation of some rooted robust subgraphs of the initial neighbor graph in the evolution of the agents with a small velocity. This preservation in turn gives rise to the conservativeness of our results. Although it is difficult to know the degree of the conservativeness, compared with the existing results of $\tau = 0$ and $r_i \equiv r$ in the literature where the initial neighbor graph itself is preserved in the evolution, our results have a clear advantage, as observed below.

We now consider the special case of $\tau = 0$ and $r_i \equiv r$, i.e. the classical Vicsek model, which is well-studied in the literature. By taking $\rho = \rho_1$, where ρ_1 is defined in (11.10), Theorem 11.1 reduces to the following result.

Corollary 11.1 *For the linear Vicsek model (11.1) and (11.2), let $r_i \equiv r$ and $\tau_{ij}(t) \equiv 0$ for all $i,j \leq k$ and $t \geq 0$. Let $\theta_i(0) \in [0,2\pi)$ for all $i \leq k$. Then the agents under the delay-free linear Vicsek model reach synchronization exponentially fast provided that*

$$v \cdot D(0) \leq \frac{r - \max\limits_{i,j} d_{ij}(0)}{2dk^d}, \tag{11.11}$$

where d is the smallest depth of $\mathcal{G}(0)$.

Remark 11.4 Note that when $r_i \equiv r$, the neighbor graph is undirected. In this case, we have $d \leq k/2$ (see e.g. Aouchiche et al. [2009]). Then a sufficient condition for (11.11) to hold is

$$v \cdot D(0) \leq \frac{r - \max\limits_{i,j} d_{ij}(0)}{k^{\frac{k}{2}+1}}.$$

A synchronization condition derived in Liu and Guo [2008] is given by

$$v \cdot 2\pi \leq \frac{r - \max\limits_{i,j} d_{ij}(0)}{k^k}. \tag{11.12}$$

It is easy to see that our condition in Corollary 11.1 is much less conservative than (11.12).

11.5 Synchronization of Delayed Nonlinear Vicsek Model

The result for the delayed nonlinear Vicsek model (11.1) and (11.3) is stated as follows.

Theorem 11.2 *Consider the delayed nonlinear Vicsek model (11.1) and (11.3). Let $\theta_i(s) \in (-\frac{\pi}{2}, \frac{\pi}{2})$ for all $-\tau \le s \le 0$ and $i \le k$. Then the agents reach synchronization exponentially fast provided that*

$$v \cdot \Delta(0) \le \sup_{\rho>0} \frac{\zeta\rho}{\alpha}, \tag{11.13}$$

where α is defined as in Theorem 11.1, and

$$\Delta(0) := \max_{\substack{1 \le i,j \le k \\ -\tau \le s \le 0}} (\tan\theta_i(s) - \tan\theta_j(s)),$$

$$\zeta := \frac{1}{2}\left(\frac{\cos|\theta(0)|_M}{k\cos|\theta(0)|_m}\right)^{\alpha}$$

with $|\theta(0)|_M = \max_{\substack{1 \le i \le k \\ -\tau \le s \le 0}} |\theta_i(s)|$, $|\theta(0)|_m = \min_{\substack{1 \le i \le k \\ -\tau \le s \le 0}} |\theta_i(s)|$.

Let

$$\Delta(t) := \tan\overline{\theta}(t) - \tan\underline{\theta}(t),$$

where $\overline{\theta}(t)$ and $\underline{\theta}(t)$ are defined in (11.5). Note that (11.3) is equivalent to

$$\tan\theta_i(t+1) = \sum_{j\in\mathcal{N}_i(t)} \frac{\cos\theta_j(t-\tau_{ij}(t))}{\sum_{p\in\mathcal{N}_i(t)}\cos\theta_p(t-\tau_{ip}(t))}\tan\theta_j(t-\tau_{ij}(t)). \tag{11.14}$$

We next show some preliminary results toward the proof.

Lemma 11.4 *Consider the delayed nonlinear Vicsek model (11.1) and (11.3). Let $\theta_i(s) \in (-\frac{\pi}{2}, \frac{\pi}{2})$ for all $-\tau \le s \le 0$ and $i \le k$. Then $\tan\overline{\theta}(t)$ is nonincreasing and $\tan\underline{\theta}(t)$ is nondecreasing.*

Proof: By (11.14) and our assumption, for all $i \le k$, we have on the one hand that

$$\begin{aligned}\tan\theta_i(1) &= \sum_{j\in\mathcal{N}_i(0)} \frac{\cos\theta_j(-\tau_{ij}(0))}{\sum_{p\in\mathcal{N}_i(0)}\cos\theta_p(-\tau_{ip}(0))}\tan\theta_j(-\tau_{ij}(0)) \\ &\le \tan\overline{\theta}(0),\end{aligned}$$

and on the other hand that $\tan\theta_i(1) \geq \tan\underline{\theta}(0)$. This implies that

$$\underline{\theta}(0) \leq \theta_i(1) \leq \overline{\theta}(0)$$

for all $i \leq k$. It thus follows that

$$\overline{\theta}(1) \leq \overline{\theta}(0), \quad \underline{\theta}(1) \geq \underline{\theta}(0).$$

Inductively, we can show that $\overline{\theta}(t+1) \leq \overline{\theta}(t)$ and $\underline{\theta}(t+1) \geq \underline{\theta}(t)$ for all $t \in \mathbb{N}$. The statements now follow. □

Lemma 11.5 *Let the hypothesis of Lemma 11.4 be satisfied. Assume that, for $t_2 \geq t_1 \geq 0$, there exists a constant $c > 0$ such that*

$$\tan\theta_i(t_2) \leq \tan\overline{\theta}(t_1) - c.$$

Then, for all $t \geq t_2$,

$$\tan\theta_i(t) \leq \tan\overline{\theta}(t_1) - c\left(\frac{\cos|\theta(0)|_M}{k\cos|\theta(0)|_m}\right)^{t-t_2}.$$

Proof: It follows from (11.14) and Lemma 11.4 that

$$\begin{aligned}
&\tan\theta_i(t_2+1)\\
&= \sum_{j\in\mathcal{N}_i(t_2)} \frac{\cos\theta_j(t_2-\tau_{ij}(t_2))}{\sum_{p\in\mathcal{N}_i(t_2)}\cos\theta_p(t_2-\tau_{ip}(t_2))}\tan\theta_j(t_2-\tau_{ij}(t_2))\\
&= \frac{\cos\theta_i(t_2)\tan\theta_i(t_2) - \sum_{j\neq i\in\mathcal{N}_i(t_2)}\cos\theta_j(t_2-\tau_{ij}(t_2))\tan\theta_j(t_2-\tau_{ij}(t_2))}{\sum_{p\in\mathcal{N}_i(t_2)}\cos\theta_p(t_2-\tau_{ip}(t_2))}\\
&\leq \frac{\cos\theta_i(t_2)\left(\tan\overline{\theta}(t_1)-c\right)}{\sum_{p\in\mathcal{N}_i(t_2)}\cos\theta_p(t_2-\tau_{ip}(t_2))} - \frac{\sum_{j\neq i\in\mathcal{N}_i(t_2)}\cos\theta_j(t_2-\tau_{ij}(t_2))\tan\overline{\theta}(t_1)}{\sum_{p\in\mathcal{N}_i(t_2)}\cos\theta_p(t_2-\tau_{ip}(t_2))}\\
&\leq \tan\overline{\theta}(t_1) - c\frac{\cos|\theta(0)|_M}{k\cos|\theta(0)|_m},
\end{aligned}$$

where the first inequality is from our assumption. We now can finish the proof by induction. □

Lemma 11.6 *Let the hypothesis of Lemma 11.5 be satisfied with $i \in \mathcal{N}_j(t_2+\tau)$. Then, it holds that*

$$\tan\theta_j(t_2+\tau+1) \leq \tan\overline{\theta}(t_1) - c\left(\frac{\cos|\theta(0)|_M}{k\cos|\theta(0)|_m}\right)^{\tau+1}.$$

Proof: Note that

$$\begin{aligned}
&\tan\theta_j(t_2+\tau+1)\\
&=\sum_{\ell\in N_j(t_2+\tau)}\frac{\cos\theta_\ell(t_2+\tau-\tau_{j\ell}(t_2+\tau))}{\sum_{p\in N_j(t_2+\tau)}\cos\theta_p(t_2+\tau-\tau_{jp}(t_2+\tau))}\tan\theta_\ell(t_2+\tau-\tau_{j\ell}(t_2+\tau))\\
&=\frac{\cos\theta_i(t_2+\tau-\tau_{ji}(t_2+\tau))}{\sum_{p\in N_j(t_2+\tau)}\cos\theta_p(t_2+\tau-\tau_{jp}(t_2+\tau))}\tan\theta_i(t_2+\tau-\tau_{ji}(t_2+\tau))\\
&\quad+\sum_{\ell\neq i\in N_j(t_2+\tau)}\frac{\cos\theta_\ell(t_2+\tau-\tau_{j\ell}(t_2+\tau))}{\sum_{p\in N_j(t_2+\tau)}\cos\theta_p(t_2+\tau-\tau_{jp}(t_2+\tau))}\tan\theta_\ell(t_2+\tau-\tau_{j\ell}(t_2+\tau))\\
&\le\frac{\cos\theta_i(t_2+\tau-\tau_{ji}(t_2+\tau))}{\sum_{p\in N_j(t_2+\tau)}\cos\theta_p(t_2+\tau-\tau_{jp}(t_2+\tau))}\left(\tan\overline{\theta}(t_1)-c\left(\frac{\cos|\theta(0)|_M}{k\cos|\theta(0)|_m}\right)^{\tau}\right)\\
&\quad+\frac{\sum_{\ell\neq i\in N_j(t_2+\tau)}\cos\theta_\ell(t_2+\tau-\tau_{j\ell}(t_2+\tau))}{\sum_{p\in N_j(t_2+\tau)}\cos\theta_p(t_2+\tau-\tau_{jp}(t_2+\tau))}\tan\overline{\theta}(t_1)\\
&\le\tan\overline{\theta}(t_1)-c\left(\frac{\cos|\theta(0)|_M}{k\cos|\theta(0)|_m}\right)^{\tau+1},
\end{aligned}$$

where the first inequality is from Lemma 11.5. □

Now we are ready to prove Theorem 11.2.

Proof of Theorem 11.2: It follows from Lemma 11.1 that $\theta_i(t)\in(-\frac{\pi}{2},\frac{\pi}{2})$ for all $t\ge 0$ and $i\le k$. Similar to the proof of Theorem 11.1, using Lemma 11.5 and Proposition 11.6, we can establish that for all $t\in\mathbb{N}$,

$$\Delta(t\alpha)\le(1-\zeta)^t\Delta(0)$$

from which the statement of Theorem 11.2 follows. To avoid repetitions, we omit the details. □

Remark 11.5 In Theorem 11.2, the initial headings are restricted to $(-\frac{\pi}{2},\frac{\pi}{2})$ since synchronization may never happen when they are relaxed to $[-\pi,\pi)$. One can refer to some counterexamples given in Liu and Guo [2008].

We now consider the special case of $\tau=0$ and $r_i\equiv r$. By taking $\rho=\rho_1$ defined in (11.10), Theorem 11.2 reduce to the following results.

Corollary 11.2 *For the nonlinear Vicsek model (11.1) and (11.3), let $r_i\equiv r$ and $\tau_{ij}(t)\equiv 0$ for all $i,j\le k$ and $t\ge 0$. Let $\theta_i(0)\in(-\frac{\pi}{2},\frac{\pi}{2})$ for all $i\le k$. Then the agents under the delay-free nonlinear Vicsek model reach synchronization exponentially fast provided that*

$$v\cdot\Delta(0)\le\frac{r-\max\limits_{i,j}d_{ij}(0)}{2dk^d}\cdot\left(\frac{\cos|\theta(0)|_M}{\cos|\theta(0)|_m}\right)^d,\qquad(11.15)$$

where d is the smallest depth of $\mathcal{G}(0)$.

Remark 11.6 When $r_i \equiv r$, the neighbor graph is undirected, i.e. $d \leq k/2$. Then a sufficient condition for (11.15) to hold is given as

$$v \cdot \Delta(0) \leq \frac{r - \max_{i,j} d_{ij}(0)}{k^{\frac{k}{2}+1}} \cdot \left(\frac{\cos |\theta(0)|_M}{\cos |\theta(0)|_m} \right)^{\frac{k}{2}}.$$

Then the critical absolute velocity is given by

$$v_1 = \frac{r - \max_{i,j} d_{ij}(0)}{\Delta(0)} \cdot \frac{1}{k^{\frac{k}{2}+1}} \cdot \left(\frac{\cos |\theta(0)|_M}{\cos |\theta(0)|_m} \right)^{\frac{k}{2}}.$$

The corresponding critical absolute velocities derived in Liu and Guo [2008] and Chen et al. [2016] are, respectively, given by

$$v_2 = \frac{r - \max_{i,j} d_{ij}(0)}{\Delta(0)} \cdot \frac{1}{k^k} \cdot \left(\cos |\theta(0)|_M \right)^k$$

and

$$v_3 = \frac{r - \max_{i,j} d_{ij}(0)}{\Delta(0)} \cdot \frac{1}{(k-1)k^{k-1}} \cdot \left(\cos |\theta(0)|_M \right)^{k-1}.$$

It is easy to verify that

$$\frac{v_1}{v_2} > k^{\frac{k}{2}-1}, \quad \frac{v_1}{v_3} > (k-1)k^{\frac{k}{2}-2}.$$

Therefore, Corollary 2 greatly improves the corresponding results in Liu and Guo [2008] and Chen et al. [2016].

11.6 Simulations

In this section, we provide some simulations to illustrate our results.

Consider a population of 10 agents in the plane. In all simulations in the section, we assume that the initial positions and heading angles of the agents are given as

$$(x_i(s), y_i(s)) = \begin{cases} (0,0) & i = 1 \\ \left(\cos \frac{i+1}{10}\pi, \sin \frac{i+1}{10}\pi \right) & i = 2, \dots, 5 \\ \left(\cos \frac{j+7}{10}\pi, \sin \frac{j+7}{10}\pi \right) & i = 6, \dots, 10 \end{cases}$$

and

$$\begin{aligned} \theta(s) &:= \left[\theta_1(s) \quad \dots \quad \theta_{10}(s) \right]' \\ &= [0.414,\ 0.408,\ 0.108,\ 0.414, 0.410, \\ &\quad -0.417, -0.195, -0.288, -0.128,\ 0.052]'\pi, \end{aligned}$$

for $-\tau \leq s \leq 0$. For simplicity, let $\tau_{ij}(\cdot) \equiv \tau$ for $i \neq j$ and $r_i \equiv 1.5$ (to compare with the known results). Note that when $0 < \rho \leq 0.5$, the corresponding $\mathcal{H}_\rho$ is rooted with $d_\rho = 1$.

We carry out the simulations for Theorem 11.2. To this end, we first take $\tau = 0$ and compute the critical absolute velocity by (11.13), i.e.

$$\overline{v} = \frac{1}{\Delta(0)} \sup_{\rho>0} \frac{\zeta\rho}{\alpha} \approx 0.00089.$$

Let $v = 0.0008 \leq \overline{v}$. Then Figure 11.1 shows the convergence of the heading angles of the agents. However, as we have pointed out, Theorem 11.2 only provides a sufficient condition but not a necessary one. By a large number of simulations, we discover that the synchronization occurs when v is less than 0.07248 and fails otherwise. Therefore, our result is conservative. It is worth noting that the critical velocities given by Liu and Guo [2008] and Chen et al. [2016] are, respectively, $v_2 \approx 8.8 \times 10^{-18}$ and $v_3 \approx 3.8 \times 10^{-17}$, which are more conservative. When $v = 0.08$, the evolutions of the heading angles of the agents are shown in Figure 11.2 which shows that the agents cannot achieve synchronization.

Next, take $\tau = 1$. The critical absolute velocity provided by (11.13) is 2.02×10^{-7}, while the simulations indicate that the synchronization occurs provided that v is less than 0.03991 and fails otherwise. Figure 11.3 shows the convergence of the

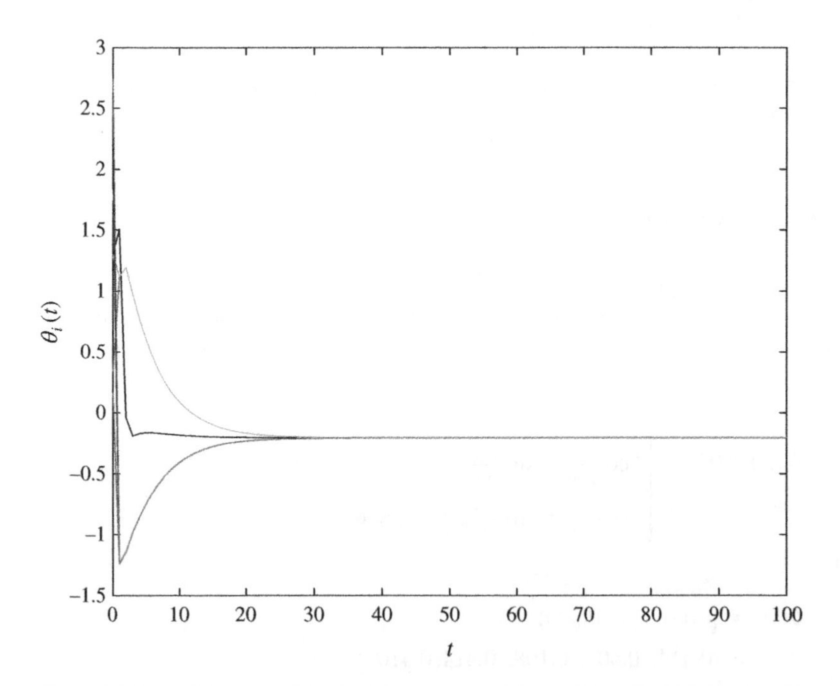

Figure 11.1 Trajectories of the heading angles with $\tau = 0, v = 0.0008$. Source: Zheng et al. (2017)/with permission of IEEE.

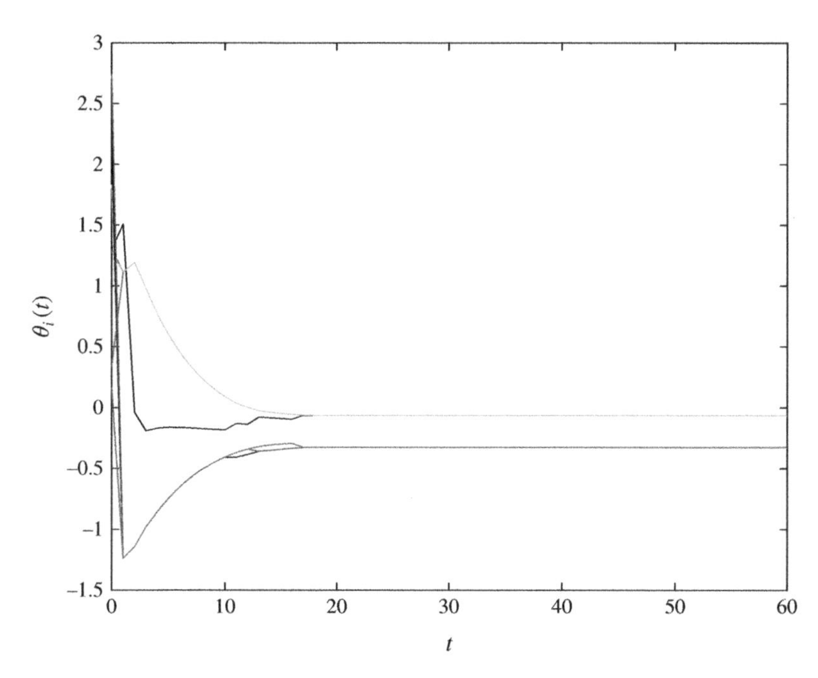

Figure 11.2 Trajectories of the heading angles with $\tau = 0, \upsilon = 0.08$. Source: Zheng et al. (2017)/with permission of IEEE.

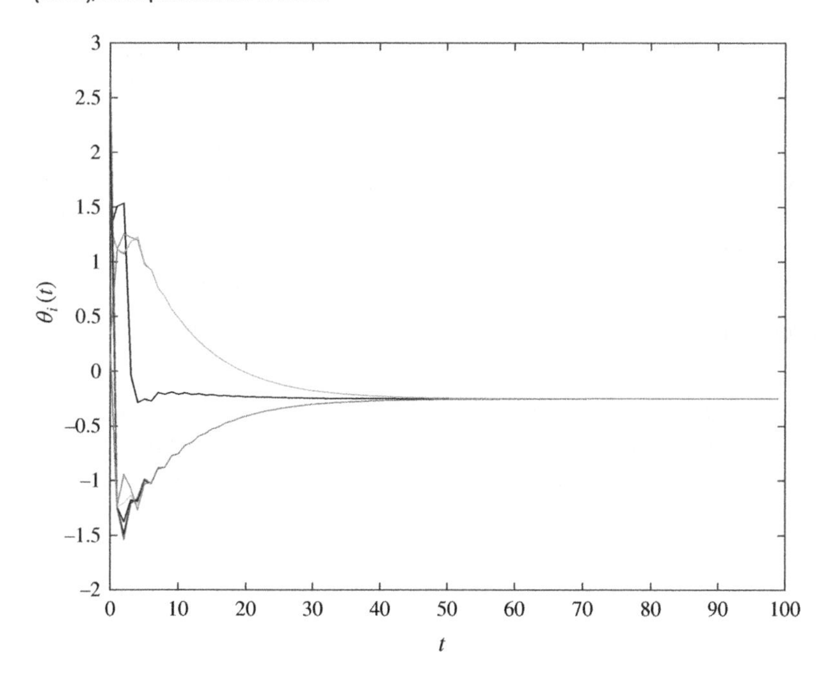

Figure 11.3 Trajectories of the heading angles with $\tau = 1, \upsilon = 0.0008$. Source: Zheng et al. (2017)/with permission of IEEE.

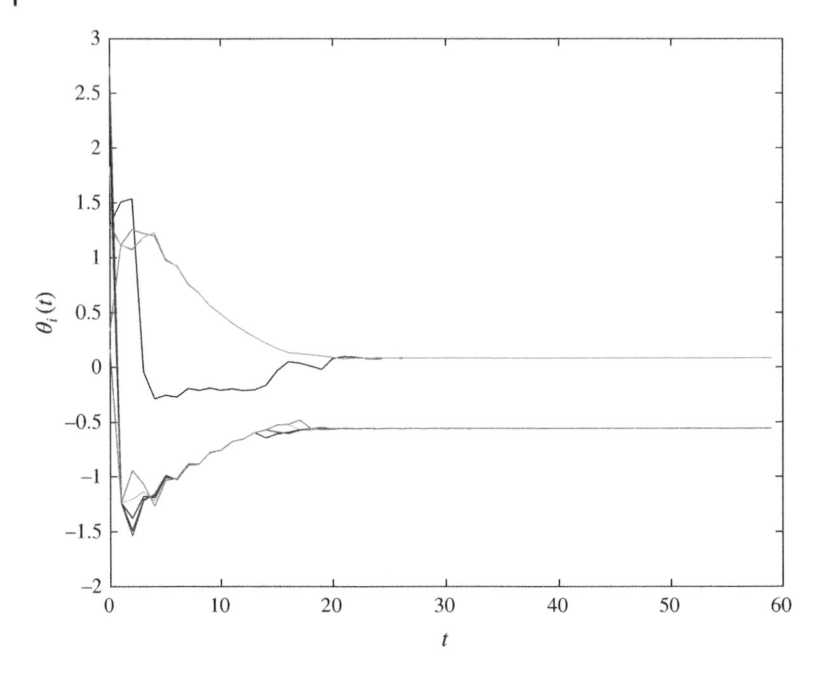

Figure 11.4 Trajectories of the heading angles with $\tau = 1, \upsilon = 0.05$. Source: Zheng et al. (2017)/with permission of IEEE.

heading angles of the agents when $\upsilon = 0.0008$. Figure 11.4 depicts the nonsynchronization behavior when $\upsilon = 0.05$. Note that without delay, the synchronization is still ensured under this absolute velocity.

At last take $\tau = 2$. The critical absolute velocity provided by (11.13) is 8.27×10^{-11}, while the simulations indicate that the synchronization occurs when υ is less than 0.02773 and fails otherwise. Figure 11.5 shows the convergence of the heading angles of the agents when $\upsilon = 0.0008$.

Comparing these simulations, we have the following observations:

(1) A delay can decelerate the synchronization and even destroy it.
(2) Our results are more conservative with a larger delay.

The conservativeness stems from the fact that Theorem 11.2 actually guarantees the preservation of some rooted robust subgraph of the initial neighbor digraph. It will be of interest to establish some tight conditions to ensure that the union of the neighbor digraphs over each bounded time interval is rooted.

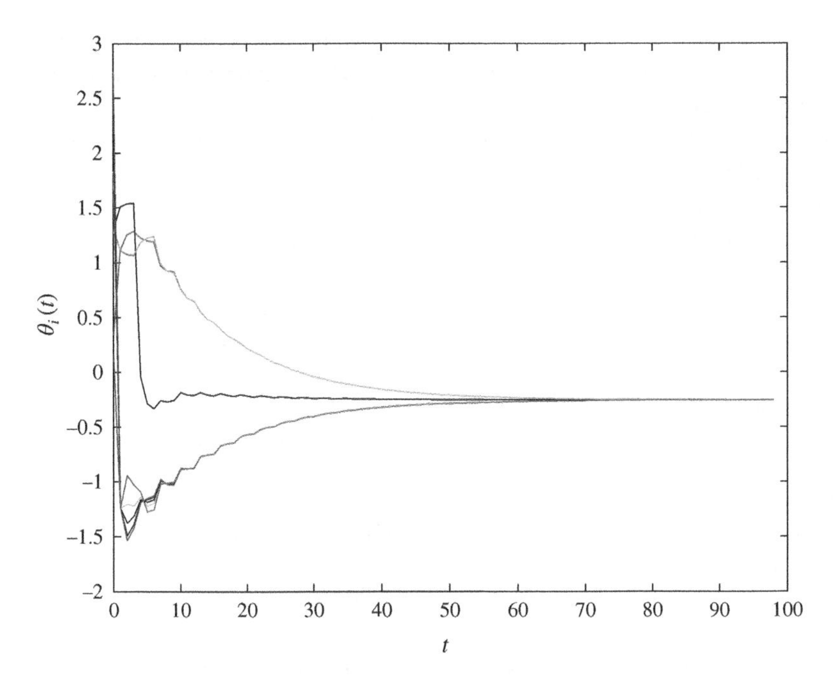

Figure 11.5 Trajectories of the heading angles with $\tau = 2, v = 0.0008$. Source: Zheng et al. (2017)/with permission of IEEE.

11.7 Conclusions

We have investigated the effect of communication delays on the linear and non-linear Vicsek models in this chapter. It has been observed that the Vicsek model is sensitive to communication delays in the sense that large delay can destroy the synchronization behavior. Some delay-dependent synchronization conditions on the initial state, which can be easily verified, have been established. In particular, for the case without delays, our results greatly improve those in the literature.

Bibliography

M. Aouchiche, O. Favaron, and P. Hansen. Variable neighborhood search for extremal graphs. 22. Extending bounds for independence to upper irredundance. *Discrete Applied Mathematics*, 157(17):3497–3510, 2009.

D. P. Bertsekas and J. N. Tsitsiklis. Comments on “coordination of groups of mobile autonomous agents using nearest neighbor rules”. *IEEE Transactions on Automatic Control*, 52(5):968–969, 2007.

Y. Cao, W. Yu, W. Ren, and G. Chen. An overview of recent progress in the study of distributed multi-agent coordination. *IEEE Transactions on Industrial Informatics*, 9(1):427–438, 2012.

G. Chen, Z. Liu, and L. Guo. The smallest possible interaction radius for flock synchronization. *SIAM Journal on Control and Optimization*, 50(4):1950–1970, 2012.

Y. Chen, D. W. C. Ho, J. Lü, and Z. Lin. Convergence rate for discrete-time multiagent systems with time-varying delays and general coupling coefficients. *IEEE Transactions on Neural Networks and Learning Systems*, 27(1):178–189, 2016.

A. Jadbabaie, J. Lin, and A. S. Morse. Coordination of groups of mobile autonomous agents using nearest neighbor rules. *IEEE Transactions on Automatic Control*, 48(6):988–1001, 2003.

Z. Liu. Consensus of a group of mobile agents in three dimensions. *Automatica*, 50(6):1684–1690, 2014.

Z. Liu and L. Guo. Connectivity and synchronization of vicsek model. *Science in China Series F: Information Sciences*, 51(7):848–858, 2008.

L. Moreau. Stability of multiagent systems with time-dependent communication links. *IEEE Transactions on Automatic Control*, 50(2):169–182, 2005.

W. Ren and R. W. Beard. Consensus seeking in multiagent systems under dynamically changing interaction topologies. *IEEE Transactions on Automatic Control*, 50(5):655–661, 2005.

T. Vicsek, A. Czirók, E. Ben-Jacob, I. Cohen, and O. Shochet. Novel type of phase transition in a system of self-driven particles. *Physical Review Letters*, 75(6):1226, 1995.

J. Zhu, J. Lu, and X. Yu. Flocking of multi-agent non-holonomic systems with proximity graphs. *IEEE Transactions on Circuits and Systems I: Regular Papers*, 60(1):199–210, 2013.

Index

Control over Communication Networks: Modeling, Analysis, and Design of Networked Control Systems and Multi-Agent Systems over Imperfect Communication Channels, First Edition.
Jianying Zheng, Liang Xu, Qinglei Hu, and Lihua Xie.

Books in the IEEE Press Series on Control Systems Theory and Applications

Series Editor: Maria Domenica Di Benedetto, University of l'Aquila, Italy

The series publishes monographs, edited volumes, and textbooks which are geared for control scientists and engineers, as well as those working in various areas of applied mathematics such as optimization, game theory, and operations.

1. *Autonomous Road Vehicle Path Planning and Tracking Control*
   Levent Güvenç, Bilin Aksun-Güvenç, Sheng Zhu, and Sükrü Yaren Gelbal
2. *Embedded Control for Mobile Robotic Applications*
   Leena Vachhani, Pranjal Vyas, and Arunkumar G. K.
3. *Merging Optimization and Control in Power Systems: Physical and Cyber Restrictions in Distributed Frequency Control and Beyond*
   Feng Liu, Zhaojian Wang, Changhong Zhao, and Peng Yang
4. *Model-Based Reinforcement Learning: From Data to Continuous Actions with a Python-based Toolbox*
   Milad Farsi and Jun Liu
5. *Disturbance Observer for Advanced Motion Control with MATLAB/Simulink*
   Akira Shimada
6. *Control over Communication Networks: Modeling, Analysis, and Design of Networked Control Systems and Multi-Agent Systems over Imperfect Communication Channels*
   Jianying Zheng, Liang Xu, Qinglei Hu, and Lihua Xie

Printed and bound by CPI Group (UK) Ltd, Croydon, CR0 4YY
16/04/2025
14658594-0001